TRANSITIONS TO SUSTAINABILITY: THEORETICAL DEBATES FOR A CHANGING PLANET

DAVID HUMPHREYS AND SPENCER S. STOBER, EDITORS

TRANSITIONS TO SUSTAINABILITY: THEORETICAL DEBATES FOR A CHANGING PLANET

DAVID HUMPHREYS AND SPENCER S. STOBER, EDITORS

First published in 2014 in Champaign, Illinois, USA
by Common Ground Publishing LLC
as part of On Sustainability book series

Library of Congress Cataloging-in-Publication Data

Transitions to sustainability : theoretical debates for a changing planet / David Humphreys and
Spencer S. Stober, editors.
      pages cm
  Includes bibliographical references and index.
   ISBN 978-1-61229-726-2 (pbk : alk. paper) -- ISBN 978-1-61229-545-9 (pdf)
   1. Sustainability--Philosophy. 2. Sustainable living. 3. Environmentalism. I. Humphreys, David,
1957- II. Stober, Spencer S.

  GE195.T73 2014
  338.9'27--dc23

                             2014025521

Cover image photo credit: Phillip Kalantzis-Cope

This book is for:

Anna Humphreys
Spencer Troy Stober
Krystal Renée Stober
Spencer Lochlyn Stober
Timothy Humphreys
Holly Renée Greene
Gary Franklin Greene
Zeniah Crystale Greene
Jenson Lynham

...and for all future generations

Table of Contents

Acknowledgements

A work such as this requires a great deal of support from others. We extend thanks to all of the contributing authors and colleagues who helped to make this book possible. We specifically thank members of the Common Ground Publishing team in particular Jamie Burns, Ian Nelk, and Phillip Kalantzis-Cope for their assistance throughout this project.

Common Ground's *International Conference on Environmental, Cultural, Economic and Social Sustainability* is an annual event held in January of each year to bring together scholars and practitioners working on sustainability. Most of the papers in this volume were first presented at one of those conferences and subsequently published in an academic journal published by Common Ground. We thank all the authors for writing such original and inspiring chapters and working so productively with us during the editing process.

We thank our families for their support and patience during the completion of the manuscript. It is our hope that this book will contribute to ongoing efforts to sustain the environment for our children, for our grandchildren, and for all those generations who have yet to tread this Earth.

Spencer S. Stober, Reading, Pennsylvania, United States
David Humphreys, Bridge, Kent, United Kingdom

August 2014

Chapter 1: The Quest for Sustainability: Mapping the Terrain

David Humphreys and Spencer S. Stober

The concept of sustainability is derived from the Latin *sustinere* (*tenere*: to hold; sus: *up*). In English, the word brings together the words "sustain" and "ability." In both languages the concept implies the maintenance of certain conditions or capabilities over time. Different academic disciplines bring different understanding and perspectives to the idea. To environmental scientists, a sustainable ecosystem is one that can endure and remain ecologically diverse and productive over time. A sustainable ecosystem may change within certain parameters but has the capacity to recover from disturbances, for example, from fire or from pest attack. Examples of ecosystems that have proved sustainable over time include forests, marine ecosystems and wetlands. In general, the more ecologically diverse an ecosystem is – that is, the greater the variety of species that live within and constitute the ecosystem – the more resilient the ecosystem and the greater its ability to recover from disturbances. While early understandings of resilience emphasized the return of an ecosystem to a more or less stable condition, later theories recognize the capability of ecosystems to change and adapt in response to broader environmental change and human interferences.[1]

Economists define sustainability in terms of natural capital, namely the totality of the ecological resources that the Earth provides. On this view, human activity that does not lead to the depletion of natural capital is sustainable. This approach is relatively uncontroversial when applied to renewable natural resources, such as fisheries and timber, and since the 1980s concepts such as sustainable forest management and sustainable water management have become benchmarks for guiding human use of renewable resources so that the resource base is maintained over time. However, the concept is more controversial when applied to non-renewable resources that are mined from the Earth, such as metals and fossils fuels. There is also disagreement whether the depletion of non-renewable resources is acceptable when it is replaced by human-made capital,

---

[1] Clark 2013.

1

such as transport infrastructure, medical equipment, telecommunications, water purification plants, and so on. *Weak sustainability* permits substitution between different forms of capital – natural and human-made – and is prepared to accept a dwindling of natural resources if this is compensated for by the production of alternative forms of capital. So deforestation would be permitted to build a school house, or even a major city, as the deforested land is being put to alternative uses that are valued by people. Proponents of *strong sustainability*, however, would either oppose forest clearance or respond that if deforestation is taking place in one place it should be compensated for by reforestation elsewhere so that the sum total of the planet's natural forest capital is maintained.[2] Both positions have something to offer, and the extent to which each will apply in practice will vary from case to case. For example, while non-renewable resources cannot be replaced their extraction may, under carefully-defined and managed circumstances, enable the generation of new technologies such as carbon capture devices and low carbon energy production that can contribute to the solution of existing environmental problems. It is equally clear, however, that the continued depletion of renewable resources (such as forests) and non-renewable resources (such as coal and oil) will threaten the long-term ecological viability of the planet.

Substituting natural capital with human made capital cannot, therefore, continue indefinitely without exceeding the carrying capacity of the planet. Carrying capacity may be defined as the population of a species that can be supported over time without degrading the ecosystem on which the species depends. The concept of carrying capacity can be applied over different spatial scales. It is central to Hardin's model of the tragedy of local commons such as grazing land.[3] The concept also applies to the tragedy of the global commons, in particular the oceans and the atmosphere. The carrying capacity of the atmosphere is the quantity of greenhouse gas emissions that can be naturally removed from the atmosphere by carbon sinks, primarily forests and the oceans, so that anthropogenic climate change does not take place. There is some disagreement between climate scientists on what the precise carrying capacity of the atmosphere is, but it seems clear that human activities are exceeding it, leading to anthropogenic climate change.

There are, therefore, limits to the disturbances that an ecosystem can withstand before the ecosystem undergoes a permanent change to another state. For example, the loss of a keystone species or a top predator within an ecosystem will lead to a fundamental change of species diversity so that the ecosystem reaches a new equilibrium. In the worst cases an ecosystem may disappear, providing the conditions for the emergence of another ecosystem (as when deforestation provides the conditions for the emergence of deserts).

The idea that there are limits to the disturbances that an ecosystem can withstand before it changes was central to the 1972 report *The Limits to Growth* produced by the Club of Rome, an interdisciplinary think tank which analyzed the relationship between the global ecosystem and human activities according to five key variables: population, industrial activity, pollution, food production and resource production.[4] As the title implies, the report argued that there are limits to

---

[2] Blowers and Glasbergen 1995, 175.
[3] Hardin 1968.
[4] Meadows et al. 1972.

human activities, and in particular to resource intensive economic growth. For this reason, arguably, the report received only limited support from political leaders. A contemporary argument in which the idea of limits is central comes from Johan Röckstrom and colleagues at the Stockholm Resilience Centre. They focus on different types of planetary boundaries within which there is a safe operating space for human activities. These boundaries include climate change, ozone depletion, ocean acidification, land use, biodiversity loss, freshwater use, the nitrogen cycle and the phosphorous cycle. Röckstrom and colleagues found that the thresholds for three of these boundaries, namely climate change, biodiversity loss and anthropogenic interference with the nitrogen cycle, have already been exceeded.[5]

Scientists have played a vital role in discovering and generating knowledge on these issues. Problems such as the thinning of the ozone layer and anthropogenic climate change were discovered through scientific research. Scientists will continue to play a major role in identifying the causes of environmental degradation, and in disentangling the natural causes from the social. Long before the evolution of humans the history of the Earth was characterized by environmental changes, often sudden and dramatic. Over the very long time horizons of geological change species evolved then became extinct, the planet warmed then cooled and entire ecosystems evolved and changed before dying out. These processes of dynamic change are seized on by those who question whether contemporary global environmental change is anything to be concerned about. The planet has always changed, it is sometimes suggested, so why should we be concerned now? This argument is made in particular by those who deny that anthropogenic climate change is taking place.

It is useful to draw a distinction here between climate change deniers and sceptics. The distinction is an important one. A sceptic is one who searches impartially and objectively for the truth but who is yet to be convinced that the available scientific evidence is sufficient to support a particular claim. But sceptics are always open to persuasion should the threshold of proof that they seek be reached. So while throughout much of the last century sceptical climate change scientists questioned the evidence that human greenhouse gas emissions were driving climate change their numbers have dwindled to just a small handful since the 1970s as the scientific evidence for anthropogenic climate change has hardened. A denier, by contrast, has no real interest in the scientific evidence except, perhaps, in the selective cherry picking of publications that confirm a preordained view. Denial is a political endeavour rather than one based on scientific knowledge or peer-reviewed evidence. Climate change denial has been supported by some important political lobbies, including Exxon-Mobil, Washington-based think tanks such as the American Enterprise Institute and certain precincts in the US Republican Party. These vocal and powerful groups have ensured that while the view that recent climate change is due entirely to natural processes has no scientific basis it remains very influential politically, occupying a disproportionate share of media space compared to the large and growing body of scientific evidence that anthropogenic climate change is taking place, and that it will get much worse. Confronting the climate change deniers

[5] Röckstrom et al. 2009.

with the latest evidence is unlikely to achieve much. Whereas sceptics can be persuaded with new scientific findings, deniers cannot because for them the science is not their main concern. Deniers, it would appear, deny because the policies needed to address climate change – such as intervention in the energy sector and increased taxation to fund the transition to a low carbon economy – run counter both to the interests of their business constituents and to their ideological beliefs on the need for light touch regulation and low taxes.[6]

The causes of and solutions to environmental issues are inextricably bound up in broader social, political and economic processes. Anthropogenic climate change caused by carbon emissions from fossil fuel burning is affecting patterns of agricultural production around the world, with the productivity of some previously fertile lands declining due to drought leading people to seek out new life opportunities through migration, both from the global North to the global South and from rural to urban areas. Changes in diets and food consumption around the world may increase pressure on some ecosystems. For example, increased meat consumption in China has been met in part by beef exports from Brazil, which has increased deforestation pressures in parts of the Amazon to free land for cattle grazing.[7]

The concept of globalization draws attention to the increasing interconnection of different parts of the world. Globalization may be defined as the intensification of social relations and social networks on a worldwide scale so that events in one geographical place are affected and shaped by events in places some distance away.[8] Processes of globalization present challenges for sustainability. Globalization can generate environmental degradation, for example when investment decisions in one place lead to mining, river pollution or deforestation in another part of the world. Often, however, the environmental consequences of globalization are unplanned and unanticipated. Furthermore, globalized processes do not affect all parts of the world equally but are spatially and temporally differentiated, pressing down more or less firmly in some places and at some times rather than others.

The concept of sustainable development seeks to draw together the different strands – social, cultural, political and economic – of sustainability. However, the concept has been criticized for being so vague as to be almost meaningless. Environmentalists tend to focus on the "sustainable" part of the term, arguing for environmental conservation, while those who are less sympathetic to environmental concerns insist that economic growth is essential if development is to be achieved. To some critics the term lacks definitional precision to the extent that it cannot be operationalized as a strategy, while to others it is an oxymoron as development is, it is claimed, inherently unsustainable, relying as it does on the conversion of natural habitats to land uses desired by humans. However, the concept has attracted widespread support from both the global South and the global North and has been endorsed by the United Nations General Assembly. It now forms an integral component of UN environmental discourse and has been adopted as a policy goal by the World Bank, the World Trade Organization and a number of international organizations in the United Nations system including the

---

[6] Humphreys 2009.
[7] Union of Concerned Scientists 2011.
[8] Giddens 1990.

United Nations Environment Programme (UNEP), the United Nations Development Programme (UNDP) and the Food and Agriculture Organization. The seventh of the eight UN Millennium Development Goals is "ensure environmental sustainability." One of the targets for this goal is to "Integrate the principles of sustainable development into country policies and programs and reverse the loss of environmental resources."[9]

Like other grand, overarching ideas that guide human activity, such as freedom and justice, sustainability is a contestable concept. Indeed the concepts of freedom and justice are often invoked in debates on sustainability. The liberal notion of freedom advances both positive and negative notions.[10] A negative freedom is a "freedom from" something that is seen as undesirable, such as crime, state surveillance, poverty or exploitation. A positive freedom is a "freedom to" do something, such as to travel, protest and worship and, more generally, to realize one's potential as a human being. Positive freedoms are often framed as rights, for example, the rights of free assembly and freedom of speech. Both types of freedom are relevant to sustainability which advances negative freedoms, principally the "freedom" from environmental degradation and pollution, and positive freedoms, such as future generations' "freedom to" to enjoy nature and to inherit the natural capital that previous generations have enjoyed. On this view, the challenge of achieving sustainability is that realizing positive freedoms for future generations requires limiting some freedoms for the present generation with some people asked to limit, or even to forgo, certain activities (such as fossil fuel burning, forest clearance and industrial production) so that future generations can enjoy positive freedoms.

Achieving sustainability thus relies on the goodwill of the present generation towards future generations. Because it is not in the self-interests of the present generation to consider the interest of other generation's, sustainability is often framed as a matter of justice. Justice is a moral notion of rightness based on ideas such as fairness and equity. The principle of equity concerns fairness between different actors. These actors may be individuals, states and generations. *Intragenerational equity* is the principle of fairness over space, that is, between different actors of the present generation. The principle holds that no actor should impose an unfair or undue burden on other actors within the present generation. *Intergenerational equity*, by contrast, applies over time and focuses on the rights of future generations; the present generation should not degrade the environment so that environmental harms are passed on to future generations.

Intragenerational equity and intergenerational equity are interlinked concepts in debates on sustainability. The point is often made that environmental degradation is caused in large measure by an unjust distribution of the world's environmental and economic resources. For example, Piet Terhal has distinguished between two connected dimensions of inequity – poverty and affluence – each of which generates different environmental problems and patterns of unsustainability.[11] Poverty and affluence are different dimensions of inequality of access; to natural resources, to life opportunities and to political influence. Poverty may be caused by limited employment opportunities, low per

---

[9] UN Millennium Development Project 2005, xix.
[10] Berlin 1958.
[11] Terhal 1992.

capita income or the displacement of rural peoples from their customary lands. Poverty, possibly accompanied by localized demographic pressures, may lead to increasing stress on natural resources for basic needs fulfilment. One example might include forest loss to provide fuelwood for local people leading to land degradation which, in turn, generates further poverty. Affluence too can generate its own distinct patterns of environmental unsustainability. It is associated with high per capita incomes and high levels of investment in technologies that are resource-intensive leading to pollution (for example, manufacturing plants and passenger jet aircraft) and resource extraction (for example, mining equipment). Affluence can generate widespread environmental despoliation and the degradation of the global commons. Terhal argues that the unsustainability of poverty and the unsustainability of affluence are interlinked and that patterns of affluence in one place may generate patterns of poverty in other places. He cautions that the temptation should be avoided of trying to address poverty by pursuing the affluence-led models of economic development in the more economically developed countries that are as unsustainable as poverty.

This suggests that achieving intragenerational equity and intergenerational equity are interrelated challenges. A fairer distribution of the world's resources will both avoid unsustainable extremes of poverty and affluence and enable the present generation to pass on the planet in a better state to its descendants. A fair distribution of resources in the present generation is thus the route to environmental conservation for future generations. This point is often made in international negotiations when delegates from the developing countries seek to link environmental and economic issues, asserting that they cannot agree to environmental conservation commitments unless the developed countries of the global North agree to a global redistribution of economic resources through financial and technology transfers and external debt relief. However, the developed countries have shied away from such commitments making modest, and often conditional, transfers of financial aid that stop well short of the demands of the global South.

The challenge of realizing sustainability is compounded by two further trends. The first relates to globalization. Tighter interconnectivity in the global system, with different parts of the globe now more strongly interdependent than ever before, is occurring during a period when, it is becoming clear, traditional sources of authority in world politics, primarily the state and international organizations, are unable to implement the measures necessary to arrest global environmental degradation. The second trend is that there is increasing evidence that processes of global environmental change will continue irrespective of any action that humans may take. Even if anthropogenic greenhouse gas emissions were to cease overnight further climate change would continue from emissions in the recent past and the positive feedbacks they will trigger through, for example, the melting of the polar ice caps and methane releases from thawing permafrost. Some scientists now suggest that human interference in the biosphere will be so lasting that the planet has entered a new geological epoch, the Anthropocene in which the dominant force for planetary change is our species. Atmospheric scientist Paul Crutzen is credited with popularizing the term Anthropocene.[12] Like

---

[12] Crutzen 2002.

earlier geological epochs, the Anthropocene will leave its trace in the geological record. Evidence of the Anthropocene that may be excavated by the people of the future (should humans survive) will include plastics, the fossils of extinct species (some of which will contain traces of industrially-manufactured persistent organic pollutants) and high carbon dioxide concentrations in layers of polar ice.

Traditional state-centric politics are failing to respond to global environmental change at the very historical moment when the problem has assumed a greater importance than ever before. This suggests the need for a fundamental transformation of global politics. The state and intergovernmental actors will play a role in any new governance mechanisms that will emerge but will not be the sole actors, with citizens' groups, community networks, scientists, knowledge networks and other actors collaborating. A "grand design" for a new global governance to address all environmental problems is unlikely, and probably unfeasible. Instead a dynamic mode of governance that that can respond with agility to new and emerging priorities for action is necessary, one that can react speedily to changing circumstances and design and implement innovative solutions, often locally-situated and context specific, that are robust and resilient while also flexible and adaptive.

The Focus of This Volume

The contributions assembled in this volume are critically guided, empirically grounded and theoretically diverse. All make a contribution to contemporary debates on sustainability and how it can be defined and operationalized. All authors in this volume share the belief that academics have a responsibility to contribute to public debate on the most pressing public issues of the day. In the twenty-first century no issue is more urgent than the degradation of the global biosphere. In order to consider how a new politics of sustainability may be constructed the contributions focus on six overarching and interconnected themes: culture, systems, business, art, rights and citizenship.

Culture

Culture is an important consideration when seeking ways to sustain the environment for future generations. Our many cultures have evolved in concert with the natural environment (and the emerging human-built environment) and this dynamic process has produced a variety of human practices and values that impact the environment. Daniel Quinn's award winning novel, *Ishmael*,[13] is a compelling dialog between an ape (Ishmael, as the teacher) and a human (Mr. Partridge, as the student). Ishmael challenges Mr. Partridge to reconsider his human-centered values and practices – to reconnect with the mother of all cultures, nature and her laws. Ishmael argues that it is our species' ability to put "Mother Culture" out of our minds that causes some of us to be "takers" (to exhaust natural resources) and others of us to be "leavers" (to replenish and live in harmony with the Earth). The first section of this edited volume considers culture and opens with a quote by Daniel Quinn where he describes our species'

---

[13] Quinn 1992; Winner of the Turner Tomorrow Fellowship.

as carelessly building a penthouse on Earth.[14] Why is our species so willing to grow with reckless abandon on Earth? Why do some of us take more from the environment than nature can replenish. Paul Derby (in Chapter 2) argues that it is our "culture of denial" that enables us to do this, and the title of his paper is a bold statement: "Not All Cultures Should Be Preserved."

Human cultures have dramatically altered the natural landscape to support a diverse array of urban habits on Earth. These human-built environments may work to distance us from the natural environment and thus alter our sense of place as a species on Earth. Therefore, urban landscape design is an important consideration if we are to nurture our relationship with what Ishmehl referred to as our "Mother Culture." Alan Derbyshire, in "The Contributory Components of Viable Eco-Diverse Landscapes" (Chapter 3), makes the case for including natural environments in urban design. Planning for ecological diversity in urban spaces requires active participation by urban dwellers in concert with design professionals. Planning for eco-diverse landscapes requires trans-disciplinary perspectives, including the study of ecology within urban settings. If our urban cultures are to survive, then we need to find ways to replenish nature and live in harmony with the Earth—what Ishmael might have referred to as "urban leavers" and not "urban takers."

Religious perspectives are an important consideration when examining cultural aspects of sustainability. How might Ishmael describe religion? These are his words to Mr. Partridge: "Mother Culture, whose voice has been in your ear since the day of your birth, has given you an explanation of *how things came to be this way*."[15] Ishmael's words could very well be describing how religious worldviews inform our relationship with the natural world. Humans may be unique among species in our ability to envision ourselves as apart from nature and to identify with a spiritual world beyond the natural word. Thus, religion is an important consideration for environmental sustainability. Chapter 4 by Dora Marinova, Amzad Hossain, and Popie Hossain Rhaman, on "Islamic Insights on Sustainability" illustrates the important role that religion plays in transitions for sustainability. The influence of religious perspectives on our interactions with the environment cannot be denied, and these values and practices are transmitted from generation to generation. The final paper in the section on culture is by Spencer S. Stober and entitled "Environmental Memes: Form, Function, and Reasons for Optimism" (Chapter 5). Stober describes how environmental memes function to create guiding principles in the minds of people, by which we judge our interactions with the environment. Both religious and secular perspectives on the environment are the source of environmental memes. The concept of stewardship is promoted by many religions, and "going green" is an emerging secular perspective. Additional memes discussed by Stober include Gaia and Pachamama.

Cultures, and the different values, ideas and social practices that they propagate, play an essential role in defining the social world in which we live. In different ways cultural forces shape the economic, social and political systems

---

[14] Quinn 2010, 13-14.
[15] Quinn 1992, 40.

that govern the planet. The theme of systems is the second organizing theme of this volume.

Systems

A system is a set of interconnected components which together form an integrated whole. A system is characterized by interactions and interrelationships that give it a structure and properties that cannot be explained solely by analyzing the component parts that make up the system. The study of sustainability requires engagement with systems as diverse as ecosystems (comprising the species in an ecological community and the non-living components of its environment such as rocks and water), political systems (which at the national level is usually defined in terms of a legislature, executives and judiciary as well as pressure groups and social movements), the international state system (comprising governments and intergovernmental organizations) and economic systems (comprising businesses, markets and banks).

It is now increasingly clear that under globalization all these different systems exist not in isolation but are inextricably interconnected, making up what we call a single "global system." For example, the global financial crisis of 2008 led to massive public sector bailout of the banking sector, in turn leading to public sector deficits and austerity policies which have resulted in less public sector money available for funding environmental protection and renewable energy programs. The financial crisis also led to decreased speculation on financial instruments such as derivatives and increased speculation on food prices. The resulting increased volatility of international food prices led many governments and food businesses to seek the acquisition of more agricultural land in order to provide a steady supply of essential foodstuffs at predictable prices. This prompted an increase in "land grabs" in many countries of the global South, especially in Africa, with village communities moved, sometimes with force, from their traditional lands which the national government had sold or leased to agriculture corporations. Examples of countries where this has happened include Senegal and Mali.[16] In some cases food insecurity, which has myriad causes in addition to those identified above, has led to food riots which in turn has prompted political instability. Food riots were a direct cause in the 2010 uprisings in Tunisia and Egypt, the so-called Arab Spring, leading to regime change in some countries and political instability and violence in others.

The ways in which different systems interact is central to Glen Kuecker's chapter "The Perfect Storm: Catastrophic Collapse in the 21st Century" (Chapter 6). Kuecker argues that while most systems are able to adapt to disturbances in ways that enable the system to evolve and survive, such adaptive capacity has important limitations. In particular, our reliance on technology has led to superficial solutions when deep rooted structural change is necessary. The "technofix" has now reached its limits for most problems, and certainly for global systemic problems such as climate change. The current global system is now producing unintended and poorly understood complexities and feedbacks that are increasingly leading to catastrophic outcomes. Different crises – such as

---

[16] Clark and Raghuram 2013; Pearce 2012.

hurricanes, the spread of disease, financial instability and social and political upheaval – are now tightly coupled with what Kuecker calls "an historically unprecedented degree of intensity." This convergence, he argues, is leading to a "Perfect Storm."

While, as Kuecker cautions, the current period of global systemic change undoubtedly comes with risks and uncertainties it does also open up political space that offers the possibility of more positive and hopeful change. In Chapter 7 Kimberly Porter Martin builds on cultural considerations in the first section of this volume when arguing that culture is a central dimension of sustainability. Anthropological theory has long recognized the complex integration of beliefs, customs and behaviors that constitute cultural systems. Martin brings together cultural theory and systems theory to provide some innovative insights on sustainability based on research into hunting and gathering societies. Her chapter examines how such societies self-organize, learn and adapt to changes to their environment. She concludes that hunting and gathering societies have a "cultural infrastructure" based on leadership, trust, flexibility and knowledge of their environments that enables the emergence of organizational forms that enhance sustainability.

In Chapter 8 David Grierson draws from the work of Thomas Kuhn to explore the possibility that we are currently undergoing a paradigm shift from a (reductionist) mechanistic paradigm to an (holistic) ecological paradigm through a series of discontinuous revolutionary breaks from our earlier values, beliefs and experiences. He suggests that a transition to a more ecological worldview leading to a radical redefinition of the relationship between society, nature and technology is not only possible but is now underway. He traces the start of the paradigm shift to the twentieth century and the development and growing scientific acceptance of Lovelock's Gaia theory. The challenge of sustainable development, he argues, is "between those who believe that technology can resolve all of our problems and those who know it cannot." Grierson's arguments on technology suggest that some major changes are needed in the role in the economy of business. This is the third theme of this volume.

Business

A common argument made in the study of international political economy is that the most powerful actors are no longer states but transnational business corporations. If the world's economies are measured in terms of either the Gross Domestic Product of countries or the annual turnover of business corporations, then approximately half of the world's largest economies are corporations (Roach 2007). Major business corporations in the oil and energy markets control vast technological infrastructures and are responsible for significant greenhouse gas emissions. Some of the world's large timber corporations are behind the extensive deforestation in the tropics. The more powerful businesses have economic assets that dwarf those of some of the countries in which they operate, making it difficult for the host government, which often depends on business for export earnings, to enforce environmental and employment regulation. Governments thus have a limited ability to pursue any agenda that is not first endorsed by transnational business leaders. While nominally the state remains the dominant rule maker through intergovernmental organizations, most of the important decisions that drive environmental degradation are taking place in privately owned businesses.

Business thus has a central role to play in political responses to environmental degradation, both as a prime driver of many environmental problems as well as potentially (although not necessarily) an active participant in generating future solutions. In Chapter 9 Dariush Rafinejad and Robert C Carlson detect some reasons for optimism on the future role that business can play in tackling global environmental degradation. First, however, they argue that it is necessary for business to examine often unquestioned assumptions on economic growth, the focus on short-term gains and the detachment from local communities. They make the case for sustainable product development intended to provide the sustainable growth of human welfare rather than the sustained growth of shareholder value. They critically examine four business models (the value-based model, the regulatory model, the entrepreneurship model and the natural capitalism model), arguing that none has led to sustainability due to fundamental shortcomings that have failed to address the root causes of environmental decline. They conclude by arguing for a new economy model that both behaves in a similar way to an ecological system while also being integrated with natural ecosystems.

In Chapter 10 Michael L McIntyre and Steven A Murphy argue that the business corporation should be seen as an essentially social organization. It needs, therefore, to be asked what the key role of the corporation in society is, and what it should be. They argue that corporations either need to make a compelling case that the current form of corporate organization is itself compatible with the sustainability goals of society, or be prepared to change in order to meet those goals. While the current corporate form has proved itself a powerful engine of economic growth, the social context within which the corporation evolved has since changed with the realization that the global environment is under threat. In order for the corporation to meet the goal of sustainability McIntyre and Murphy recommend a 3R economy (Recycle, Renew, Re-birth).

The relationship between the corporation as an organization and the broader political context in which business operates is explored further by David Humphreys in Chapter 11. He argues that the dominant model for business responses to social and environmental problems, namely corporate social responsibility (CSR), is flawed as it offers an agent-led response to a systemic problem. Furthermore, the idea of CSR accepts the idea of an enfeebled state: that is, the state can no longer pass legislation and regulation to govern the activities of transnational corporations under global neoliberalism and instead must rely on the corporations themselves taking voluntary measures. This, Humphreys argues, is a contradiction in terms as, first, the corporation's primary fiduciary responsibility is to its shareholders, not to broader society and, second, realizing the public interest cannot be left to private business organizations. The chapter concludes by arguing for a nested model of corporate accountability to publicly-accountable authorities with final decision-making authority resting with local communities.

Art

Art enriches our lives and sustains human cultures. For example, the painting *The Hay Wain* by the artist John Constable depicts a horse drawn farm cart crossing a stream. It is often invoked as a scene of idyllic rural beauty and tranquility in early nineteenth century England. Other paintings may portray a more disturbing image. Art historians have speculated whether the red sky in Edvard Munch's famous painting the *The Scream* was inspired by the eruption of Krakatoa in 1883 which led to the spread of atmospheric aerosols and particulate matter around the globe leading to spectacular sunsets in many parts of the world.

We have four reasons for selecting articles to illustrate the role that art can play in our efforts to sustain the environment for future generations. First, art can challenge us to reflect on how we view the natural world. The section on art opens with a pencil drawing entitled "Displaced Enchantment" by Spencer S. Stober (co-editor). This drawing challenges us to reflect on our relationship with nature. Some of us may attempt to understand the environment in material terms, via reasoning and scientific methods that suggest humans are of nature and subject to natural constraints. Others of us may embody nature through experiences that defy material explanation. This perspective enabled our human hunter-gather ancestors to sustain themselves in the natural world, and some human cultures to this day. Others of us may accept on faith that nature requires our stewardship because we are privileged among other creatures. Does this perspective work to distance our species from nature, our more primitive and mystical relationship with nature? Are we ignoring material evidence that humans cannot continue to encroach on nature's capacity to nurture life on Earth? The drawing "Displaced Enchantment" challenges us to consider these questions.

Second, art is an important component in the cultural dimension of sustainability. Art may enrich and sustain the human aspects of a culture, but it also endorses human interactions with the environment that are necessary for survival. Alisa Moldavanova's article entitled "Sustainability, Aesthetics, and Future Generations: Towards a Dimensional Model of Arts' Impact on Sustainability" (Chapter 12), provides a compelling argument that art is an essential consideration for a sustainable future. Third, art is a powerful medium by which humans can share experiences. Art can be, as Elizabeth More Graff and Wolfram Hoefer suggest, "A Call to Experience the Earth Collectively" (Chapter 13). Graff and Hoefer studied the "Earth From Above" exhibit by Yann Arthus-Bertrand, and they provide us with several compelling images so that we may share a bird's eye view of the Earth. The fourth and final reason discussed here, is the possibility that art is necessary to enrich, nurture, and repair our relationship with the natural world. Jade Wildy, in "Sculpting Sustainability: Art's Interaction with Ecology" (Chapter 14), describes several examples of how environmental artists are experimenting with techniques to make aesthetics part of their efforts to sustain the environment. These works are reasons for optimism.

Rights

The focus of the fifth section of this book is rights. Sustainability, it was noted above, is often expressed in terms of freedoms, both freedom from harm (negative freedoms) and freedom to do something (positive freedoms). Positive freedoms may be seen as entitlements or rights. In domestic legal systems rights may be granted by states to people or to organizational forms created by people, such as business corporations. Similarly, in international law intergovernmental organizations may grant, or recognize, rights. Examples include international human rights law and international private law which seeks to regulate interactions between, and uphold the rights of, private sector actors, such as businesses. A body of international law has also developed on the rights of the state. Over the last 30 years several important legal principles have crystallized in international environmental law, such as the polluter pays principle (polluters should bear the costs of cleaning up any pollution they have emitted in order to restore environmental quality), the precautionary principle (lack of scientific certainty should not be used as an excuse for postponing measures to prevent environmental degradation) and common but differentiated responsibilities (all states have a responsibility to prevent environmental degradation and to restore environmental quality, but these responsibilities are differentiated according to the different historical contributions to these problems).

In Chapter 15 David Humphreys surveys five sets of rights which, he argues, are central to the study of sustainability. The first four are the rights of states (in particular the right to sovereignty over natural resources), human rights (both the rights of individuals as well as collective rights, such as the rights of indigenous peoples), property rights (or, more accurately, the rights of individuals, groups or organizations to a particular piece of property, such as land, that is recognized as binding in law) and corporate rights (such as the right to trade or invest internationally). However, the bulk of the chapter focuses on the recognition by two countries in South America (Ecuador and Bolivia), and of some sub-federal public bodies in the United States, of a fifth set of rights; the rights of nature. Humphreys argues that while the emergence of what has been termed an "Earth jurisprudence" should be seen as positive, it is not a panacea for environmental degradation. Even in Ecuador and Bolivia, the first two countries to recognize rights of nature, conflicts remain on the extent to which a strong conservationist ethos based on respecting nature's rights may come at the expense of using natural resources for community-driven or national development.

In Chapter 16, Jennifer E Michaels examines the work of the Indian scientist and campaigner Vandana Shiva who has been active in promoting human and community rights throughout the global South. Michaels outlines the evolution of Shiva's work as an activist and scholar, in particular her opposition to corporate globalization, which she sees as a war against the Earth, people, democracy and freedom. Shiva condemns corporate ownership of intellectual property rights of seeds and genetic material as the colonization of life, and she considers carbon trading a mechanism that legitimizes continued atmospheric pollution. Shiva insists on the relationship between cultural diversity and biological diversity.

The remaining two chapters in this section, by Spencer S Stober and Christine Dellert, examine the emergence of the rights of nature in Ecuador (Chapters 17 and 18 respectively). Stober traces the cultural and sociopolitical factors that led to the recognition of rights of nature in Ecuador in the 2008 constitution to the prior recognition of rights for indigenous peoples in the 1998 constitution. This is similar to developments elsewhere in South America where local leaders have emerged seeking to reclaim a more intimate relationship with nature such as Chico Mendes in Brazil. Stober argues that the recognition of the rights of nature in Ecuador was enabled by the way that the indigenous Ecuadorians view citizenship, with the emphasis very much on a collective view of citizenship rather than the individual notion of citizenship found in liberal democracies. Like Stober, Dellert considers the ramifications of the recognition of rights of nature in Ecuador for how citizenship is conceived. She argues that while the constitution has transformative potential there is as yet "little evidence to suggest a decommodification of nature in a country so heavily dependent on its exploitation." She argues for a reframing and expansion of ecological citizenship in order to promote more effective decision-making about the future of our planet and its inhabitants. The concept of citizenship forms the sixth and final theme of this volume.

Citizenship

To a large extent how we view our role as citizens is shaped by our values, in other words our social preferences and moral-based beliefs about what is good or desirable, and the actions we take in pursuit of the sort of world we want to inhabit. Values are central to the study of sustainability, and of society in general. Different actors may value the environment in different ways. Environmental values may be defined in relation to the valuer (the person or people who value) or the valued (the species or things that people value). This has given rise to the familiar distinction between instrumental value (the value that a species, ecosystem or natural feature has for humans) and intrinsic value (the value that nature has in in and of itself, irrespective of how people may value it). Another familiar distinction is between light greens, who tend to see environmental degradation as an accidental condition of modernity that can be addressed through changes to industrial production such as technological refinement, and deep greens, who see environmental problems as an integral and unavoidable condition of modernity, with industrial production and modern technology part of the problem rather than the solution.

Our personal values shape how we view citizenship, and in particular the responsibilities we have to other people and to nature. Should we wait for those who hold political and economic power to take action to address environmental degradation, or should we take action as citizens with a view to changing the broader social structures within which we live and act? Or do we have a responsibility to act irrespective of what others do? A useful distinction is drawn between environmental and ecological citizenship by Andrew Dobson. Environmental citizenship is defined with respect to the relationship between the state and the citizen. Environmental citizens give their consent to the state to define and uphold environmental rights and to make environmental policy in their interest; citizens who claim environmental rights have a duty to respect the environmental rights of other citizens. The obvious problem with this notion of citizenship is that it relies on the state, and no state acting in isolation can safeguard the global environment.

Dobson argues against a right-based notion of citizenship, arguing that citizenship should also embrace obligations and in an ecologically fragile and interdependent world there is no good reason why obligations should cease at national borders. He argues for ecological citizenship which is based on the fair usage of global ecological space. Ecological space may be defined in terms of, for example, water consumption, greenhouse gas emissions and land used for living and agriculture. There is a limit to the ecological space that the world's population can consume before environmental degradation ensues, and that space should be fairly divided between the world's populations. Dobson argues that any citizen who occupies an unsustainable amount of ecological space, or who imposes upon the ecological space of others, has an obligation to reduce their consumption of ecological space.[17] On this notion of citizenship obligations are unequally divided between the people of the world. Because the wealthier populations of the world occupy more of their fair share of ecological space, they have a moral responsibility to consume less. Poorer communities have no such obligation.

Globalization, it was suggested above, can present challenges for sustainability but it can also generate alternatives. Innovative experiences and solutions developed in one place can be rapidly disseminated through social networks, such as, for example, the Transition Towns network.[18] Globalization, therefore, may not only reinforce and perpetuate the status quo; it may also promote disruptions to it. As Samuel Alexander argues in Chapter 19, disruption brings with it alternatives. He develops the concept of *disruptive innovation* to denote the rapid and far-reaching social changes that may be catalyzed by the sudden emergence of a new social movement, technology or business model, or even a confluence of such phenomena. He examines the role that disruptive innovation may play in prompting the transition to a low carbon world. Such a transition, he notes, need not necessarily lead to an homogenous globalized alternative. Different citizens, he suggests, will make different choices on energy technology depending on their social and cultural context. This opens up the possibility that "a low carbon world can indeed emerge from the grassroots – if only we choose it."

Alexander and Sabina Lautensach examine how environmental education may enable learners to develop and adopt more sustainable ways of living as engaged global citizens (Chapter 20). An ethics of responsible citizenship, they suggest, has always been a part of a school education. They consider how school education can incorporate teaching on sustainability. They insist that ideology can never be absent from the classroom and that an environmental education should seek a comparative evaluation of the merits of different ideologies from an environmental perspective. Such an endeavor, they argue, would lead to a destabilization of the dominant economistic paradigm and a shift towards ecocentrism.

Elin Wihlborg and Per Assmo examine the role of the home as an arena for promoting sustainable development (Chapter 21). Global environmental problems, they suggest, are both a symptom and the cumulative result of the daily practices of people at the local level. They examine the home along three dimensions: as a physical dwelling, as a place of economic resource management,

---

[17] Dobson 2003.
[18] Bingham and Clark 2013.

and as a social and emotional space for the family. They argue for a broader notion of citizenship that includes the family. The liberal democratic notion of citizenship is flawed because it ignores relationships and lacks an ethic of care, kinship, mutuality and empathy. The "good" citizen, they argue, will always recognize her obligation to the family.

Conclusion

A recurring line of argument throughout this book is that to greater or lesser degrees most of us contribute to environmental problems. Much global environmental degradation today is caused by the aggregate of routine, ordinary and every day actions, such as turning on a light, driving a car, applying pesticides to a field or garden, and disposing of waste in a careless manner. We may knowingly buy foodstuffs that have been flown halfway around the world in plastic containers that we know are not biodegradable. These actions that we take are shaped both by the broader structures of world that we inhabit and by our agency as social actors. We all, therefore, have a role to play in addressing environmental degradation; as citizens, as consumers, and as family members. We may also have a role to play in promoting change in communities and organizations, such as our employers, and those bodies that claim to represent us such as local and national government. How may the leadership we exercise as individuals translate into organizational leadership and structural change?

Benjamin Redekop is an authority on leadership and he reminds us that "sustainability is no longer an optional element of leadership, if it ever was, but a central task."[19] This mandate calls for a new and more egalitarian perspective on leadership. Leadership is by people with people and it is therefore reasonable to expect our efforts to focus on the social and cultural aspects of environmental sustainability, but it is also important for us to reflect and dialog on how humans relate to the natural world. "Nature-centered Leadership" is a new perspective on leadership that calls for this dialog, while building an aspirational narrative for a more sustainable future.[20] Nature-centered leaders are primarily transformational in that their actions facilitate a shared vision for environmental protection, one of hope where humans are living in harmony with nature. They include visionaries such as Rachel Carson, James Lovelock, and Chico Mendes. Each of us can strive to be nature-centered in our everyday actions, and we can also be part of the dialog to seek common ground for concerted action. The collection of papers in this volume gives voice to a range of perspectives. It is our hope that you will enjoy and find the papers to be inspiring in your efforts to sustain our many cultures and the global environment for the benefit of future generations.

References

Berlin, Isiah. 1958. *Two Concepts of Liberty.* Oxford: Clarendon.
Bingham, Nick and Nigel Clark. 2013. Transition, sharing and an environmental
    imagination. In *Environment: Sharing a Dynamic Planet – Carbon,*

---

[19] Redekop 2007, 134.
[20] Stober, Brown and Cullen 2013.

Food, Consolidation, edited by Parvati Raghuram, Nick Bingham, Jane Roberts, David Humphreys, Joe Smith, Sandy Smith and Philip O'Sullivan, 361-290. Milton Keynes: Open University.

Blowers, Andrew and Pieter Glasbergen. 1995. The search for sustainable development. In *Environmental Policy in an International Context: Perspectives,* edited by Pieter Glasbergen and Andrew Blowers, 163-183. London: Arnold.

Clark, Nigel. 2013. Shaping and sustaining places. In *Environment: Sharing a Dynamic Planet – Introduction, Life, Water,* edited by David Humphreys, Nigel Clark, Sandy Smith, Petr Jehlička and Nick Bingham, 81-110. Milton Keynes: Open University.

Clark, Nigel and Parvati Raghuram. 2013. Securing food, sharing land. In *Environment: Sharing a Dynamic Planet – Introduction, Life, Water,* edited by David Humphreys, Nigel Clark, Sandy Smith, Petr Jehlička and Nick Bingham, 115-145. Milton Keynes: Open University.

Crutzen, Paul J. 2002. Geology of mankind. *Nature* 415 (6867): 23.

Dobson, Andrew. 2003. *Citizenship and the Environment.* Oxford: Oxford University Press.

Dresner, Simon. 2002. *The Principles of Sustainability.* London: Earthscan.

Giddens, Anthony. 1990. *The Consequences of Modernity.* Cambridge: Polity.

Hardin, Garrett, 1968. The tragedy of the commons. *Science* 162(3859): 1243-8.

Humphreys, David. 2009. The role of science in climate change policy. In *A Warming World* edited by David Humphreys and Andrew Blowers, 57-96. Milton Keynes: Open University.

Meadows, Donella H, Dennis L Meadows, William W Behrens and Jørgen Randers. 1972. *The Limits to Growth.* New York: Club of Rome.

Pearce, Fred. 2012. *The Landgrabbers: The New Fight Over Who Owns the Planet.* London: Random House.

Quinn, Daniel. 1992. *Ishmael: An Adventure of the Mind and Spirit.* New York, NY: Bantam Books.

Quinn, Daniel. 2010. The Danger of Human Exceptional. In *Moral Ground: Ethical Action for a Planet in Peril,* edited by Kathleen Dean Moore, and Michael P. Nelson, 9-20. San Antonio, TX: Trinity University Press.

Redekop, Benjamin. 2007. Leading into a Sustainable Future: The Current Challenge. In *Leadership: Impact, Culture, and Sustainability,* edited by Nancy S. Huber and Michael Harvey, 134-146. College Park, MD: International Leadership Association.

Roach, Broan. 2007. *Corporate Power in a Global Economy.* Medford MA: Tufts University. Available at http://www.ase.tufts.edu/Gdae/education_materials/modules/Corporate_ Power_in_a_Global_Economy.pdf, accessed June 4, 2014.

Rockström, Johan, Will Steffen, Kevin Noone, Åsa Persson, F. Stuart Chapin, Eric F. Lambin, Timothy M. Lenton, Marten Scheffer, Carl Folke, Hans Joachim Schellnhuber, Björn Nykvist, Cynthia A. de Wit, Terry Hughes, Sander van der Leeuw, Henning Rodhe, Sverker Sörlin, Peter K. Snyder, Robert Costanza, Uno Svedin, Malin Falkenmark, Louise Karlberg, Robert W. Corell, Victoria J. Fabry, James Hansen, Brian Walker, Diana Liverman, Katherine Richardson, Paul Crutzen and Jonathan A. Foley.

2009. A Safe Operating Space for Humanity. *Nature* 461: 472-5 (September 24).

Stober, Spencer S., Tracey L. Brown and Sean J. Cullen. 2013. *Natured-centered Leadership: An Aspirational Narrative. Champaign, IL*: Common Ground Publishing LLC.

Terhal, Piet. 1992. Sustainable development and cultural change. In *Environmental Economy and Sustainable Development*, edited by H.B. Opschoor, 129-142. Amsterdam: Wolters-Noordhoff.

Union of Concerned Scientists. 2011. *The Root of the Problem: What's Driving Tropical Deforestation Today?* Cambridge MA: Union of Concerned Scientists.

UN Millennium Development Project. 2005. *Investing in Development: A Practical Plan to Achieve the Millennium Development Goals.* London: Earthscan.

Part I: Culture

Editors' note: Human cultures are an important consideration when exploring transitions to sustainability. Our many cultures have evolved in concert with the environment and provide for our needs as a species – to reproduce and to thrive. Daniel Quinn offers this "penthouse" metaphor:

> We are like people living in the penthouse of a hundred-story building. Every day we go downstairs and at random knock out 150 bricks to take upstairs to increase the size of our penthouse. Since the building below consists of millions of bricks, this seems harmless enough ... for a single day. But for thirty thousand days? Eventually – inevitably – the streams of vacancy we have created in the fabric of the walls below will come together to produce a complete structural collapse. When this happens – if it is *allowed* to happen – we will join the general collapse, and our lofty position at the top of the structure will not save us.

Daniel Quinn (2010)[1]

Image 1: *At What Cost?*, 2013, by Spencer S. Stober, pencil drawing, 9x12 inches. Permission to reprint has been granted by Spencer S. Stober (co-editor).

---

[1] Quinn, Daniel. 2010. The Danger of Human Exceptionalism. In *Moral Ground: Ethical Action for a Planet in Peril*, edited by Kathleen Dean Moore and Michael P. Nelson, 9-14. San Antonio, TX: Trinity University Press. This quotation can be found on pages 13-14.

Chapter 2: Not All Cultures Should Be Preserved: The Culture of Denial, Its Effects on Sustainability and What Should Be Done About It

Paul Derby

What if the values and behaviors of one culture are detrimental to the natural environment or the existence of other cultures? This paper addresses this question through a discussion of a growing denial of global warming among people of the United States of America. It is argued that the emergent Culture of Denial results from threats to a dominant Western ideology caused by challenges from the sustainability movement and the current global economic recession. It is further argued that this dominant ideology of the so-called developed societies promotes and legitimizes cultural values that are unsustainable, thus harmful to the natural environment and other cultures. A case study of a lake community in United States demonstrates the power these cultural values have on the decision-making of the people of the community and the negative effects their decisions have had on the natural environment. The paper concludes by suggesting ways to challenge the Culture of Denial and change the ideology of unsustainability that threatens the natural and human environments of our shared ecosystem.

A Growing Denial of Global Warming in the United States

On December 10, 2009 the editor of a weekly newspaper in my hometown wrote an article expounding his angst about "this global warming issue." [1] While begrudgingly acknowledging that greenhouse gas emissions appear to be increasing and so too global warming, [2] the editor questioned the "apparent consensus" from the scientific community and felt the issue "more politically-

---

[1] Frost 2009, 7.
[2] The term "global warming" is used here rather than climate change for two reasons: (1) this is the term used in the editorial and (2) the phrase "climate change," although more accurate, is represented by proponents on both sides of the issue as purposely confusing rhetoric.

charged" than necessary.[3] Not surprisingly, the editorial was timed to coincide with the 2009 United Nations Climate Change Conference (Copenhagen Summit) and President Obama's highly anticipated announcement that would have committed the major industrialized nations to an agreement to significantly reduce greenhouse gas emissions by 2050.[4] As we now know, the Copenhagen Accord was not passed and holds no legal and little political clout. In the next week's "Letters to the Editor" section, there was a flurry of "local skepticism" about so-called "global warmth," with opinions ranging from a lament of a perceived absence of serious intellectual and scientific debate to the outright dismissal of global warming as a hoax.[5] This piqued my interest, because in the last few months it seems everyone I talked to opined that either the current climate change is a natural cycle or that global warming was just an overstated scare by politicians and environmentalists.

I wondered if this was a local phenomenon, since I live in a very conservative region of the United States, or a societal trend, and if so, what is driving it and what does it mean for the sustainability movement? As it turned out, this is a trend amongst people of the United States. Two recent opinion polls bear this out. A December 2009 Harris poll indicates that only 51% percent of Americans now believe that the release of carbon dioxide and other gasses into the atmosphere will lead to global warming. This is down significantly from 71 percent in 2007 and the lowest percentage since Harris began asking the question 12 years ago.[6] A second poll analyzed by The Pew Resource Center in October 2009 revealed nearly the same results, again showing sharp decreases of those believing that the Earth is warming. Even more disturbing, the survey found that only 36 of 100 Americans believe that there is solid evidence that climate change is being caused by human activity, and, what I found personally troubling, that level of education – college graduates, some college, and high school or less – made no difference in attitudes for adults in the United States.[7] Finally, for those who did acknowledge that global warming is a problem, a growing percentage believe that technologies are being developed and will be in place in time to keep greenhouse gas emissions within acceptable limits.

It is clear that there is a growing disbelief amongst Americans about the validity of global warming, but what is the scientific evidence? While not every scientist agrees,[8] the overwhelming majority in the scientific community has reached the consensus that global warming is real and human activity is the predominant reason for it. The Intergovernmental Panel on Climate Change (IPCC), made up of thousands of scientists from 194 countries from across the political and cultural spectrums, is the largest consortium of scientist ever assembled.[9] This body concluded in its report that

---

[3] Frost 2009.

[4] China Rejects UK Claims it Hindered Copenhagen Talks, *BBC World News*, December 22, 2009.

[5] Global Warmth 2009, December 17. *The Chronicle Newspaper* 30 (1315). 1,17.

[6] Hall 2009.

[7] Fewer Americans See Solid Evidence of Global Warming 2009.

[8] CATO Climate Change Ad 2008.

[9] IPCC 2007.

Global atmospheric concentrations of carbon dioxide, methane and nitrous oxide have increased markedly as a result of human activities since 1750 and now far exceed pre-industrial values determined from ice cores spanning many thousands of years. The global increases in carbon dioxide concentration are due primarily to fossil fuel use and land use change, while those of methane and nitrous oxide are primarily due to agriculture.[10]

Additionally, most scientific voices warn that humans need to act now in order "to cut emissions around 80 percent by 2050"[11] to keep the level of carbon dioxide below 450 parts per million, which is generally accepted as the "point of no return." However, with growing disbelief among the general public and the inability of political commitment among the industrialized nations, the chances of sufficient cuts in CO_2 emissions by 2050 appear unrealistic.

So, why is it that given the substantial evidence from the scientific community to the contrary a growing number of people in the United States either don't believe the science or choose to ignore it? It is argued here that many people of United States have constructed a Culture of Denial about global warming in response to threats to their ideology and way of life.

What is this Culture of Denial?

Denial appears to be a common enough occurrence among human beings, but what is denial and why would humans deny apparently real problems that may, in reality, threaten their survival? Denial can be defined as "any of a group of mental processes that enables the mind to reach compromise solutions to conflicts that it is unable to resolve."[12] Perhaps the main theory of human denial derives from Freud's theory of personality, which conceptualizes denial as a basic human defense mechanism. In short, denial is a mental, usually unconscious, response that "involves concealing from oneself internal drives or feelings that threaten to lower self-esteem or provoke anxiety."[13] For example, on a recent trip to Ecuador my wife found the twisty, fog-filled and often too close to the precipice road up the mountains to the city of Cuenca very frightening. The rest of us in the minivan assured her that this driver had driven through the clouds along this switch backed and washed out road "at least a thousand times before." In truth, none of us had any idea if this was the driver's first or five thousandth time up this road or if the weather this day was any better or worse than any other. Further, none of us cared to ask. It was simply more comforting to deny the perhaps very real possibility that it may actually be a life-threatening event. From this perspective, the denial of global warming can be explained as the human defense mechanism to quell the anxiety associated with the perceived or real threat to our well-being.

---

[10] IPCC 2007, 2. Note that 80 percent of greenhouse gas emissions come from the burning of fossil fuels, and most of the remaining from deforestation and other changes in land use.
[11] The Carbon Bathtub 2009, 26.
[12] Defense Mechanisms 2014.
[13] Heffner 2002.

A second theory of human denial comes from my own field of anthropology, specifically from physical anthropology that examines human biological evolution. Within evolution theory, denial can be understood as a genetically encoded survival adaptation. That is, it may have been (and may be) an evolutionary advantage for humans, as a species, to be able to deny certain thoughts or behaviors. Because humans are rational beings – i.e. we have the ability to reflect upon (think about, ponder) what we have done – certain acts such as having to kill other life in order to eat would have produced internal conflicts. Thus, it would be an advantage for the reasoning human to conceptualize or rationalize killing for food as not murder but supper. Further, humans are often faced with incongruous human actions such as deceptions (when is it okay to lie), disloyalty (infidelity), even rather unscrupulous acts of persuasion by some (politicians) for the so-called good of the whole. Such acts are routinely rationalized by reasonable humans as necessary or even courteous; yet they are, in fact, inherently contradictory and often harmful to others, to the natural environment, or to ourselves. From this perspective, the denial of global warming can be explained as the innate reaction to rationalize anthropogenic behaviors that bring harm to others or the natural environments.

Both Freud's defense mechanism and human evolution are theories of nature (instincts and genetics). This begs two nagging question: Is denial human nature? If so, what does this imply about the human ability to recognize and remedy unsustainable human-caused actions such as global warming? Denial is probably both a psychological and biological quality of the human animal. However, such explanations leave some gaping holes. If denial is human nature, how can we explain differences among people within the same society or across cultures? How is it that millions of Americans, and perhaps billions of others in the world, don't deny that anthropogenic global warming is actually happening? Further, how can it be that people could and do change their minds over time? The answer to these seeming contradictions resides in the other part of the rational human being: culture – those shared ideas, values, and behaviors that people construct in order to adapt to the natural environment and/or to adapt the natural environment to the fulfill their wants and needs. For humans, it appears that culture trumps nature, or at the very least allows humans to be reasonable decision-makers. Thus, while denial may be a psychological defense against internal conflict and a rational justification of human-created incongruous acts, it is the culturally constructed value systems in which these thoughts and acts are embedded that establish the context for behaviors and subsequent justification and denials.

As Casimir wrote in *Culture and the Changing Environment*, "the causes of environmental problems… are related to culture-specific human behaviors, values, norms, needs and wants."[14] More so, our reactions to them are based in culture rather than unconscious repression or biological reactions. This is important to the sustainability movement, because it is human constructed ideas, values and behaviors that got us into situations of unsustainability such as anthropogenic global warming, then humans can reconstruct (change) their cultural values and behaviors to get us out.

---

[14] Casimir 2008, 3.

The Culture of Denial as a Reaction to Threats upon the Dominant Western Ideology

Nevertheless, the question remains as to why a Culture of Denial about global warming is emerging at this time? It is argued here that the growing Culture of Denial about global warming results from threats to a dominant Western ideology. Ideology – "the body of doctrine, myth, belief, etc., that guides an individual, social movement, institution, class, or large group"[15] – is itself a culturally constructed value system or worldview – a basic set of shared and cohesive values, ideas and associated behaviors that constructs and legitimizes a perception or understanding of the world and one's or one's culture's place within it.[16] This dominant Western or Northern, as it is often referred to in recent literature, ideology is a cultural value system that is promoted and legitimized by the power structures of the industrialized nations of North America and Europe, and of the nations of China, Russia, and India. It is important in this context to note that these nations are the leading producers of greenhouse gasses and thus the primary causing agents of global warming.[17] Further, many scholars[18] have criticized the Western world for not only promoting and legitimizing unsustainable practices within their own nations but also for the globalization of this worldview in, as Bowers stated, "a messianic drive to spread progress around the world."[19]

Following the work of Bowers,[20] the cultural values of this dominant Western ideology can be identified as: 1) the individual is the basic social unit, 2) an anthropocentric view of the world (i.e. constructed knowledge and values derive from a human perspective and the fulfillment of human needs), 3) the separation between human (secular) and the spiritual (sacred), 4) social or culture change is understood in terms of "progress" which is achieved through economic and technological "development," 5) the past as banality and backwardness, and thus an inhibitor to progress, and 6) machines (technology) "serve as the analog for understanding life processes."[21] "Progress" and "development" are placed in quotes to point out that this worldview presupposes that change should be understood as advancement and in terms of an ever growing economic system based upon consumption and the commodification of nature. However, as others have demonstrated,[22] the Earth will not be able to sustain a global economy with the consumption rates of the Western world, and the commodification of nature has resulted in the scarcity of natural resources and the degradation of what remains. Simply put, cultural values based upon "modern progress leads to environmental degradation."[23]

The Culture of Denial is a social reaction to perceived or real threats to the dominant Western ideology, which is being threatened by two factors. The first

---

[15] Ideology 2014.
[16] Geertz 1973; Kawagley 1995; Derby 2008.
[17] The world's 12 largest GHG emitters CBCNEWS, World, December 14, 2009.
[18] Casimir 2008; Shiva 2008; Nadeau 2006; Uhl 2004; Shiva 2003; Speth 2003; Derby 2009.
[19] Bowers 1997, 8.
[20] Bowers 1997.
[21] Bowers 1997, 8.
[22] Brown 2008; McKibben 2007; Merchant 2005.
[23] Bowers 1997, 3.

threat results from the realities of global warming and the resultant sustainability movement. Since the industrial age was powered by the burning of fossil fuels, the manner in which humans have "progressed" has been called into question by the environmental degradation resulting from global warming. As well, global warming has made humans realize that what they do to nature does matter, and therefore the actions of humans can no longer be logically separated from the effects of those acts. Further, ideas of social progress or cultural achievement through technological and economic development have been seriously called into question, because the very technologies and machines that were thought to be the means to progress and prosperity were in large part the causes of degradation of the human and natural environments. The second threat to the dominant ideology came with the crash of the global economy in 2009. The failing global economy, based on a Western capitalist economic system, has been a very real and intimate threat to the way of life for many. This economic crisis was a trigger that brought to a head the fragility of their worldview. In response to this threat, the industrialized nations have attempted to reestablish their dominance by refueling the economic and political systems that drive their ideology, thus the bailouts and reorganization of financial institutions. As a result, the societal and political reaction to these threats has been a growing denial of global warming – the poster child of the sustainability movement. The challenge to the sustainability movement, then, will be to find ways to promote the values of sustainability as a means to alleviate the real or perceived threats, and to establish itself as the solution rather than a threat.

A Culture of Denial in Action: A Case Study of Glen Lake

A case study of a small lake community in the United States demonstrates the tremendous power the cultural values of the dominant ideology have on the decision-making of the people and the negative effects their decisions have had on the natural environment in which they live. The case study herein follows from my dissertation research work on the ecological and cultural histories of this small, freshwater lake known as Glen Lake located in upstate New York, US.[24] It should also be noted that my wife and I are members of this community; we maintain a home at Glen Lake and I was president of the lake's homeowners' association from 2002-2009.

Like many lakes and ponds in North American, Glen Lake is threatened by a variety of environmental problems such as decreasing water quality due to erosion from over development of the shoreline, nutrient loading from septic leaching, fertilizer use, stormwater runoff, and the introduction of non-native, invasive species. All of these environmental threats are greatly exacerbated cultural eutrophication – the rapid acceleration of the degradation of a freshwater system due to human land and water use activities.[25]

The biggest threat to the lake's ecology in recent years has come from the introduction and unchecked growth of Eurasian watermilfoil, an invasive, non-native aquatic plant that can devastate a lake's ecology by taking over the habitat

---

[24] Derby 2008.
[25] Diet for a Small Lake: A New Yorkers Guide to Lake Management 1990.

.of native plants and depleting oxygen levels, thus interfering with the fishery and other organisms in the food chain. Additionally, left uncontrolled, milfoil can grow so dense as to inhibit recreational uses, diminish the aesthetic quality, and markedly decrease riparian owners' property values. Eurasian watermilfoil (Myriophyllum spicatum) was most likely introduced into North America from the ballast waters of ships bound from Europe and Asia in the mid-twentieth century.[26] The non-native species rapidly spread from lake to lake throughout North America by plant fragments stuck on boats and trailers, transported in fishermen bait buckets, or carried by migrating waterfowl. Once introduced, the growth and spread of this invasive plant was spurred through human land and water use patterns, specifically nutrient loading of phosphorous and nitrogen from lawn fertilizers and farm runoff, and erosion of shorelines and stormwater runoff from clear cutting and development of the watershed for cabins, homes, and tourist businesses. To combat these environmental and social threats, the lake community's homeowner's association – the Glen Lake Protective Association – considered two management approaches. The first sought to control the invasive plant by non-human means, with machines and chemicals. The second sought changes in human behaviors at and near the lake to lessen cultural eutrophication; this strategy also included managing the plants by hand pulling and minimizing sources of reintroduction. In practice, both approaches were tried.

To control the growth of the milfoil, the homeowner's association used chemical herbicides three times between 2005 and 2009 at a cost of more than $200,000. While providing seasonal relief for the summer months, this management strategy has not and likely will not sustain long-term control. To change human behaviors, the association introduced several programs: One, the association asked residents to voluntarily restrict fertilizer use on lawns within 200 feet of the lake. Two, the association wanted to set up boat cleaning and inspection stations at public launches to keep new sources of the plant (and other unwanted things) out of the lake. Three, a grant was secured to reimburse homeowners one-half of the cost of a septic system pump out. Four, the association asked for volunteers to hand pull areas of regrowth of milfoil after the herbicide application. Five, the Town was asked to institute new building and zoning regulations that would require buffers of natural plants within 20 feet of the shoreline and reduced footprints for new construction within the previously designated "critical environmental zone" within 100 feet of the water's edge.

Changing behaviors proved extremely difficult and politically frustrating. For example, even though the evidence of phosphorous loading from lawn fertilizers at lakes is overwhelming[27] and information about this was continually provided to lake residents at meetings and in monthly newsletters, few refrained from using them, including some of the officers of the lake's protective association. Similarly, less than one-third of the residents participated in the septic cleanout program. Many who refused stated that they feared that if their systems were failing they would be reported to the environmental conservation agency. This was not true and repeatedly explained to lake residents. Perhaps even more disturbing, one might ask why someone with a failing septic would not want to

[26] Myriophyllum Spicatum 2014.
[27] Phosphorous Fertilizers and Water Pollution 2014.

correct it. As well, plans for the boat cleaning stations were never completed, not enough people volunteered to do hand pulling, and although the Town was willing to implement restrictions of fertilizer uses and more stringent building codes within the critical environmental zones, the new regulations met with such resistance that no significant changes were passed.

Why would the community choose the path of chemicals and machines over a longer-term human-centered management strategy to lessen cultural eutrophication? The answer resides in the values and behaviors of the Glen Lake community; values and behaviors which are informed by an unsustainable Western worldview. In short, the people of the community do not perceive their behaviors as detrimental to the ecology of the lake, even though there is clear evidence to the contrary. Thus, as with the growing American cultural denial of global warming, the Glen Lake community has constructed a Culture of Denial in reaction to threats to their way of life.

By applying the values of unsustainability of the dominant Western ideology as shown above by Bowers,[28] this case study exhibits a clear picture of a Culture of Denial shaped by resistance to the ideological and an unsustainable worldview held by the Glen Lake community. The first cultural value of unsustainability identified by Bower of the dominant Western ideology – the individual is the basic social unit – can be demonstrated in the community by the libertarian ideology about the legal and moral sanctity of what one can do with his or her piece of private property. At Glen Lake these values are evidenced by the continued use of fertilizers in order to have manicured lawns and landscaping in one's yard, as well as by the prolongation of untended or failing septic systems. Such behaviors demonstrate self-interest and reveal a basic lack of understanding that the actions of individuals can have negative environmental and social effects beyond one's property lines.

Bower's second value of Western unsustainability – an anthropocentric view of the world – can readily be seen in the attitudes of most lake residents who believe that the lake is an object for the pleasure of the people in order to fish, boat upon, swim in, or just to look at. The lake is there for human entertainment, or a weed-free lake is important to sustain high property values. These individualized and anthropocentric values lead into Bower's third point of an unsustainable ideology – the separation between human (secular) and the spiritual (sacred). While lake residents certainly enjoy the lake, it is not understood as sacred – i.e. something regarded with great respect or reverence. The lake is not perceived as an entity on par with human needs and wants. The lake is a profane object and an economic commodity to be preserved not for itself but to achieve or sustain human, secular wants and needs.

The values and lifeways demonstrated by Glen Lakers are products of an European worldview imported with colonialism. These are demonstrated in the fourth and fifth points by Bowers: social or culture change is understood in terms of "progress" which is achieved through economic and technological "development"; and the past is understood as banality and backwardness, and thus an inhibitor to "progress." After the sanctity of private property, the most important cultural value for Glen Lake residents relates to ideas of progress in

---

[28] Bowers 1997.

terms of social, economic, and technological development. Glen Lake is the most intensely developed landscape in the Town of Queensbury. There are 288 households along 5.2 miles of shoreline, with an average property lot size of 50 x 75 feet (0.086 acres or less than 1/10th of an acre), all within a designated "critical environmental area."[29] Yet the shoreline is zoned Waterfront Residential One Acre, meaning it would require a minimum lot size of one acre to build a one family residential home today. Only one percent (3 of 288 homes) meet that zoning requirement. This is because nearly every home, most of which have been converted in recent years from small seasonal cabins to year round homes, are considered by the Town to be pre-existing, non-conforming lots. The Town allows this because Glen Lake is held up as an example of social, economic achievement and progress. It is the wealthiest neighborhood in the Town with more than $100 million of assessed, thus taxable, value. As a result, building rules and zoning regulations are routinely bent or overlooked in the name of progress or community development. These values of progress and development are driven by an essential dismissal of the past, which is viewed an antithetical to progress. The idea of looking to the past as an example of a healthier, more sustainable human-nature relationship carries little merit for the community today. Case in point, the archaeological and protohistorical evidence of indigenous populations before Europeans came upon the landscape shows that Glen Lake was a shared waterscape by as many as seven Native American cultures that used the lake for seasonal occupation for subsistence fishing and hunting. However, no single indigenous culture claimed the lake as its own, no fences or property markers were placed to section it off, and no wars were fought to claim it as part of any culture's real estate. However, by the mid-eighteenth century the lake was known to colonial British troops as French Pond, and the lake was understood as a marker for a borderline dispute between the French and English during the war to establish it as colonial property. Further, once settled as a part of the English colony of New York, the very first act was to survey the landscape by rod and chain and split it up into 102 parcels of private property to be developed by pioneer settlers, who were given the land to imprint upon it their cultural ideology. This culturally constructed idea of private property is one of the hallmark ideological differences between indigenous and colonial populations in the Americas.[30] Along with the idea of nature as private property came imported European cultural values of the objects from nature as commodities for human use and the dominion of human over nature. It is this imposition of a radically different cultural ideology regarding human's relationship with nature that we see in the justification of the acts and beliefs of the residents of Glen Lake today.

Bowers' final aspect of a dominant, unsustainable ideology – understanding machines as an analogy of life processes – is readily observed in the choices of lake management. The health of Glen Lake, and most lakes in North America, is managed as the majority of people in the United States manage their healthcare, with drugs and machines. The lake itself is not perceived as an entity that is alive, although the things that live in it are alive and generally considered a nuisance, but as a machine that, like the human body (also understood as a machine),

---

[29] Glen Lake Watershed Management Plan 1998.
[30] Merchant 2002; Cronon 2003.

should be made or kept healthy with drugs – at the lake that is with herbicides to kill the problem – or with machines such as a weed harvesting, benthic barriers, and dredging. The use of chemicals to control nuisance plants and the large amounts of money spent to do this demonstrate the community's ready acceptance of technology as the appropriate approach to manage the lake's health.

Thus, the people of Glen Lake do not wish to see that their behaviors negatively affect the ecology of the lake. Individuals, by and large, do not believe that their fertilizers do much harm, or that their failing septic system could create enough pollution to be a detriment to the water quality. And as with the global warming issue, most people of the Glen Lake community choose to believe that even if their activities are harming the lake, new technologies or better drugs and machines will be in place in time to remedy them. The message is clear enough: I should not have to change my behaviors, because my behaviors are not causing significant or lasting harm. The same faulty reasoning is used to deny global warming; thus the emergence of a Culture of Denial of the ideas and behaviors that are contradictory to the worldview of the people of the Glen Lake community.

Conclusion: Ending the Culture of Denial and Changing the Ideology of Unsustainability

This paper has shown that many people of the United States are in denial about the negative effects their behaviors and attitudes have upon the natural environment. It is also argued that this Culture of Denial results from perceived or real threats to the dominant Western ideology caused by environmental problems such as global warming and the current economic crisis.

To effectively combat these attitudes will require people of the United States to take an honest and critical look at their ideological values and associated behaviors. It is our job as advocates for sustainability to make them take this look at themselves and their worldview. Rather than turn a deaf ear to the rhetoric of denial, more of us need to stop contributing to the Culture of Denial with our silence. While it may be uncomfortable to confront others, especially in casual situations, to avoid it out of courtesy or tolerance is contributing to it. As well, it our obligation to become better educated about the scientific evidence of global warming and other issues in order to challenge the deniers and better teach about the risks denial can have on the natural environment and our very survival. Also, we need to be able to articulate the inconsistencies in the dominant Western ideology that promotes and legitimizes an ideology of unsustainability.

Finally, it seems to me that the sustainability movement has lost momentum over the last few years. While this in part is due to problems associated with the poor global economy, the movement is also is being eroded by multinational corporations who are co-opting the movement by portraying themselves as green, responsible global citizens, when in actuality few are doing very little. Again, it is our job to challenge these marketing ploys and teach about the realities of corporate persuasion. Further, we need to keep the issues in front of the public and hold corporations and governments responsible for misinformation and misdirected marketing. As well, we need to re-energize our movement with fresh ideas and approaches. The Inconvenient Truth[31] that once drove the movement has been too easily replaced with convenient denial. To change this, we need to recognize that sustainability must be a global social, economic, cultural, political, and environmental movement; we need to bring diverse voices from diverse cultures into the global sustainability dialogue; we need to provide examples of cultures and ideologies that do promote values of sustainability, and, in the end, we need to realize that the global sustainability movement is one of global culture change. As Robert Uhl reminds us:

> Culture is not static. It changes and evolves as a result of the things that human beings do – the ways that we behave, the stories that we tell, the conversations that we have, and the values we espouse during our lifetimes. Each day, humans participate in the process of making culture. In this vein, the efforts throughout the planet to give birth to a new story – one grounded in sustainability – can be viewed as a grand social-change movement.[32]

References

Bowers, Chet A. 1997. *The Culture of Denial: Why the Environmental Movement Needs a Strategy for Reforming Universities and Public Schools.* Albany, New York: SUNY Press.

Brown, Lester R. 2008. *Plan B3.0: Mobilizing to Save Civilization.* New York: W.W. Norton and Company.

Casimir, Michael J. 2008. The Mutual Dynamics of Cultural and Environmental Change: An Introductory Essay. In *Culture and the Changing Environment: Uncertainty, Cognition and Risk Management in Cross-cultural Perspective,* edited by Michael J. Casimir, 1-58. New York: Berghahn Books.

CATO Climate Change Ad. 2008. With all due respect Mr. President, that is not true. Letter to the President (US) dated November 19, 2008. Available at http://www.sustainableoregon.com/Cato%20Climate%20Change%20Ad .pdf, accessed March 29, 2014.

Cronon, William. 2003. *Changes in the Land: Indians, Colonists, and the Ecology of New England.* New York: Hill and Wang.

Defense Mechanisms. 2014. *Encyclopedia Britannica* Online. Available at http://www.britannica.com/EBchecked/topic/155704/defense-mechanism, accessed March 15, 2014.

Derby, Paul. 2009. The Role of Anthropology in the Global Sustainability Movement: Maintaining the Small While Transforming the Large. *International Journal of Environmental, Cultural, Economic and Social Sustainability* 5(3): 209-216.

Derby, Paul. 2008. *Indian Trails, Military Roads, and Waterwheels: Cultural and Ecological Transformations at Glen Lake, New York,* Doctoral

---

[31] Gore 2006.
[32] Uhl 2004, 255.

Dissertation, Syracuse University, ProQuest Publishing, UMI Dissertations Publishing. UMI number 3323048.

Diet for a Small Lake: A New Yorker's Guide to Lake Management. 1990. New York: New York State Department of Environmental Conservation, and the Federation of Lake Associations, Inc.

Fewer Americans See Solid Evidence of Global Warming. 2009. The Pew Resource Center for People and the Press, October 22, 2009. Available at http://www.people-press.org/2009/10/22/fewer-americans-see-solid-evidence-of-global-warming/, accessed March 15, 2014.

Frost, Mark. 2009. The Inside Scoop: The Global Warming Debate. *The Chronicle Newspaper* 30 (1314): 7.

Geertz, Clifford. 1973. *The Interpretation of Cultures*. New York: Basic Books.

Glen Lake Watershed Management Plan. 1998. Unpublished manuscript, Glen Lake Technical Advisory Committee, New York.

Gore, Al. 2006. *An Inconvenient Truth*. London: Bloomsbury

Hall, Alyssa. 2009. Big Drop in Those Who Believe that Global Warming is Coming. The Harris Poll Press Release, December 2, 2009. Available at http://www.harrisinteractive.com/vault/Harris-Interactive-Poll-Research-Global-Warming-2009-12.pdf, accessed March 15, 2014.

Heffner, Dr. Christopher L. 2002. Freud's Ego Defense Mechanism AllPsych Online, Heffner Media Group, August 21. Available at http://allpsych.com/personalitysynopsis/index.html, accessed March 15, 2014.

Ideology. 2014. Dictionary.com. Available at http://dictionary.reference.com/browse/ideology, accessed March 15, 2014.

IPCC. 2007. Summary for Policymakers. In *Climate Change 2007: The Physical Science Basis. Contribution of Working Group 1 to the Fourth Assessment Report of the Intergovernmental Panel on Climate Change*, edited by S. Solomon, D. Qin, M. Manning, Z. Chen, M. Marquis, K.B. Averyt, M.Tignor and H.L. Miller. New York, NY: Cambridge University Press.

Kawagley, A. Oscar. 1995. *A Yupiaq Worldview: A Pathway to Ecology and Spirit*. Prospect Heights, Il: Waveland Press, Inc.

McKibben, Bill. 2007. *Deep Economy: The Wealth of Community and the Durable Future*. New York: Henry Holt and Company, Inc.

Merchant, Carolyn. 2005. *Radical Ecology: The Search for a Livable World*. New York: Routledge.

Merchant, Carolyn. 2002. *The Columbia Guide to American Environmental History*. New York: Columbia University Press.

Myriophyllum Spicatum. 2014. Center for Aquatic and Invasive Plants, University of Florida IFAS. Available at http://plants.ifas.ufl.edu/node/278, accessed March 15, 2014.

Nadeau, Robert L. 2006. *The Environmental Endgame: Mainstream Economics, Ecological Disaster, and Human Survival*. New Brunswick, New Jersey: Rutgers Press.

Not all scientists buy into global warming. 2009. *The Chronicle* 30(1315): 1, 17.

Phosphorous Fertilizers and Water Pollution. 2014. PlantTalk™ published by Colorado State University Extension, Denver Botanic Gardens, and Green Industries of Colorado, Inc. Available at http://www.ext.colostate.edu/PTLK/1620.html, accessed March 15, 2014.

Shiva, Vandana. 2003. The Myths of Globalization Exposed: Advancing toward Living Democracy. In *Worlds Apart: Globalization and the Environment* (Yale School of Forestry and Environmental Studies), edited by James Gustave Speth, 141-154. Washington: Island Books.

Shiva, Vandana. 2008. Working Together for Sustainability On Campus and Beyond. Keynote Lecture at 2nd Biennial Conference of the Association for the Advancement of Sustainability in Higher Education, November 11.

Speth, James Gustave, ed. 2003. *Worlds Apart: Globalization and the Environment* (Yale School of Forestry and Environmental Studies). Washington: Island Press.

The Carbon Bathtub. 2009. *National Geographic* December: 26-29.

Uhl, Christopher. 2004. *Developing Ecological Consciousness: Path to a Sustainable World*. Lanham: Rowman and Littlefield Publishers, Inc.

Chapter 3: The Contributory Components of Viable Eco-diverse Landscapes

Alan Derbyshire

Introduction

It is estimated that population growth within urbanity will rise exponentially throughout the present century.[1] If current migration trends from rural to urban areas continue urbanity will become the primary habitat for humans. The implications for sustainable development within urbanity is obvious; the global ecological footprint for cities is disproportionately large in relation to their share of land cover. This reality frames the broader narrative of the sustainable city and the context for the rationalization of ecologically diverse urban landscapes as consequential to sustainable city living.

Exponential rise in United Kingdom city center living is a relatively new phenomenon. Regional cities such as Manchester, Birmingham, and Leeds have undergone significant redevelopment and revitalization over the past decade. Traditional industrial infrastructures within these cities have been replaced or refurbished by architectural structures that "speak" of aspiration and status. The dramatic increase in dwellings within metropolitan centers reflects the commercial pragmatism of developers, but also the perception amongst governmental stakeholders and agencies that higher density living is a more sustainable recourse for urban habitation.

This study reviews the role of the urban landscape against the backdrop of this significant increase in population and considers the function of the urban landscape in engendering emotional, cultural and social well-being. Accordingly it assesses the issues connected with sustainable development of urban landscapes and considers four critical subject areas.

The issue of increasing urban population densities and how the establishment of "placeness" contributes to the emotional and cultural well-being of increasing human thresholds is fundamental to sustainability. The association of landscapes and existential notions such comfort, prosperity and emotional security are well

---

[1] United Nations 2004, 4.

documented and closely associated with the concept of place. Placeness is an abstract conceptual foundational element in the structure that is the landscape. The built environment design professions play a key role in establishing these material attachments to the places we frequent on the way to work, home and recreation areas. These shared spaces can lift or depress the spirit, make us feel closer to or detached from nature; in other words they are central to the habitability, and consequently the sustainability, of urbanity.

Ecological diversity within the urban landscape as well as being a factor in energy conservation and reducing carbon emissions, is fundamental to establishing the notion of placeness. How this is effectively and creatively implemented is a source of dispute within the built environment professions. The case for a closer relationship between design and science is examined as a potential model for a "new paradigm" in designed ecological applications within the urban landscape.

Contemporary methods of implementing ecologically diverse urban landscapes have reflected a reliance on formal, modernist and post-modernist design methodologies amongst designers. This approach has arguably contributed to what have been described as "corporate blandscapes."[2] Non-standard strategies for a more sustainable employment of urban landscaping processes could be characterized as radical. However, they offer a more nature inspired approach to design, a more sustainable natural processes driven approach to urban ecological development.

The role of universities and design pedagogy in determining the direction of sustainable aware design graduates is considered and discussed. Progress has been made in the tuition of sustainable design principles. However the prospect of a viable trans-disciplinary approach to sustainable urbanity is hampered by the perceived failure to impart a more holistic understanding of sustainable development amongst design graduates. If ecological diversity is to be creatively implemented it is arguably increasingly necessary to have a deeper understanding of scientific disciplines.

The Significance of Place to Higher Population Densities

Urban communities are primarily identified by a high concentration of people and activities within a space commensurate with the surrounding regions. The compact nature of human interaction and dwelling within cities is associated with high levels of culture and civilized expressions of social behavior.[3] The arguments for high-density urban living and relationship to sustainable objectives are well established and advocated in many UK land use planning policy and guidance documents. For example, 60 percent of all new development in the UK is destined for brown field re-use in urban areas, transport and housing policy guidance also strongly advocates intensifying developments.[4]

---

[2] Osmond 2002, 99.
[3] Lozano 1990, 157.
[4] Williams 2000, 30.

Much of the debate relating to the sustainability of urbanity focuses on energy efficiency and transport or energy efficiency and urban form.[5] However, the concentration on energy efficiency is unlikely to stimulate a wider understanding of the notions of sustainable urbanity or act as a catalyst for the development of guidelines for designers, planners and governmental agencies.[6] Energy issues need to be balanced against socio-economic and environmental objectives.[7] The human component of the urban framework and the collective response to higher densities is a principle element in the objective of realistic sustainability. To generate the human interactions that make urban activities and functions viable, it is necessary to have certain densities or thresholds of people in a given area. Therefore a greater number of people, it is argued, manifestly increase the potential for the number and variety of activities and consequently contributes to the richness of a community. According to Lozano: "Urbanity is based on density."[8]

The dramatic increase in city center living in the UK has been relatively sudden. The population in Manchester city center in 1990 was approximately 1,300 rising to an estimated 20,000 in 2010.[9] The significant increase in city dwelling is a marker for sustainable urbanity in terms of population density and management of energy efficiency. However, it is also a source of concern when the cultural and emotional well-being of the community is considered.

Landscapes play an important part in establishing and reflecting cultural values, and as such are influential in creating attachments to local environments. With increasing population densities and cultural diversity within cities, the question of how urban landscapes contribute to the livability of urbanity become more germane. Creating a sense of ownership and attachment to these shared spaces is a marker to their sustainability. In the absence of this sense of attachment or guardianship people are less inclined to care about or look after shared public places.[10] The sense of ownership of public places and spaces is fundamental to their sustainability; places need to be cared for in order to be maintained. Creating a sense of place is an essential element in the process of constructing a sense of community and belonging.

The concepts and design rationale relating to the establishment of place is the subject of much debate. The creation of a sense of place is a principle objective within the built environment design professions when attempting to re-negotiate the character of a site. Places are perceived to include psychological, social and human activities rooted within a physical setting.[11] Placeness in its broadest sense is the collection of symbolic meanings and the collective or individual attachment with a spatial setting. We attribute meaning to landscapes and consequently become attached to the meanings according to Lynch.[12] The more established approach to developing conceptual perceptions of place within the built

---

[5] Frey 1999, 37.
[6] Breheny 1992, 83.
[7] Frey 1999.
[8] Lozano 1990, 316.
[8] Manchester.gov 2009.
[10] Nassaur 1997, 75; Nassuar 1995.
[11] Brandenburg and Carroll, 1995; Relph 1997, 214.
[12] Lynch 1960, 113.

environment design professions is rooted in the phenomenological tradition, which is less open to positivistic hypothesis testing and interpretation, [13] and therefore by its nature hard to define.

The origins of the perception of places as phenomena of experiences are well documented. [14] It is Christian Norberg-Schulz's writings that have directly informed place related design strategies within the design professions. Schulz draws on works such as Martin Heidegger's (1971) seminal essay "Building, Dwelling, Thinking" and Edmund Husserl's (1936) book *The Crisis of European Sciences and Transcendental Phenomenology* to urge a return to place-based design. Norberg-Schulz reasons that the universality of modernism militates against a more accessible or popular meaning and the communicative role of design is being lost. To him the designer should strive to "concretize" the physical character and essence of places and make visual references to their differences. The celebration of this difference is what gives individual places their character and this correspondingly cultivates notions of ownership and care. [15]

Environmental phenomenology as such now provides a theoretical context for concepts such as materiality, texture and sensory experience. The character and placeness of the urban landscape is in effect fabricated from the configuration of material, ecological and spatial elements. The origins of these components are central to establishing a sense of place, but are also fundamental elements in the establishment of sustainable urbanity. The question of how the designed application of diverse ecologies in particular is successfully accomplished is one more ingredient in the "concretization" of place. The increase in urban densities underlines the fundamental requirement of places within urbanity. If the city is to be more consequential than a functional, energy efficient metropolis, in effect a more livable environment for humanity, the establishment of place is vital.

Urban Form, Cultural Attitudes

The process of establishing culturally vibrant and livable urban environments, in particular the issues associated with the implementation of policy relating to ecological principles within urban communities, is far from straightforward. Top down "expert driven" sustainable indicators and policy that ignore communal experience and networks help contribute to unintended outcomes, not least of all the alienation of local communities from design and planning processes. The friction between "expert" and community driven models of sustainable urban indicators is well documented and in general it is recognized that integrating the two approaches could lead to more productive outcomes. [16] However, establishing urban design models that both recognize and incorporate scientific/environmental theory and policy and also embrace local vernacular traditions can be problematic. Little is known about the practicalities involved with integrating cultural expertise with established expert led sustainable initiatives, or if "standard" sustainable practices even chime with local values and understandings of sustainability. [17]

---

[12] Stedman 2002.

[14] Husserl 1936; Heidegger 1971; Norberg-Schulz 1980; Relph 1997.

[13] Husserl 1936, 355; Heidegger 1971, 11; Norberg-Schulz 1980, 18.

[14] Reed et al. 2006.

[15] Turcu 2012.

Accordingly, central to establishing the notion of urban resilience is the ability to include communal participation in on-going debates involved with urban development. Not only is this good for creating genuine understanding of ecological principles at various rates of scale, but also as a foundational element in alleviating the homogenizing nature of globalism, as distinctive cultural perspectives are fed into design and implementation processes.

The perception of identity and culture as markers of sustainable health has largely been neglected by established theoretical benchmarks of sustainable development. However, a distinctive identity is central to the long-term habitability and cultural evolution of urban environments. For example, utilizing local materials is a benchmark of sustainable practice; their embodied energy costs are lower. Nevertheless, perhaps of greater significance is that local materials are more recognizable by indigenous communities and as a result help to embed emotional attachments to the places they are employed. As discussed in the previous section of this study, the links between ownership, conservation and preservation are essential features in establishing a sense of place and community. The rationale for a more "culturally constructed space" is proposed by Seghezzo, arguing that the concept of place should be a more prominent element of the sustainability paradigm.[18]

The relationship between cultural identity and rapidly evolving global and local influences over urban spatial characteristics is a potential source of discord or opportunity. The challenge for stakeholders is to conceptualize regional identity and expression as a bold and inclusive cultural process. The connection between cultural conventions and regional environments lie at the heart of vernacular architectural theory and philosophical reasoning. Vernacular architecture, it can be argued, addresses many of the associated issues of cultural and societal adjustment of the landscape in response to evolving economic and environmental factors. The dynamic nature of vernacular forms and how such forms occur are largely determined by local and external factors. It is this ability to respond to vigorous change in an incremental manner that suggests a vernacular response to architectural and spatial development is integral to the objective of preserving cultural identities.

Developers are eager to promote and market the sustainable and cultural character of their development projects based on claims of vernacular heritage. However, more often than not this amounts to nothing more than flirting with tradition and for most part is a superficial imitation of traditional forms based on commercial factors.[19] Connecting the marketing of commercial developments with misleading claims of vernacular heritage is standard green-washing practice, but more importantly is an obstacle to the desire of communities to retain their own identities. Vernacular traditions should not be associated with anachronistic notions of vanishing cultures and societies, but is an essential feature of culturally sympathetic sustainable architecture.[20] It could also be argued that all urban form needs to be vernacular in order to be sustainable, as historically vernacular traditions have responded to climate, nature, indigenous materials and culture.[21]

---

[16] Seghezzo 2009.
[17] Vuksanovic 2006.
[20] Oliver 2003, 14.
[19] Rau and Schierle 1995.

In essence, what the debate into vernacular form as a marker for sustainability highlights is the need to fully account for cultural variables in urban design frameworks. There is broad agreement that local communities should play a greater participatory role in establishing sustainable indicators, though in practice this generally remains an anomaly.

The Case for Ecological Urban Sustainability and the Design of Nature

The arguments for more imaginative designed applications of ecology within urbanized areas are not new. The notion that science can be the principle source for the design of landscapes was regarded as a radical innovation in the 1960s and 1970s, but is now viewed as a model for the sustainable landscape. Integrating landscape ecology principles within the urban landscape (i.e. combining urban morphology with ecological functioning) is not a straightforward undertaking. Assimilating ecological principles with architecture, planning and design in order to achieve urban sustainability is still in the early stages and as many challenges as opportunities are expected.[22]

The underlying rationale for this paradigm change centers on the perceived failures of the formal designed application of ecology and the need for more profound response to the issues surrounding population growth. For instance, urbanized areas cover about 2% of the Earth's land surface, but account for 78% of carbon emissions, 60% of residential water use and 76% of wood used for industrial purposes.[23] Moreover, altering land cover patterns and therefore surface radiation regimes and energy balance can affect local and regional climatic conditions. Impervious surfaces within cities can significantly alter the flow paths of surface water, potentially leading to contamination through the overloading of sewerage systems.[24] Urbanization often leads to the fragmentation and loss of natural habitats as well as the introduction of exotic species.[25]

With the increase population growth within urban centers these problems are estimated to "accelerate explosively"; it is against this that the need for the increase in ecology within urban centers should be gauged. The scientific justification for eco-diversity as a basis for sustainability within urbanity is well researched and persuasive, particularly the rationale for landscape ecology. Roadside plants, for example, help reduce carbon monoxide levels, and wall-climbing vines can reduce summer temperatures on a street by 5 percent. The emerging field of sustainability science provides a scientific basis for sustainable development and also focuses on the dynamic relationship between nature and society. Its place-based and solution-driven approach to complex regional and global environmental, social and economic issues provide a basis for a trans-disciplinary inspired process in establishing designed urban landscapes.[26]

---

[22] Ahern 2005, 30.

[23] Brown 2001, 188.

[22] Wu 2008.

[23] Hope et al. 2003.

[24] Wu, 2008; Wu and Hobbs, 2002; Hough, 1984; Kates, et al. 2001; Parris and Kates, 2003; Clark, Dickson; 2003.

Within the field of landscape architecture, environmentalism has been a source of division. For instance, various polarized rhetorical positions have been expressed on the preservation, ecological and integrative nature of environmentalism. Nonetheless, in reality individual landscape architects express numerous perspectives on environmental issues and for the most part the professional practice does not support the rhetoric.[27] The dominant ongoing challenge to the designer is the mediation of practices between the disparate dialogues relating to "art" and "environmentalism." One of the more fundamental issues connected with the "debate" is the disconnection between site analysis and design expression or, in other words, between environmental values and form generation, from the perspective of the Landscape Architect, according to Meyer.[28] Biologist Robert Cook reasons that a "new paradigm" has emerged due to the dynamic change in the underlying assumptions that support an understanding of the natural world.[29] Cook argues that a better understanding of ecology helps the landscape designer in two ways:

1. Design projects usually involve intervention and rearrangement of the land, a biological understanding of the consequences would help predict and control the outcome of the intervention.
2. The narrative of the ecology and feelings provoked by an ecological perspective could serve as an aesthetic challenge, and therefore as an inspirational approach, to the design process.

The process of adopting the perceived paradigm shift necessary to re-negotiating sustainable ecological principles to the urban landscape is at the heart of the matter. The task of challenging what landscape designer Michael Hough refers to as "the fixed mould of aesthetic convention" is not straightforward.[30] Hough argues that the tradition of adapting historic "artistic philosophies" to urban landscape settings ignores the "working vernacular landscape of town and country." The landscapes of nature, with their origins in poverty and necessity hold significant lessons in the search for a more sustainable urban form. For example, according to Hough, the natural urban plants of forgotten places within the city such as cracks in the pavement, walls, rooftops, or wherever a foothold can be gained. This natural process driven ecological approach to landscaping urbanity runs counter to well-established pedigreed landscapes of mown turf and flowerbeds, and ultimately frames the dilemma of the "new paradigm" approach to urban landscaping.

Design Practice and Alternative Perspectives

The creation of the urban landscape is a trans-disciplinary activity, defined by the contributions of architects, urban designers, landscape designers and other associated (governmental and commercial) stakeholders. Over the course of the

---

[27] Nadenicek and Hastings, 2000, 133.
[28] Meyer 2000, 187.
[29] Cook 2000, 115.
[30] Hough 1995, 9.

last half-century, the substantial development of urban spaces has been determined by the influences of modernist and postmodernist design philosophies. Modernism responded to the challenge of creating social order within mass societies through notions of structural honesty and functionality. Modernism is also viewed as the antithesis of place-based design theory and is more concerned with the architectural object than the requirements of the site.[31] Post modernism, by contrast sought to address the wider needs other than the rigid adherence to form and function. In other words, there was an attempt to address the relationship between human experience and architecture, a regard for architecture within the context of society.[32]

Postmodernism advocated a revitalization of vernacular architecture that responded to social, economic and functional circumstances, the connection with the local, engendered placeness. Whilst this view of postmodernism strengthened the notional connection with humanity and the desire to avoid placelessness and stifling minimalism, postmodernism gave rise to equally impassioned analysis. Some looked beyond the focus on aesthetics and corresponding notions of playfulness and superficiality and observe something more sinister than a stylistic fad. Whereas modernism intended to shock its audience by using new materials and breaking with the past, postmodernism uses familiar and borrowed elements from older styles such as arches, columns and pilasters, ergo more recognizable and appealing to consumers. Given this populist disposition it is reasonable to assume that postmodern design lacking contextualization may promote opportunities for profit making and consumption. Frederic Jameson contends that postmodernism represented "the cultural logic of late capitalism."[33] In addition to this, despite its attempt to counter negative aspects of modernism, postmodern urbanism's preoccupation with irony and surface treatment make it correspondingly guilty of neglecting the human element.[34] Urban design has often reflected this dichotomy and the lack of clearly defined contextualization within the urban landscape. The placelesness of the postmodern shopping complex is indistinguishable from the austerity of modernist functionality. Recent applications of ecology within new "re-vitalized" developments within UK cities such as Manchester do nothing to dispel this contention. The designed applications of ecology within these developments owe their origins to the utilitarian modernist tradition. Planting is regimented: mown grass in boxes as depicted in Figure 3.1 and trees planted in rows as depicted in Figure 3.2. There is little or no interaction or connection with the surrounding architecture, or pretensions of creating a sense of care or ownership from their inhabitants.

---

[31] Ellin 1996, 155.
[30] Huxtable 1981.
[33] Jameson 1991, 239.
[32] Ellin, 1996.

Figure 3.1: Landscaped Area; Spinningfields, Manchester, UK[35]

Figure 3.2: Regimented Planting; Spinningfields, Manchester, UK[36]

Champions of a more nature-driven approach to landscape design would reason that the presumed relationship between the origins of vernacular architecture, and place-based design, could be extrapolated to urban ecological applications. Both, the built vernacular and the fortuitous landscape have evolved in response to minimum input from authority. Hough argues that purposeful design has done more to generate placelessness than to promote a sense of place.[37]

There is no distinction in Hough's thesis between modernism and postmodernism. The creation of place requires considered intent from the designer. The sustainability of designed placeness is dependent on its regional identity, in that it is connected to local ecological values and principles. This has

---

[35] Photograph by the author.
[36] Photograph by the author.
[37] Hough 1995, 94.

obvious implications for all of the built environment professions and their desire for viable sustainable urban landscapes. Architects for example are designing and employing ecological features such as living walls and green roofs within buildings, both as energy conservation mechanisms and as a perceptible connection with the landscape. It is the role of the landscape in influencing the holistic anatomy of the urban landscape that is fundamentally characterized within the emergent notion of "landscape urbanism".

The origins of the landscape urbanism "agenda" are founded in the recognition that the landscape can act as a model for urbanism. As Corner has observed, the heightened awareness of environmentalism and global concerns with ecology and sustainability has given rise to the reappearance of the landscape within the larger cultural imagination.[38] Constituents within the built environment design professions such as architects, designers, urban designers and planners are beginning to move to a shared form of hybrid practice where the landscape is a formative element. This is a radical departure from the "established" view of nature as being detached from the city. He also reasons that the notion of the ecology of the city and the cultural, political and social implications associated with this symmetrical existence with the natural world, are yet to be fully understood. According to Corner it is landscape urbanism's promise that "the development of a space time ecology that treats all forces and agents working in the urban field and considers them as continuous networks of inter-relationships."[39]

Academics such as Alan Berger[40] argue that the emergence of landscape urbanism is a reaction to the polarizing arguments of pro- and anti-urbanization ideologies. Berger reasons that the unorthodox approach to landscape study and practice advocated by landscape urbanism is not exclusive of contemporary design disciplines. A more radical approach can be formulated through existing knowledge structures, such as rethinking the nature of place within small patches of urban spaces of what he terms "drosscape." Drosscapes are the de-industrialized (or post-industrial) areas within cities, the horizontal spaces separating environment form architecture. Berger attaches no value system to these spaces; they are neither bad nor good, but in need of new conceptualization. The challenge for the urban design professions is to channel existing knowledge structures in creating a radical agenda for the renegotiation of placeness within such spaces.

One of the more progressive and creative examples of the more direct interaction between ecology and the urban landscape is the "High Line" in New York City. Originally an iron clad raised freight train track built 30ft. above the street in the 1930s until it's disuse in 1980, the "High Line" has been transformed into a public park. Within the park rail tracks, fortuitous ecologies and native grasses harmoniously co-exist with engineered sections and designed paving systems (Figures 3.3 and 3.4 respectively). The perceived radical complexion of the development and construction was embraced by the city and the risk appears to have paid off.[41] For the first time in recent history crickets can be heard in

---

[38] Corner 2006, 23.
[39] Corner 2006, 23.
[40] Berger 2006, 236.
[41] Martin 2009, 38.

lower Manhattan, the public are responding positively, nature is no longer "far away." The "High Line" can be viewed as a potential benchmark for future ecologically diverse urban renewal developments, a genuine contribution to the realization of the "new paradigm" of eco-diverse urban landscapes.

Figure 3.3: Paved Landscaped Area; "High Line" in New York City[42]

Figure 3.4: Detail of Landscaping; "High Line" in New York City [43]

---

[42] Photograph by the Author.
[43] Photograph by the Author.

Landscape urbanism attempts to address the seemingly inextricable global condition of revitalizing urbanity with sustainable, ecological and place-based initiatives; radical perhaps, but arguably a more considered sustainable alternative vision of urbanity.

Design, Ecology, and Education

Higher education is one of the largest industries in the world and the specialists within most industries were educated at university, it is the mother of all industries.[44] Naturally, it has a critical role to play in the development of design professionals and their awareness of sustainable matters. Universities are therefore key to the training of environmentally informed architects, designers, urban designers and planners. However, the pedagogical strategy for implementing sustainable principles within course curricula has been constrained by the absence of a shared understanding of notional sustainable principles and objectives. Many academics contend that the concept of sustainability is still evolving and there is as yet no single framework or conceptualization of sustainability or sustainable development. [45] Nevertheless, there are more proactive signals being advanced in the form of commitments from university presidents and vice chancellors to the principles of sustainable development.[46] If these commitments are to be more meaningful than rhetoric, and more of a formative characteristic of design pedagogy, then academics need to demonstrate a willingness to facilitate viable sustainable principles within course structures.

Currently, sustainability as a concept can be broadly agreed on an abstract level amongst academics, but when this is translated into practical application within course curricula, support may dwindle. If this conceptual ambiguity is a hallmark of academic attitudes and commitments it is perhaps of little wonder that students perceive sustainability as primarily a narrow environmental issue. More nuanced associations between the environment, diversity, participation and inclusion are rarely made.[47] These are essential features in the assimilation of sustainable development within urbanity, particularly in the establishment of placeness. The fundamental understanding and interconnection between notions of place, diversity and inclusion are innately incidental to sustainable urbanity and as such should be considered within design processes.

The environmentalist and academic David Orr advocates the embedding of holistic ecological/sustainable principles at a preliminary level within educational systems. He argues for the integration of the study of natural systems within education from primary/kindergarten level to PhD and for an educational system that values ecological intelligence:

> To teach economics, for example, without reference to the laws of thermodynamics or ecology is to teach a fundamentally important

---

[44] McGonigle and Starke 2006.
[45] Hopkins and McKeown 2002; Sauve 1996; Selby 2006.
[46] Wright 2004, 7; Tilbury 2004, 97.
[47] Gray-Donald et al. 2007.

ecological lesson: that physics and ecology have nothing to do with the economy. It just happens to be dead wrong.[48]

The demarcation between core subject areas and the contextualization of ecological and sustainable concepts within design pedagogies and subject curricula is central to the matter. Whereas the teachings of place-based design philosophies are integral to the built environment design pedagogies, the ecological and sustainable characteristics of place are less so. The boundary between design and the science of ecological applications within design curricula needs to be more intimate. If future design professionals are to fully engage in the development of sustainable urbanity, the reality is that they will need to work effectively with ecologists and sustainability scientists. The promotion of a trans-disciplinary approach to the convergence of "science" and "art", ecology and humanity is vital in providing a more comprehensive understanding of urban sustainability.[49]

In order for design graduates to be sensitive to the ecological narrative and to view this as an effective addition to the design process, there needs to be a structural acknowledgement and contextualization of biological/ecological matters. The "new paradigm" objective is dependent on the commitment of designers to demonstrate more than a rhetorical appreciation of the significance of the ecological landscape. University design schools have a central role in framing the issues and contextualizing the structure of science within urban design pedagogy. Good design is sustainable, designed nature is fundamental to urban sustainability. The issues facing urban landscapes have their origins and solutions in university design departments.

Conclusion

As population densities grow within cities, the capacity of the landscape to inspire and connect humanity with nature will act as a fundamental element in the livability of urbanity. Cities have been viewed as a major source of environmental problems, but conversely they have a major role to play as an avenue to regional and global sustainability. How this is achieved is ultimately down to the design philosophies and commitment of architects, designers, planners, and associated stakeholders.

Urban sustainability requires active participation from humans. The interconnection of city dwellers and the urban landscape is part of the broader process of achieving genuine sustainability. Built environment design professionals also live or work in cities, their stake in the habitability of urbanity is as distinct as any other inhabitant. As a species humans are natural "ecosystems engineers"; it is in our nature to create and maintain our habitats.[50] As stakeholders of urbanity we have a fundamental choice to make. Do we continue to ignore or eradicate natural ecosystems replacing them with concrete, bricks and unfamiliar ecologies, replacing "biological complexity with artificial regularity."

---

[48] Orr 1994, 12.
[49] Wu 2008.
[50] Jones et al. 1994.

or take an alternative route forward? If the paradigm shift necessary for the realignment from the actuality of "ecology in cities" to the notional "ecology of cities" is to materialize, then rhetorical expressions of environmental principles need to be substituted with practical applications.[51]

New paradigm actualization is doable. Ecologists are beginning to view urbanity as a source of legitimate study and the perception that ecologists are hostile to architects, developers and planners is no longer accurate. It is argued that a trans-disciplinary approach to the development of urban landscapes is an essential prerequisite to establishing the sustainable city. However, stakeholders need to embed relevant cultural traditions and technologies as an integral component part of any prospective trans-disciplinary design model.

Universities have an integral part to play in the training of ecologically aware design graduates, cognizant and capable of implementing designed ecological applications. The designer's ability to be at ease with the broader scientific implications of the ecology of the urban landscape is central to sustainable viability, but the designed urban landscape still needs to function aesthetically. The establishment of the ecology of place is dependent on the architect and designers adeptness in marrying science and art harmoniously, a sensory connection to nature. Benchmark projects such as the "High Line" are a source for optimism, but in order for such projects to have a wider impact, the arguments for their realization need to be explicit. Ultimately the only realistic authority the architect, designer and planner have to influence decision-making processes is the vigor of the argument. The nature of the argument lies at the heart of the matter. The practice of paying lip service to the "new paradigm' is instrumental in effectively diminishing the principle of eco-urbanity; the inevitable consequence of such conventions is the maintenance of the status quo.

References

Ahern, Jack. 2005. Integration of Landscape Ecology and Landscape Architecture: An Evolutionary and Reciprocal Process. In *Issues and Perspectives in Landscape Ecology*, edited by John Wiens, and Michael Moss, 307-315. Cambridge, MA: Cambridge University Press.

Berger, Alan. 2006. *Drosscape*. New York, NY: Princeton Architectural Press.

Brandenburgh, Andrea and Matthew Carroll. 1995. Your place or mine? The Effect of Place Creation on Environmental Values and Landscape Meanings. *Society and Natural Resources* 8 (5): 381-398.

Breheny, Michael. 1992. *Sustainable Development and Urban Form*. London: Pion Limited.

Brown, Lester. 2001. *Eco-Economy: Building an Economy for the Earth*. New York: W.W. Norton & Co.

Clark, William, and Nancy Dickson. 2003. Sustainability Science: The Emerging Research Program. *Proceedings of the National Academy of Sciences* 100: 8059-8061.

Cook, Robert. 2000. Do Landscapes Learn? Ecology's "New Paradigm" and Design. In *Environmentalism in Landscape Architecture*, edited by

---

[51] Wu 2008.

Michael Conan. Dumbarton Oaks Colloquium on the History of Landscape Architecture 22, Washington DC.

Corner, J. 2006. Terra Fluxus. In *The Landscape Urban Reader*, edited by Charles Waldheim, 23-33. New York: Princeton Architectural Press.

Ellin, Nan. 1996. *Postmodern Urbanism*. New York: Princeton Architectural Press.

Frey, Hilderbrand. 1999. *Designing the City: Towards a More Sustainable Urban Form*. New York, Routledge.

Gray-Donald, James, Debby R. E. Cotton, Martyn F. Warren, Ian Bailey, and Fumiyo Kagawa. 2007. Researching Sustainability in Higher Education: From Local to Global. *International Journal of Environmental, Cultural, Economic and Social Sustainability* 3 (5): 205-216.

Heidegger, Martin. 1971. Building Dwelling Thinking. In *Poetry Language Thought*, translated by Albert Hofstadter, 1-11. New York, NY: Harper Colophon Books.

Hope, Diane, Corinna Gries, Weixing Zhu, William F. Fagan, Charles L. Redman, Nancy B. Grimm, Amy L. Nelson, Chris Martin, and Ann Kinzig. 2003. Socioeconomics Drive Urban Plant Diversity. *Proceedings of the National Academy of Sciences* 100: 8788-8792.

Hopkins, Charles and Rosalyn McKeown. 2002. Education for Sustainable Development an International Perspective. In *Education and Sustainability: Responding to the Global Challenge*, edited by Daniella Tilbury, Robert Stevenson, John Fien, and Danie Schreuder, 13-24. Cambridge: International Union for Conservation of Nature.

Hough, Michael. 1995. *Cities and Natural Process: A Basis for Sustainability*. New York: Routledge.

Husserl, Edmund. 1936. *The Crisis of European Sciences and Transcendental Phenomenology*. Evanston IL: Northwestern University Press.

Huxtable, Ada. 1981. The Troubled State of Modern Architecture. *Architectural Record* 169 (3): 72-79.

Jameson, Frederic. 1991. *Postmodernism, or, The Cultural Logic of Late Capitalism*. Durham, NC: Duke University Press.

Jones, Clive, John Lawton, Moshe Shachak. 1994. Organisms as Ecosystem Engineers. *Oikos* 60 (3): 373-386.

Kates, Robert, William C. Clark, Robert Corell, J. Michael Hall, Carlo C. Jaeger, Ian Lowe, James J. McCarthy, Hans Joachim Schellnhuber, Bert Bolin, Nancy M. Dickson, Sylvie Faucheux, Gilberto C. Gallopin, Arnulf Grübler, Brian Huntley, Jill Jäger, Narpat S. Jodha, Roger E. Kasperson, Akin Mabogunje, Pamela Matson, Harold Mooney, Berrien Moore III, Timothy O'Riordan, Uno Svedin. 2001. Sustainability Science. *Science* 292 (5517): 641-642.

Lozano, Eduardo. 1990. *The Crossroad and the Wall*. New York, NY: Cambridge University Press.

Lynch, Kevin. 1960. *The Image of the City*. Cambridge, MA: MIT Press.

Manchester.gov. 2007. Manchester City Centre Population Estimate Planning Area. Available at http://www.manchester.gov.uk/downloads/A15_CityCentrePopEst_Oct0 7_1_.pdf, accessed on 3rd April 2014.

Martin, Guy. 2009. New York's Hanging Gardens. *The Observer Magazine,* November 8, 38.

McGonigle, Michael, and Justine Starke. 2006. *Planet U: Sustaining the World, Reinventing the University.* Gabriola Island: New Society Publishers.

Meyer, Elizabeth. 2000. The Post Earth Day Conundrum: Translating Environmental Values into Landscape Design. In *Environmentalism Landscape Architecture,* edited by Michael Conan. Dumbarton Oaks Colloquium on the History of Landscape Architecture 22, Washington DC.

Nadenicek, Daniel and Catherine Hastings. 2000. Environmental Rhetoric, Environmental Sophism: The Words and Work of Landscape Architecture. In *Environmentalism Landscape Architecture,* edited by Michael Conan. Dumbarton Oaks Colloquium on the History of Landscape Architecture 22, Washington DC.

Nassauer-Iverson, Joan. 1995. Culture and Changing Landscape Structure. *Landscape Ecology* 10 (4): 229-237.

Nassauer-Iverson, Joan. 1997. *Placing Nature, Culture and Landscape Ecology.* Washington DC: Island Press.

Norberg-Schulz, Christian. 1980. *Genius Loci Towards a Phenomenology of Architecture.* New York, NY: Rizzoli International Publications Inc.

Oliver, Paul. 2003. *Dwellings: The Vernacular House World Wide.* London: Phaidon Press.

Orr, David. 1994. *Earth in Mind: On Education, Environment, and the Human Spirit.* London: Island Press.

Osmond, P. 2002. The Sustainable Landscape. In *Design for Sustainability,* edited by Janis Birkeland, 99-103. London: Earthscan.

Parris, Thomas and Robert Kates. 2003. Characterizing and Measuring Sustainable Development. *Annual Review of Environment and Resources* 28: 559-586.

Rao, Srinivas and Goetz Schierle. 1994. Sustainability: The Essence of Vernacular, ACSA Conference, Milwaukee, WI. Available at http://www-classes.usc.edu/architecture/structures/papers/GGS-Rau.pdf, accessed March 30, 2014.

Reed, Mark, Evan Fraser and Andrew Doughill. 2006. An Adaptive Learning Process for Developing and Applying Sustainability Indicators with Local Communities. *Ecological Economics* 59 (4): 406-418.

Relph, Edward. 1997. Sense of Place. In *Ten Geographic Ideas That Changed the World,* edited by Susan Hanson, 205-227. New Brunswick, NJ: Rutgers University Press.

Sauve, Lucie. 1996. Environmental Education and Sustainable Development a Future Appraisal. *Canadian Journal of Environmental Education* 1: 7-34.

Seghezzo, Lucas. 2009. The Five Dimensions of Sustainability. *Environmental Politics* 18(4): 539-556.

Selby, David. 2006. The Firm and Shaky Ground of Education for Sustainable Development. *Journal of Geography in Higher Education* 30 (2): 353-67.

Stedman, Richard. 2002. Towards a Social Psychology of Place: Predicting Behaviour from Place-based Cognitions, Attitude, and Identity. *Environment and Behaviour* 34 (5): 561-581.

Tilbury, Daniella. 2004. Environmental Education for Sustainability: A Force for Change. In *Higher Education and the Challenge of Sustainability*, edited by Peter Blaze Corcoran and Arjen E.J. Wals, 97-112. Dordrecht: Kluwer Academic Publishers.

Turcu, Catalina. 2012. Re-thinking Sustainable Indicators: Local Perspectives of Urban Sustainability. *Journal of Environmental Planning and Management* 56 (5): 695-719.

United Nations. 2004. *World Population in 2300*. New York, NY: United Nations.

Vuksanovic, Dusan. 2006. Architectural Atlas of Montenegro. Available at http://www.gov.me/files/1247231246.pdf, accessed 5th April 2014.

Williams, Katie, Elizabeth Burton, Mike Jenks. 2000. *Achieving Sustainable Urban Form*. London: Spon Press.

Wright, Tara. 2004. The Evolution of Sustainability Declarations in Higher Education. In *Higher Education and the Challenge of Sustainability*, edited by Peter Blaze Corcoran and Arjen E.J. Wals, 7-19. Dordrecht: Kluwer Academic Publishers.

Wu, Jianguo. 2008. Making The Case for Landscape Ecology. *Landscape Journal* 27(1): 41-50.

Wu, Jianguo and Richard Hobbs. 2002. Key Issues and Research Policies in Landscape Ecology: An Idiosyncratic Synthesis. *Landscape Ecology* 17 (4): 355-365.

Chapter 4: Islamic Insights on Sustainability

Dora Marinova, Amzad Hossain, and Popie Hossain Rhaman

Introduction

Religions, like ligaments, describe our ties to the members of our family, our community, the larger human community, the environment and the transcendent.[1] This chapter argues that Islam as a religion contains a range of guidelines and prescriptions which encourage its followers to pursue a sustainable life on Earth. For example, Islam urges people to live in a way that does not destroy finite natural resources and to maintain modest consumption. Islam also emphasizes one of the main principles of sustainability, namely that one should contribute to the common and not just to the private good.[2]

Islam is the dominant religion of 49 countries around the world where at least half of the population identify themselves as Muslim. However, despite the deeply rooted sustainability principles in this religion, the lifestyle adopted by its followers varies widely. This generates a great disparity between countries, such as the United Arab Emirates (UAE), Qatar, Bahrain, Kuwait and Saudi Arabia, on the one hand, and Bangladesh, Afghanistan, Palestinian territories, Pakistan and Yemen on the other. For example, the ecological footprint of the UAE is the largest in the world compared to that of Bangladesh, which has one of the smallest (see Table 4.1). Similarly, a citizen of Saudi Arabia uses around 6.7 times more hectares per person than their counterpart in Pakistan although the share of Muslim population within both these countries is similar at around 97 percent.

Islam as a combination of law, religion and morality[3] provides a comprehensive framework for the individual, family, community and governments to follow. However, as Islamic societies changed over the course of their history, so did the interpretations of the Quran and its teachings.[4] What we

---

[1] Smith 2005,103.
[2] Mulder 2006.
[3] Dahl 1997.
[4] Samani and Marinova 2007.

witness around the world nowadays are the differences in lifestyles and value systems resulting in outcomes that are drastically contrary to Islamic norms and expectations. This chapter therefore is an attempt to go back to the basics of the Islamic religion and outline the foundations in the traditions of the Prophet. It also uses evidence from Bangladesh, one of the happiest nations in the world, as an example of a country whose population has one of the lightest ecological footprints in the world but is still exceeding its domestic biocapacity.

Table 4.1: Muslim country data from the Pew Research Center (2011) and the Global Footprint Network (2010).

	Country	Total population	% Muslim population (estimated)	Ecological footprint ha/person	Within domestic biocapacity
1	UAE	3,577,000	76.0%	10.68	NO
2	Qatar	1,168,000	77.5%	10.51	NO
3	Bahrain	655,000	81.2%	10.04	NO
4	Kuwait	2,636,000	86.4%	6.32	NO
5	Saudi Arabia	25,493,000	97.1%	5.13	NO
6	Oman	2,547,000	87.7%	4.99	NO
7	Malaysia	17,139,000	61.4%	4.86	NO
8	Kazakhstan	8,887,000	56.4%	4.54	NO
9	Turkmenistan	4,830,000	93.3%	3.93	NO
10	Gambia	1,669,000	95.3%	3.45	NO
11	Libya	6,325,000	96.6%	3.05	NO
12	Lebanon	2,542,000	59.7%	2.90	NO
13	Turkey	74,660,000	98.6%	2.70	NO
14	Iran	74,819,000	99.7%	2.68	NO
15	Mauritania	3,338,000	99.2%	2.61	YES
16	Niger	15,627,000	98.3%	2.35	NO
17	Jordan	6,397,000	98.8%	2.05	NO
18	Mali	12,316,000	92.4%	1.93	YES
19	Albania	2,601,000	82.1%	1.91	NO
20	Tunisia	10,349,000	99.8%	1.90	NO
21	Azerbaijan	8,795,000	98.4%	1.87	NO
22	Uzbekistan	26,833,000	96.5%	1.74	NO
23	Chad	6,404,000	55.7%	1.73	YES
24	Sudan (including South Sudan)	30,855,000	71.4%	1.73	YES

25	Guinea	8,693,000	84.2%	1.67	YES
26	Egypt	80,024,000	94.7%	1.66	NO
27	Algeria	34,780,000	98.2%	1.59	NO
28	Syria	20,895,000	92.8%	1.52	NO
29	Comoros	679,000	98.3%	1.42	NO
30	Somalia	9,231,000	98.6%	1.42	NO
31	Iraq	31,108,000	98.9%	1.35	NO
32	Burkina Faso	9,600,000	58.9%	1.32	NO
33	Kyrgyzstan	4,927,000	88.8%	1.25	YES
34	Morocco	32,381,000	99.9%	1.22	NO
35	Indonesia	204,847,000	88.1%	1.21	YES
36	Senegal	12,333,000	95.9%	1.09	YES
37	Sierra Leone	4,171,000	71.5%	1.05	YES
38	Tajikistan	7,006,000	99.0%	1.00	NO
39	Yemen	24,023,000	99.0%	0.94	NO
40	Pakistan	178,097,000	96.4%	0.77	NO
41	Palestinian territories	4,298,000	97.5%	0.74	NO
42	Afghanistan	29,047,000	99.8%	0.62	NO
43	Bangladesh	148,607,000	90.4%	0.62	NO
44	Brunei	211,000	51.9%	n/a	n/a
45	Djibouti	853,000	97.0%	n/a	n/a
46	Kosovo	2,104,000	91.7%	n/a	n/a
47	Maldives	309,000	98.4%	n/a	n/a
48	Mayotte	197,000	98.8%	n/a	n/a
49	Western Sahara	528,000	99.6%	n/a	n/a
	The World	6,892,319,000	23.4%	2.70	NO

Note: "NO" denotes difference larger than 2; "n/a" denotes information not available.

The Quran and the Tradition

Islam stresses the accomplishment of the pre-requisites in accordance with the guidelines of the Quran and the Tradition. The Quran is comprised of 114 chapters and 6,666 verses.[5] The Tradition – a commentary for implementing the Quranic guidelines – is comprised of the sayings and doings (sunna) of Prophet Muhammad. However, in this analysis the commentary of the Prophet's

---

[5] The chapters and verses of the Quran are referred as 'chapter number: verse(s) number', e.g. 1:2 or 23:12–22.

successors and other prophets of the past, Sufi saints and philosophers including the Baul-philosophers[6] of Bangladesh as well as proverbs, idioms and values wisdom are included in conformity with the spirit of the Quran.[7] Thus the Quran is the primary source for understanding the Islamic view of sustainability and the Tradition helps sustainable implementation, accomplishment and management.

Every verse of the Quran has four kinds of meaning: an exoteric sense (zahir, sharia), an inner sense (batin, marifa)[8], a limit (hadd) and a lookout point (muttala). The exoteric sense is the recitation, the inner sense is understanding, the limit is what the verse permits and prohibits, and the lookout point is the elevated place of the heart (qalb) what was intended by it as understood from God. The knowledge of the exoteric sense is public knowledge (ilm amm) while the understanding of its inner sense, and what was intended by it, is private – khass.[9] For understanding sustainability we use the terms Sharia and Sufism to recognize respectively the exoteric and inner senses of the Quranic verses.

The Quran addresses God with 99 attributive names. Some of them relate to sustainability: Praise be to Allah, the Cherisher and Sustainer of the worlds (1:2); And He is the Provider (5:58); God maintains all things (4:85). This suggests that sustainability is intrinsic to the Quranic revelations for "a people who understand" how God as the Creator-Sustainer of nature has provided all things to be self-sustained. The Quran repeats "people who understand" many times in relation to individual, social, economic and ecological aspects of sustainability. It offers this description: "And in the earth there are tracts side by side and gardens of grapes and corn and palm trees having one root and (others) having distinct roots – they are watered with one water, and We make some of them excel others in fruit; most surely there are signs in this for a people who understand" (13:4).[10]

The "people who understand" such as Ulama (scholars) and Sufis (philosophers) are revered by Muslims as living guides in the absence of the Prophet. Ulama act as guides in implementing the Sharia (a system of learning the divine law, way of belief and practice) and Sufis act as guides for people towards spiritual development for achieving perfection in doing things. It is envisaged that in order to understand God's signs in nature[11] one is required to follow the guidelines of Ulama and Sufis as they are seen as a people of Islamic values:

---

[6] The Baul-philosophers are mostly unlettered, yet full of poetic, musical, and philosophical talent. They are seen as being at the root of Bengali culture. Bauls are unique in socio-religious syncretisation (Hossain 1995).

[7] Traditional Islam refers to both the Quranic revelation and the prophetic sunna (sayings and doings) in themselves as well as the subsequent life and activity of the Islamic community, whether it be law, philosophy, art, mysticism, politics or social life, which can be seen as a historical commentary upon, and continuity of, the original revelation. As for modernism, it refers to a worldview, with a whole body of ideas and their resultant institutions.

[8] Marifa is a key Islamic concept; the core ingredient of Islamic spirituality and the opposite of normative Islam (Karamustafa 2006, xiii).

[9] Sands 2006, 9.

[10] See also 2:164, 5:99, 29:35, 30:24, 30:28, 40:3 and 45:5.

[11] The term "sign" in relation to sustainability has been revealed in the Quran in numerous places in various contexts. For example: "And Allah sends down rain from the skies, and gives therewith life to the earth after its death: verily in this is a Sign for those who listen" (16:65).

peaceful and submissive, charitable, modest, spiritual and responsible. They possess divine justice and human order with Islamic worldview that embraces a seamless web of ideas beginning with the belief in a single God.

This belief permeates various aspects of the Quranic teaching, from creation and the nature of the universe to ethics, social relations, and commercial and constitutional matters.[12] Islam is very strict about maintaining justice in every respect of economic, social and ecological affairs. The Quran repeatedly states that all of God's creations, even fruits, vegetation, and the sea and its contents, are signs for those who ponder about sustainability.[13]

Besides the Quranic contents, Islamic culture is also rich with proverbs, Sufi tales and folk songs which convey a sense of values that belong to the realm of oral and folk traditions. They comment on the universal human conditions in proverbial style. Examples include: "The door of the carpenter is always broken"; "The shoemaker goes barefoot"; "In every piece of tender meat there is a bone"; "Take the bad with the good"; "The most important thing about a house is its neighbors"; "You bring someone to solve your problem and then they become the problem"; "Every time has its knowledge"; "When you are losing your patience, try a little more patience"; "And tomorrow there will be apricots."[14]

Some Islamic proverbs appear as verses of the Quran: "Those who seek to make wealth or status a means to happiness will come to know in the end how futile and fruitless their efforts were. And truly you have come unto Us alone (without wealth, companions or anything else) as We created you the first time. You have left behind all that which We had bestowed on you" (6:94). "Did you think that We had created you in play (without any purpose)?" (3:115). "It is not for the sun to overtake the moon, nor does the night outstrip the day. They all float, each in an orbit" (36:40). "We cause to grow wonderful gardens full of beauty and delight" (27:60). Hence, both the Quran and the Tradition help constitute the Islamic culture and many features of it are closely linked to values that support sustainability.

The Concept of Sustainability in Islam

According to David Suzuki, one of the world's leading environmentalists, there is much to be learned from the wisdom of cultures formed by the cosmic religions if we are to have a future on the planet.[15] Islam is an example of a cosmic religion that deals with sustainability, rather vividly and deeply. It exhorts its followers to look at nature and admire it as God's creation, to nurture it, and to share it with other human and non-human beings, especially in terms of provisions such as food and shelter. The Quran regards the human as God's steward (Khalifa) on Earth, possessing an innate disposition or capacity (fitrah) to know God and act righteously.[16] This implies that human mannerism and behaviour are directed by envisaging peaceful intervention for survival which allows full measure of justice to natural resources.

---

[12] al-Hibri 2006, 238.
[13] al-Hibri, 2006, 243; See (55:11-13, 19-23, 55:5-25; 16:64-69; 29:43; 23:12-22).
[14] Fluehr-Lobban 2004, 74.
[15] Thornhill 2000, 108.
[16] Barr 2002, 121.

The Quran reveals that sustainability exists as a "measurable" entity through the visible state of the ecosystems' health, as a tree is visible through its roots, stems, branches, leaves, fruits and seeds. It says: "We send down water from the sky according to (due) measure, and We cause it to soak in the soil; and We certainly are able to drain it off (with ease)" (23:18); "With it We grow for you gardens of date-palms and vines: in them have ye abundant fruits: and of them ye eat (and have enjoyment)" (23:19); "And We have provided therein means of subsistence, – for you and for those (other creatures) for whose sustenance ye are not responsible" (15:20).

Islam warns against over-extraction of depletable natural resources and over-consumption. According to the Tradition: "Your wealth is not wide enough to encompass the people, so let your cheerful countenance and beautiful character surround them". One of the ancestors in faith said: "A little humility suffices for a lot of action, and a little spiritual science suffices for a lot of knowledge."[17]

The Quran employs a different but related figure of thought on sustainability by requiring its followers and humankind to behave like Vicegerents (representatives, khalifa) of the Creator of the universe (2:30). Human beings are authorized to represent God on Earth as the caretakers of nature's sustainability; hence they possess intrinsic worth and also, in submission to God, are called to live in justice. In this respect Islam is understood to be a religion that stipulates the unique purpose of human existence as protecting and caring for nature.[18]

The Islamic religion also highlights ways to practice sustainability in terms of protection, restoration and conservation of nature with Sufism showing how to accomplish this in action. The Sharia literature, mysticism, sayings and doings of Sufis, including the mystic Baul singer-philosophers of Bangladesh, conceive sustainability as a reflection of a lifestyle that reinforces regeneration of ecological goods. The Bauls are environmentalists by their belief and practice; they are truly soul stirring and take the listeners closest to nature, namely the divine. They are simple, natural, unembellished and rooted in the soil[19] and advocate for a non-violent and non-destructive use of technological deployment, the way Islam does. Bauls are illiterate and do not write down their innumerable spontaneously composed songs, although the messages they convey cannot but portray them as supreme Pundits (scholars). The impact that they have on listeners in shaping behavior and values is enormous. Despite the fact that the majority of the Baul-philosophers come from a Muslim background, they display a religious indifference or neutrality, which suggests that the practice of secularism by "people who understand" can also reinforce the proper management of natural resources. The unity within the diversity is one of the most important goals of the Islamic messages.[20]

Food is a very interesting aspect of Islam which by dint of its books and traditions makes diverse regulations on consumption, including categorical specification on halal (permissible) and haram (prohibited) foods. Overall it calls for moderation – one third of the stomach is for food, one third for drink and one third for air; the latter is also left empty for spiritual exercise and yogic

[17] Renard 2004, 243.
[18] Johnson et al. 2006, 9.
[19] Hossain 1995.
[20] Mondal 1997, 48.

contemplation. It is believed that the prophet Mohammed and many of his followers practiced this. The Baul philosophers stress that the very basis for a sustainable ecological footprint is a sustainable "food-print." Because of its heavy and unsustainable production footprint, meat [21] should be considered as a celebrating dish, consumed only occasionally and well within the recommended healthy limits. [22]

The foundations of Islam encourage human behavior that is entirely in line with the sustainability concept. However, the reality is that such behavior is very rare. In 2003 the global human population's ecological footprint overshot the planet's capacity by the equivalent of 0.25 Earths. [23] Today we need 1.5 planet Earths to meet the human demand on nature if we were to remain within its regenerative capacity. [24] These ecological limits need to be recognized. Not surprisingly the current environmental crisis cannot be met without a change in human priorities, values and lifestyles. To accomplish this there is a need for a different ethos which the roots of Islam offer. [25]

Sustainability Education in Islam

According to Ahmad, education is "an attempt on the part of individuals and society to transmit to the succeeding generations their accumulated store of the knowledge of arts, values, customs and their ideals of life as a whole as well as their experiences in various fields which should help the younger generation in carrying on their activities of life effectively and successfully." [26] Education is essential for understanding the natural and social environment around us and for connecting our knowledge with our concerns about nature, ecology and other people. Since the Quran provides Muslims with an outlook towards life, [27] its principles must provide a starting point and guide Islamic education for sustainability. Moreover, Regan argues that the Muslim educator will find in the Quran the guiding principles that also help in selecting the content of the curriculum. [28] Such a planned transmission enables the new generation to acquire and assimilate, within a short span of time, the fruits of learning of thousands of years of its predecessors. This generation will also enrich the accumulated treasures of knowledge with its own experiences.

---

[21] Estimates show that about 6 kg of plant proteins are needed to produce 1 kg of animal proteins; 1 kg of beef requires 15 m^3 of water compared to only 0.4–3 m^3 of for 1 kg of wheat. See Mulder 2006, 272.

[22] Raphaely and Marinova 2014.

[23] Global Footprint Network 2008.

[24] Global Footprint Network 2010.

[25] Edwards 2006, 100.

[26] Ahmad 1990, 1.

[27] Murata and Chittick 1994, 182-184, describe the Quran as a complete code of life having in mind verses such as: "We have sent down to thee the Book as an Elucidation of all things, and as a Guidance and a Mercy, and as good news to those who submit" (16:89); "We have left nothing out from the Book" (6:38); "We have counted everything in a clear register" (36:12) and "There is nothing whose treasures are not with Us and We only send it down in a known measure" (15:12).

[28] Regan 2000, 190.

Renard explains that in Islamic education, two kinds of knowledge are seen: knowledge generated and knowledge gathered.[29] Gathered knowledge is useless if it does not include knowledge generated; just as the sun is useless when the eye is incapable of receiving illumination.[30] The Quran reveals: "When you do not know, then inquire of the people of recollection" (16:43; 21:7). This transmission of knowledge and generation of new knowledge within a constantly changing ecological and social world are essential for sustainability education. The two essential methods in Islamic education are: the general ways or the Sharia laws that show how to act to protect, restore and conserve the natural environment and relate to fellow human beings; and Sufism which provides the spirituality of how to transmit and acquire values that guide people's way of living and doing things. The learning approach common in both the Sharia and Sufi education focuses on non-coerced, task-oriented action individually or by a group of individuals. The theories of reality (Sufism) and of practice (Sharia) influence each other and relate respectively to worldviews and norms of behavior within task-oriented groups.[31]

With the growing consumption of finite natural resources worldwide, especially that underpinning the uptake of extravagant consumerism in fast developing countries,[32] the practice of the Islamic tradition and education can offer an alternative code of economic development and consumerism that puts less emphasis on material wealth. This could enable people to pursue a lifestyle that supports local and global sustainability, providing them with an educational basis for the social, economic, technological and environmental aspects of sustainability.[33]

Rediscovering Sustainability

Islam recognizes sustainability as a gift from the Sustainer. The Quran and the Tradition show a straight path in order for its sustainable maintenance, which is to live simply and gently within the nature's gifts.[34] On the other hand, the Quran refers to the people who go astray as wasteful, extravagant, corrupts, mischief and Satan.[35] The straight-path concept along with its educating and implementing guidelines depicts an icon of holistic sustainability: individual, social, economic and ecological sustainability.

---

[29] Renard 2004.

[30] Renard 2004, 205.

[31] Abraham 2006, 225.

[32] For example, while the world population grew from 2.7 billion to 6 billion within 50 years, meat production and consumption increased fivefold in the same period (Mulder 2006, 272).

[33] Huckle and Sterling 1996, 3.

[34] This is consistently stated throughout the book. Examples include 3:101; 4:48 and 66-68; 6:39; 87:126,153 and 161; 16:121-124; 22:54; 48:2 and 20; 36:60-61; and 42:13.

[35] See 2:60 and 205; 5:64; 7:31 and 56; 17:26-27; 28:4; and 3:138.

Individual Sustainability

Islam takes sustainability of an individual first as the basic unit of social, economic and ecological sustainability. It recognizes individual spiritual development as the very starting point of one's progress to individual sustainability (longevity) management. Spiritual development or spirituality "energises the soul to provide what the world (materialism) lacks."[36] Spirituality is also the source of the will to act morally where other fellow humans and other beings in nature are not affected.[37]

For individual development, Islam prescribes training from a spiritual guide (Sufi). Saeed observes that Sufis tend to be more accommodating and inclusive because they care for the heart's inner disposition. [38] They tolerate local differences, such as between Christianity and Islam, and pursue individual development drawing diverse elements from other traditions. Chittick gives an account of Sufi master Shams-I Tabrizi's influence upon Rumi who was transformed from a sober jurisprudent to an intoxicated celebrant of the mysteries of Divine Love.[39]

Edwards argues that in order to convert oneself to develop a sense of kinship with and responsibility for the creatures of Earth, for the land, atmosphere, seas and rivers that support them, one must strive to get involved in the struggle for a more just and ecologically sustainable world that can be fulfilling and meaningful.[40] This is possible by creating a way of life, a culture and educational system that provide the motives, the stimulation and the facilities for self-transformation to which the foundations of Islam can contribute.

Social Sustainability

Mondal observes that the ideal society in Islam designates a human agglomeration possessing a common faith and belief in maintaining social sustainability.[41] The virtuous living according to Islam includes alleviation of the suffering of others, rendering assistance to the needy, good neighborliness and maintenance of peace in the society. Its social system is based on equality, brotherhood[42] and justice.

Islam, similar to Christianity and other religions, calls for an interpersonal relationship with God in grace and for graceful relations with other fellow creatures [43] rejecting everything that leads to divisiveness. Islam does not recognize barriers of birth, race, color, creed, language, or even geographic barriers.[44] Its focus is on the purification of motives, the ultimate test being action

---

[36] Chittister 1998, 1.
[37] Smith and Standish 1997, 122.
[38] Saeed 2006, 9.
[39] Chittick 1983.
[40] Edwards 2006, 115.
[41] Mondal 1997, 46.
[42] The meaning implied by brotherhood (in the absence of a word such as siblinghood) is a feeling of fellowship, sympathy, compassion, kindness and consideration for others.
[43] Edwards 2006.
[44] Williams and Humphrys 2003, 155.

and behavior to appease the Sustainer. Social justice and care, peace and improved quality of life as social goals of sustainability can be enhanced by the Islamic religion.

Economic Sustainability

Economic growth, which is an increase in quantity, cannot be sustainable indefinitely on a finite planet.[45] Conversely, Costanza et al. point out that economic development, which is an improvement in the quality of life without causing an increase in the quantity of consumed resources, could be sustainable.[46] Over-extraction of finite resources for over-consumption in order to lead wasteful lifestyle is repeatedly prohibited by Islam (2:60, 2:205, 5:64, 7:56, 28:4). Islam also prohibits the generation of growth in money by charging interest or excessive profit. From a religious point of view Islam considers money just as a convenient medium of exchange. Its use as a commodity to lend or hoard for excessive profit can injuriously deprive others and is strictly prohibited. People can have surplus money to lend to those in need, but with profit sharing, not with fixed rate interest or high profit. For economic health, money ought to circulate in a community like blood. In the case of piling up of wealth, the Quran warns: "Who pileth up wealth and layeth it by, Thinking that his wealth would make him last for ever! By no means! He will be sure to be thrown into That Hell which Breaks to Pieces" (104:2-4). Living simply and without greed for more than required is the standard of maintaining economic sustainability in Islam.

Fluehr-Lobban argues that poverty is not for the sake of hardship but for the sake of managing long-term sustainability within the regenerative capacity of renewable resources.[47] In many ways attempts coming from outside Islam directed towards what is known as "poverty alleviation" clash with the fundamental Muslim values and effectively push such countries, including Bangladesh, towards unsustainability.[48]

Environmental Sustainability

Environmental sustainability exists where the natural resource base and ecosystem health are not negatively affected by development. The Islamic view of individual, social and economic development is devoid of such development activities that destroy the natural resource-base. Islam recognizes that everything God created is an essential part of the sustainability management by nature. The Torah reads: "Even things which appear to you to be superfluous in the world, such as flies, fleas and mosquitoes, they are also part of the creation of the world, and God performs His operations through the agency of all of them, even through a snake, mosquito or frogs."[49]

Mulder explains that as recently as in the 1960s, most people perceived the natural environment as infinite until Rachel Carson's book *Silent Spring* exposed

---

[45] Costanza et al. 1991.

[46] Costanza et al. 1991.

[47] Fluehr-Lobban 2004, 188.

[48] Hossain and Marinova 2005.

[49] Mulder 2006, 37.

that the opposite is the case.[50] For the followers of Islam, the Quran revealed the finiteness of natural resources more than 1400 years ago: "We have counted everything in a clear register" (36:12); "There is nothing whose treasures are not with Us and We only send it down in a known measure" (15:12). They are commanded to bear this finiteness and live simply.

Contrasting Western scientific understanding of the world with that of the Islamic belief, Fazlun and O'Brien argue that in Western science theories are always open to rejection and change while Muslim scientists explain the "natures" of created things in terms of their supernatural origin.[51] This emphasizes the sacred character of creation, since to abuse what is natural, or to show lack of respect for nature, is to show lack of respect for the Creator.

Many Muslim people have lived and are living happy and fulfilled lives that are deliberately simple and deliberately restrained. Their goal is a quality style of life based on cooperative relationships, appreciation of nature and celebration of the spirit, not on high productivity and consumption.[52]

Conclusion

For the millions of followers of Islam around the globe, its fundamentals provide guidance how to manage sustainability as a concept, understanding, educating and practicing. In fact, Islam discourages people from playing a double role of destroying first, and then being concerned about healing the natural and social world, as they have been throughout history.[53] A sustainable lifestyle requires a moral unity of humankind. Islam can be a powerful unifier as its teachings blend religious, moral and social practice into an indivisible whole for the believer-practitioner.[54]

In no way is this chapter an attempt to argue that Islam (or any religion for that matter) is the only solution to the current environmental and social crises. What it tries to convey is that sustainability is deeply rooted in Islam and the millions of its followers (representing more than 20 percent of the global population) can strongly benefit from adhering to its foundations. Those of them who have already become "people who understand" deserve the respect and acknowledgement of the global community for developing their human potential through simple lifestyle, proper food choices, aspiration for social justice and care for nature.

References

Abraham, Martin A., ed. 2006. *Sustainability Science and Engineering: Defining Principles*. Amsterdam: Elsevier.
Ahmad, Manzoor. 1990. *Islamic Education: Redefinition of Aims and Methodology*. New Delhi: Qazi Publishers & Distributors.

---

[50] Mulder 2006, 15.
[51] Fazlun and O'Brien 1992, 42.
[52] Cummings 1991, 130.
[53] Schweiker et al. 2006, 2.
[54] Regan 2000, 196.

al-Hibri, Azizah Y. 2006. Divine Justice and the Human Order: An Islamic Perspective. In *Humanity Before God: Contemporary Faces of Jewish, Christian, and Islamic Ethics*, edited by William Schweiker, Michael A. Johnson, and Kevin Jung, 238–255. Minneapolis, MN: Fortress Press.

Barr, Michael D. 2002. *Cultural Politics and Asian Values: The Tepid War*. London: Routledge.

Chittick, William. 1983. *Sufi Path of Love: The Spiritual Teachings of Rumi*. Albany, NY: State University of New York Press.

Chittister, Joan. 1998. *Heart of Flesh: A Feminist Spirituality for Women and Men*. Grand Rapids, MI: Wm. B. Eerdmans Publishing Company.

Costanza, Robert, Herman E. Daly, and Joy A. Bartholomew. 2001. Goals, Agenda and Policy Recommendations for Ecological Economics. In *Ecological Economics*, edited by Robert Costanza, 1–21. New York: Columbia University Press.

Cummings, Charles. 1991. *Eco-Spirituality: Toward a Reverent Life*. New York: Paulist Press.

Dahl, Tove S. 1997. *The Muslim Family: A Study of Women's Rights in Islam*. Oslo: Scandinavian University Press.

Edwards, Denis. 2006. *Ecology at the Heart of Faith: The Changing of Hearth That Leads to a New Way of Living on Earth*. New York: Orbis Books.

Fazlun, Khalid, and Joanne O'Brien. eds. 1992. *Islam and Ecology*. New York: Cassell Publishers.

Fluehr-Lobban, Carolyn. 2004. *Islamic Societies and Practices*. Gainesville, FL: University Press of Florida.

Global Footprint Network. 2008. *Humanity's Footprint 1961-2003*. Available at www.footprintnetwork.org/gfn_sub.php?content=global_footprint, accessed March 16, 2008.

Global Footprint Network. 2010. *Ecological Footprint Atlas 2010*. Available at http://www.footprintnetwork.org/en/index.php/GFN/page/ecological_foo tprint_atlas_2010, accessed March 31, 2014.

Hossain, Amzad. 1995. *Mazar Culture in Bangladesh*. PhD Thesis, Perth, Australia: Murdoch University.

Hossain, Amzad and Dora Marinova. 2005. Poverty Alleviation – a Push towards Unsustainability in Bangladesh? *Proceedings of the International Conference on Engaging Communities*. Brisbane: Queensland Department of Main Roads. Available at, http://www.engagingcommunities2005.org/abstracts/Hossain-Amzad-final.pdf, accessed March 31, 2014.

Huckle, John, and Stephen R. Sterling. eds. 1996. *Education for Sustainability*. London: Earthscan Publications.

Johnson, Michael A., Kevin Jung, and William Schweiker. 2006. Introduction. In *Humanity Before God: Contemporary Faces of Jewish, Christian, and Islamic Ethics*, edited by William Schweiker, Michael A. Johnson, and Kevin Jung, 1–18. Minneapolis, MN: Fortress Press.

Karamustafa, Ahmet T. 2006. Preface. In *Knowledge of God in Classical Sufism: Foundations of Islamic Mystical Theology*, edited by John Renard, xi–xiv. Mahwah, NJ: Paulist Press.

Mondal, Sekh Rahim 1997. *Educational Status of Muslims: Problems, Prospects and Priorities*. New Delhi Inter-India Publication.

Mulder, Karel. ed. 2006. *Sustainable Development for Engineers: A Handbook and Resource Guide*. Sheffield, UK: Greenleaf Publishing.

Murata, Sachiko, and William C. Chittick. 1994. *The Vision of Islam*. New York: Paragon House.

Pew Research Center. 2011. The Future of the Global Muslim Population: Projections for 2010–2030. Available at http://www.pewforum.org/2011/01/27/the-future-of-the-global-muslim-population/, accessed March 31, 2014.

Raphaely, Talia, and Dora Marinova. 2014. Flexitarianism: A More Moral Dietary Option. *International Journal of Sustainable Society* 6 (1/2): 189–211

Regan, Timothy. 2000. *Non-Western Educational Tradition: Indigenous Approaches to Educational Thought and Practice*. London: Lawrence Erlbaum Associates.

Renard, John. 2004. *Knowledge of God in Classical Sufism: Foundations of Islamic Mystical Theology*. Mahwah, NJ: Paulist Press.

Saeed, Abdullah. 2006. *Islamic Thought: An Introduction*. London: Routledge.

Samani, Shamim, and Dora Marinova. 2007. Integrating Islamic Values in Domestic Violence Mitigation. In *Proceedings of the 2007 International Women's Conference "Education, Employment and Everything... the triple layers of a woman's life"*, 127-131. Toowoomba, Australia.

Sands, Kristin Z. 2006. *Sufi Commentaries on the Quran in Classical Islam*. London: Routledge.

Schweiker, William, Michael A. Johnson, and Kevin Jung. eds. 2006. *Humanity before God: Contemporary Faces of Jewish, Christian, and Islamic Ethics*. Minneapolis, MN: Fortress Press.

Smith, II George P. 2005. *The Christian Religion and Biotechnology: A Search for Principled Decision-making*. Dordrecht, The Netherlands: Springer.

Smith, Richard, and Paul Standish. 1997. *Teaching Right and Wrong: Moral Education in the Balance*. London: Trentham Books.

Thornhill, John. 2000. *Modernity: Christianity's Estranged Child Reconstructed*. Grand Rapids, MI: W. B. Eerdmans Publishing Company.

Williams, Michael, and Graham Humphrys. eds. 2003. *Citizenship Education and Lifelong Learning: Power and Place*. New York: Nova Science Publishers.

Chapter 5: Environmental Memes: Form, Function, and Reasons for Optimism

Spencer S. Stober

Introduction

Will our species reproduce and consume energy unchecked until we exceed the world's capacity to support us? Is an irreversible "tragedy of our commons" inevitable? We are able to identify and dialogue about environmental issues, but we are also slow to garner the collective forces necessary to address the issues we identify. When Erika Rosenthal, an attorney for Earthjustice.org, returned from the United Nations Framework Convention on Climate Change (UNFCCC) negotiations in Cancun, she reported that steps had been taken to hasten the transfer of green technologies and to create a framework to compensate developing countries that preserve their tropical forests.[1] But she believes that we are still "woefully short of what science says is urgently needed, doing little more than kicking the can down the road on the hard decisions."[2] It remains to be seen whether or not the governments of the world will unite to mitigate our environmental issues. It is the opinion of this author that while we strive to solve these challenges with rational thinking and emerging technologies, we must also reconnect to, or at least acknowledge through more dialogue about, our relationship with Mother Nature.

This paper suggests that there are environmental memes that can create principles in the minds of people, and that these principles can then serve as benchmarks by which we judge and mediate our interactions with nature. What are environmental memes? Environmental memes are defined here as a special category of memes that inform our view of nature, and in so doing, they create principles in our minds that we use as benchmarks, both consciously and subconsciously, to judge and mediate our interactions with nature. This paper discusses environmental memes that have developed and how they play a role in our decision-making. "Green" and the principle of "stewardship" are memes

---

[1] UNFCCC 2010.
[2] Rosenthal 2010.

operating at the day-to-day decision-making level. "Gaia" and "Pachamama"[3] are also memes, and while they have become part of our rational decision-making process, they also have the potential to restore our relationship "with" the ecology of nature. To illustrate this reasoning process, the form and function of these environmental memes are considered. These abstractions cannot be used to demonstrate cause and effect, but they do provide a framework for dialogue concerning our species' relationship with nature, particularly the tension we feel because it is apparent that the economic growth of our societies is at Mother Nature's expense. Our day-to-day decisions as leaders and members of organizations are influenced by these environmental memes, and this may be reason for optimism. It is concluded that even though environmental memes vary greatly, there is value in their rapid proliferation, and the subsequent dialogue indicating a deep-seated concern about our relationship with nature is reason for optimism as we seek ways to restore our relationship with Mother Nature.

Memes

The term "meme" was coined in 1976 by Richard Dawkins in his famous book entitled *The Selfish Gene*. He focused on evolutionary theory from the perspective of a gene. Genetic information perpetuates when the organism carrying that information is able to reproduce and transmit the information to the next generation. It is in a gene's "self-interest" that its host organism survives, and it is for this reason that Dawkins referred to organisms as vessels which "selfish genes" use to perpetuate themselves. Dawkins suggests that our understanding of how genes replicate may serve as an analogy for the transmission of culture, a process that has evolved along with our ability to learn and imitate the actions of others. In his words:

> Examples of memes are tunes, ideas, catch-phrases, clothes fashions, ways of making pots or of building arches. Just as genes propagate themselves in the gene pool by leaping from body to body via sperms or eggs, so memes propagate themselves in the meme pool by leaping from brain to brain via a process which, in the broad sense, can be called imitation.[4]

Is it useful to compare memes to genes? Today the concept of a gene is well-defined and supported with empirical evidence, in part, because the science of genetics is now well-established and continues to yield mechanisms that explain gene function. Memetics is an emerging concept that has yet to mature in a scientific sense. If we hold memetic science up to the genetic science standard of today, then we overlook the fact that the concept of a gene and its associated mechanisms developed long after Gregor Mendel identified hereditary "factors" in the 1800s. Thus, it is reasonable for a theoretical concept to be ill-defined while in its infancy, and this appears to be the case for memes, but if we set the idea's immaturity aside, is there a more substantive reason to use the gene-meme

---

[3] The word Pachamama is sometimes written as Pacha Mama.
[4] Dawkins 2006, 192.

analogy? Even though some scholars have harsh words for the study of memes—calling it an ill-defined, misguided, misinformed, unhelpful, and unscientific field—others suggest that this body of work could become helpful in our understanding of language, mind, and culture.[5]

How do anthropologists view this gene-meme analogy? The idea that cultural transmission is analogous to the transmission of genes was not at first well-received by anthropologists.[6] They were absent from the early dialogue surrounding memes, and from their point of view the concept of memes sounded a lot like evolutionary determinism all over again. They had already examined the diffusion of "bits" or "traits" in the context of culture, and the notion that "culture is ultimately made of distinguishable units [memes] which have a life of their own" is inconsistent with their paradigm.[7] That said, anthropologist Maurice Bloch suggests that a more functional approach could open up the possibility for working across disciplines to better understand culture: "a commitment to seeing culture as existing in the process of actual people's lives, in specific places, as part of the wider ecological process of life."[8] He reminds us that "theory cannot be about what is circumscribed by disciplinary boundaries."[9] If we are to span disciplinary boundaries, then we should work to develop an appreciation for methods and emerging ideas within the disciplines of others, even though they may at first seem difficult and foreign to us. As a start and for our purposes here, let us assume that at the very least, the idea of a meme is a useful heuristic device. While the gene-meme analogy may reveal a bias for biological explanations, a functional approach to anthropology could enable us to join forces and explore "environmental memes" as principles that become part of our moral reasoning and ultimately influence our interactions with nature. The next section will examine four environmental memes, namely the green stewardship, Gaia and Pachamama memes.

Environmental Memes

The Green Meme

The English word "green" by itself is a color, but it can take on other meanings depending on the context in which it is used. For example, if someone states that their organization is "going green", then it is reasonable for us to assume that their organization is taking steps to make "good" environmental decisions when it comes to the use of energy, the handling of wastes, and other actions that impact upon ecosystems. In this sense, the word "green" is a perception that we hold regarding a need to do what is right for the environment. The word "green" has become an influential idea or catch-phrase that our brains are capable of processing with consequent actions. Memes replicate when someone acts, another

---

[5] Poulshock 2001, 68-80.
[6] Bloch 2005.
[7] Bloch 2005, 91.
[8] Bloch 2005, 16.
[9] Bloch 2005, 17.

person imitates, and the process continues.[10] Green memes, like genes, may or may not render survival value for their hosts, but when it comes to the need to act differently toward the environment, humans could be catching on to the idea that "going-green" makes sense. Is this a reason for optimism? If the "green" meme has the power to induce actions that are then imitated, if it has the power to consistently spread from brain to brain, and if those actions lead to environmentally sound decisions, then yes there could be reason for optimism.

A thorough analysis of the green meme's power to consistently induce action is beyond the scope of this paper, but viewing the "meme as a catalytic indexical"[11] suggests a mechanism for future consideration. As an indexical, the word "green" has multiple meanings depending on its context. A "green building" may be called green because it is green in color. In Canada and the United States it might be called a green building because it is certified by the Green Building Council as having energy saving-components.[12] There are thousands of certified projects world-wide; the idea is catching on as more and more people and business see the value of a certified green building. This is not to say that the green meme is making all of this happen, but the term "green" is acquiring meaning as it is associated with these types of endeavors, albeit not the same for all people and businesses. "Green" ideas are becoming a part of our conscious decision-making process that informs our actions. When it comes to being-green, some people may consider it to be the right thing to do, while others may see it as an economic savings, but nevertheless, "going-green" is taking hold as an idea with the power to reliably induce actions by people and organizations. If one simply paints their building green because it blends in with a forest, an environmentalist may judge this action to be frivolous from an environmental point of view, but for some people, this may seem like a green initiative. Depending on one's point of view regarding consumption, some companies are taking steps to "paint" their products as green so that they can blend into the "landscape" of environmentally conscious consumers—green marketing is big business. In fact, there is a growing tendency for environmental organizations such as The National Wildlife Federation, The Nature Conservancy, and others, to recognize companies such as Shell and BP for environmental stewardship. These environmental organizations may have good intentions with hopes that they can change corporate behavior, but when dealing with corporations and the United States government, their leaders may have become political realists who are willing degrade the environment and externalize environmental risks to sustain the economy.[13] Businesses are also realizing that going-green is good for business; we can only hope that the business leaders also realize it is not just about selling products. Consumers are already looking for authentic "green" products. The Greenlife organization exposes "greenwashing" business when they fail to deliver sustainable products as claimed in their advertising.[14] An expectation of authenticity is growing as the green meme replicates.

---

[10] Dennet 2009.
[11] Lissack 2004.
[12] US Green Building Council 2010.
[13] Hari 2010.
[14] The Greenlife 2014.

Peter Senge in his book *The Necessary Revolution* calls for business leaders to recognize the seriousness of our pending environmental crisis. He suggests that the situation is like a financial bubble that bursts when the laws of economics take over because of bad financial management (a metaphor that business leaders can appreciate). If we continue business as usual when dealing with the environment, a crisis will eventually occur as the laws of nature take over because we have made bad environmental decisions.[15] We can only hope that "going-green" will continue, because even with a lack of precision, the green meme may serve as a catalyst for dialogue and action concerning environmental issues.

The Stewardship Meme

The concept of "stewardship" is much older than "green" as a meme. The word "green" as a meme may have gotten its start with the formation of Greenpeace in the early 1970s. The idea of stewardship has usually been associated with the Judeo-Christian tradition and its mythic account of creation going back thousands of years. In the book of Genesis in the Bible, the first creation account has God, after having created Adam and Eve, saying: "Be fruitful, and multiply, and replenish the earth, and subdue it: and have dominion over the fish of the sea, and over the fowl of the air, and over every living thing that moveth upon the earth."[16] Traditional Christianity in particular has interpreted this verse, and others related to it, as meaning that humans have superiority over the rest of creation, and in fact that they have been commanded by God to have such a relationship over it. That is, dominion implies that there is a hierarchy of beings with humans at the top, and that it is incumbent upon humans to manage God's creation. While no Christian theologian then or now would likely maintain that this means that we can do whatever we want to do to animals and the rest of the created order, Christianity has assumed that we have a relationship to creation as God's caretakers. Thus, we are ultimately in charge. Some people have even laid the entire history of the mistreatment of animals and the devastation of nature at the feet of Christianity, and while that is too sweeping an indictment, some contemporary Christian theologians and biblical scholars have challenged the traditional understanding of stewardship as meaning dominion.

One of the shifts in thinking over the past 30 years is that some Christian theologians are emphasizing stewardship rather than dominion. Stewardship has been offered as a kind of benevolent dominion. However, some contemporary theologians want to go even one step further than stewardship, or at least understand it in a more radical way. Thus, traditional Christian theologians such as Andrew Linzey, process theologians such as John Cobb and Jay McDaniel, and feminist theologians such as Sally McFague and Rosemary Radford Ruether, have argued that we need to re-conceptualize our relationship with nature. Views differ on the nature of this relationship. Some maintain that animals have rights, while others believe that God suffers along with all of God's creatures (not just humans). Some argue that the Earth should be understood as the body of God, while others maintain that we are co-creators with God in creation. In all cases,

---

[15] Senge 2008.

[16] Bible, King James Version, Genesis 1:28.

stewardship is emerging as a necessary approach for dealing with the created order. The concept of stewardship, whether viewed as a responsibility to replenish nature because of a God-given right to use nature, or as a responsibility to co-operate with God's right to reign peace and justice, is an influential idea that our brains are capable of processing. It is a meme. We process what we believe about God's word, we act, someone imitates, and the process continues. This is a very old meme, but current efforts to conserve natural resources are modern-day examples of its enactment. Thus, a tradition of stewardship and going-green may be reasons for optimism when it comes to rational actions that people and organizations can take to mitigate the many environmental issues that we face.[17]

The Gaia Meme

"Go forth and multiply" should not be justification to rape and pillage Gaia.[18] Some of us may be offended by this statement because it challenges a Christian perspective on nature without acknowledging the human capacity for stewardship. Others of us may be offended because it personifies a pagan god (Gaia was the Greek goddess of Mother Earth). Others still may consider the entire statement to be inflammatory for a host of reasons. Memes can function in this way because they are words (or symbols) that gain meaning from the context and perceptions that we hold. In this case, the stewardship meme is not mentioned, but nonetheless the statement confronts our ability to be good stewards by reminding us that humans are capable of "rape" and "pillage"—even a pagan god deserves better.

What is the Gaia meme? Greek minds gave birth to Gaia as a goddess for Mother Earth, and she was resurrected in modern times as a name for a hypothesis proposed by James Lovelock in 1988, along with microbiologist Lynn Margulis. Lovelock briefly states the hypothesis as follows: "Organisms and their material environment evolve as a single coupled system, from which emerges the sustained self-regulation of climate and chemistry at a habitable state for whatever is the current biota."[19] This idea is difficult to verify scientifically, but for now, as Lovelock suggests, "Its greatest value lies in its metaphor of a living Earth, which reminds us that we are part of it and that human rights are constrained by the needs of our planetary partners."[20] Thus, Gaia has become a meme for a living Earth. The terminology is enchanting with the potential to spur action. For example, Andy Lechter in "Gaia Told Me to Do It," explains how the "Eco-Pagans" view nature as "sacred, sentient and crying out for protection." [21] We should probably not be hopeful that all humans will rise up to save the Earth because they have been infected with the Gaia meme, but nevertheless, this meme is part of the dialogue raising our environmental awareness with a potential for subsequent action.

---

[17] This section was prepared in consultation with Dr. Donna Yarri, Associate Professor of Theology at Alvernia University, US.
[18] Stober 2009.
[19] Lovelock 2003, 769.
[20] Lovelock 2003, 770.
[21] Letcher 2003, 61.

The Pachamama Meme

The Gaia meme is alive and well today as a scientific hypothesis and a metaphor for a living Earth. The Pachamama meme is also alive and well today with legal standing in Ecuador's Constitution. Constitutions are by their very nature anthropocentric in that they provide for the rights of people. About 50 nations have established a constitutional right for persons to a clean environment, and this number grows to more than 100 nations if general language supporting a clean environment is included.[22] Yet one country, Ecuador, recognizes "La naturaleza o Pacha Mama" (nature or Mother Earth) as worthy of respect and having rights according to their 2008 Constitution, Articles 71 to 74 in *Chapter Seven, Rights of Nature.* Article 71 of the constitution states that

> Nature, or Pacha Mama, where life is reproduced and occurs, has the right to integral respect for its existence and for the maintenance and regeneration of its life cycles, structure, functions and evolutionary processes.

> All persons, communities, peoples and nations can call upon public authorities to enforce the rights of nature. To enforce and interpret these rights, the principles set forth in the Constitution shall be observed, as appropriate.

> The State shall give incentives to natural persons and legal entities and to communities to protect nature and to promote respect for all the elements comprising an ecosystem.[23]

Is Pachamama a meme? Does she have the power to induce actions that are then imitated by others? The indigenous populations are a majority in Ecuador, and their intimate relationship with nature fueled their advocacy for Pachamama. This made Ecuador a perfect place for the rights of nature to be recognized in a constitution for the first time.[24] Today, the Pachamama Alliance continues to advocate recognition of the rights of nature in Ecuador and throughout the world.[25] What is it about Pachamama that induces us to act? Some of us may embody her spiritual aspects, to be at one with nature. Others of us may embrace her as a metaphor for a planet with intrinsic value in need of protection. Still others among us may understand ecosystems in material terms and recognize these systems as being essential to our existence. All of these and other views may be reason for optimism if they inform our day to day decision making in a positive way when we interact with the environment. In this capacity, Pachamama is an environmental meme. She informs our dialogue regarding the rights of nature, and in so doing, she has the potential to influence our actions,

---

[22] Hayward 2005, 22.
[23] Constitution of the Republic of Ecuador 2008. This English translation of the constitution uses "Pacha Mama" rather than "Pachamama." See note 3 above.
[24] Stober 2010.
[25] The Pachamama Alliance 2014.

both consciously and subconsciously. We can then judge and mediate our interactions with nature. Pachamama is similar to other memes in that she becomes part of the moral reasoning process that informs our actions; but as nature, she nurtures us as well.

Competing Memes

It is possible that competing memes operate to offset our efforts to "go green" and be good stewards of the environment. If current trends in consumption and population growth continue, by the year 2030 it will take two worlds to support all of us.[26] Clearly, environmental memes as discussed here are not slowing our patterns of consumption and population growth around world. Patterns of consumption and population growth are not easily categorized as memes, yet these patterns characterize the sociopolitical and cultural landscape.

What fuels our species' drive to thrive and flourish? Are we hosting a complex array of memes that in concert become cultural imperatives that contribute to the evolution of our species and our many cultures? [27] The mantra for policies that promote economic growth is an example of a complex concept that may compete with or confound the expression of environmental memes. Gross Domestic Product (GDP) is a common indicator of economic growth, but it does not account for environmental degradation or resource depletion.[28] This can be problematic. Some political decision makers advocate for ways to promote economic growth instead of engaging in a serious (and often difficult) discussion about the reallocation of resources according to need.[29] These approaches are a concern for two reasons. First, economic growth drives climate change and the loss of biodiversity, and the larger the economy the greater the environmental degradation.[30] Second, social injustices and poverty may disempower individuals and communities in their efforts to use resources wisely. Some political decision makers consider economic growth to be the way forward because economic growth generates revenue to support the research and development that they believe is necessary for new technologies to offset environmental degradation and the depletion of resources. Would efforts to slow economies be a better approach?

Strategies to slow economic growth are emerging with discussions around values, ethics, and preferences that fuel consumption. Efforts to slow economic growth also focus on changes in GDP. Even though the evidence suggests that GDP is a poor measure of human wellbeing and that the negative effects of this limitation are compounded by a perspective that economic growth is an unconditional requirement when developing environmental policy.[31] It remains to be seen whether policy makers can navigate free market economies to mitigate human impacts on the environment. Is there reason for optimism?

---

[26] Global Footprint Network 2014.

[27] Dawkins 2006, Dennett 2009, and Blackmore 2009, provide insight as to how humans might act as "meme machines" by hosting a complex array of memes working as a so-called "memeplex" in the evolution of our species and its many cultures.

[28] World Bank 2014.

[29] Conca et al. 2001.

[30] Rosales 2008, 1411.

[31] van den Bergh 2010.

Reasons for Optimism

Environmental memes such as Gaia and Pachamama inspire some of us to seek a relationship with nature that recognizes her inherent value above and beyond the utility of her resources. Environmental memes such as "green" and "stewardship" inspire others of us to make rational judgments to gauge our interactions with nature in light of the long-term consequences. Given these distinctly different, although not incompatible, views of how we relate to nature, should we be optimistic that environmental memes can actually bring about change that leads to a more sustainable natural world? Memes influence individual action, but how might they induce collective actions by groups of people and their organizations. To illustrate, we need to briefly examine how memes are transmitted within organizations and how leaders can facilitate this transmission for collective action.

It is important that we as leaders of organizations pay attention to the elements that define our particular organization, and that we utilize the appropriate carriers of information for that organization. In this way, memes influence leaders, who in turn spread the word and mobilize the action of their organization, and thus further spread the word with subsequent actions. Organizations are made up of regulative, normative, and cultural-cognitive elements that with associated activities and resources maintain the organization and provide meaning.[32] If the dominant element of an organization is regulative, then laws, rules, protocols and mandated specifications provide meaning and support actions. For example, Ecuador's 2008 constitution reflects a view that citizens hold regarding nature (Pachamama). A reason for optimism is the possibility that their constitution will influence the development of laws to protect nature with consequent court decisions to uphold those laws when they are violated. If the dominant element in an organization is normative, then values, roles and standards provide meaning and support actions. Members within these types of organizations may be more responsive to actions that emerge from dialogue around environmental concerns. For example, many scientists purport a value neutral standard which dismisses the study of Gaia as a living Earth, but this does not preclude a dialogue around the Gaia as a metaphor. In this capacity, Gaia as a meme fuels multiple perspectives that are not mutually exclusive. Some of us may view Gaia as a hypothesis while seeking to understand our biosphere as a system that can be understood in material terms. Others of us may view her as a metaphor for a living Earth. Still others among us may experience her in a spiritual sense. Clearly, the Gaia meme has the capacity to stimulate dialogue and induce actions in a variety of forms to improve our relationship with the natural world. If the dominant element of an organization is cultural-cognitive, then schema, identities and symbols provide meaning and support action. For example, more than four-hundred college and university presidents have signed-on to the The Talloires Declaration since its ratification in 1990, and each new signature is a symbolic act and another commitment by a university to model efforts that promote a more sustainable future.[33] Efforts include green buildings and sound environmental management practices at their respective institutions, research into

---

[32] Scott 2014.
[33] ULSF 2014.

green and related technologies, and educational programming for a sustainable future. Thus leaders and individuals are in a position to enact environmental values within an organization. Ecuador's new constitution, scientists working for a better understanding of our biosphere, and university presidents modeling environmental stewardship are positive examples. But even if their motives are not purely for environmental reasons—such as green-marketing strategies, or the stewardship of company resources to enhance the bottom-line—the environmental memes are still functioning. That said, perhaps our best lessons can be learned from Mother Nature.

Daniel Quinn's story of *Ishmael* reminds us of the possibility that we are of nature, can learn from nature, and most importantly, we must consider the possibility of a world without us. This compelling story is about Mr. Partridge's dialogue with his wise teacher, a gorilla named Ishmael. The story debunks human superiority and offers a bio-centric perspective on life through Ishmael's eyes. Mr. Partridge began to examine his own "Mother Culture." This is the culture that according to Ishmael nurtures our views and endorses our actions. Ishmael's lessons were about "takers" and "leavers" and forced Mr. Partridge to recognize his human-centered views. According to Ishmael, the so-called "civilized" humans living today have become "takers" who exhaust nature's resources. In contrast, our human ancestors, the hunter and gatherer cultures, and some cultures today, are "leavers" because the live in harmony with nature. Throughout the story Mr. Patridge was puzzled by a question that was displayed on a sign in Ishmael's den—"With man gone, will there be hope for gorilla?" The final lesson was delivered loud and clear when, upon Ishmael's death, Mr. Partridge came to realize that the most important question was on the back of the sign—"With gorilla gone, will there be hope for man?"[34] This is an ominous message given the fact that humans, more than ever, are responsible for the extinction of other species.[35] We talk about going green and being good stewards, but we focus our efforts on ways to grow our economies and develop technologies to support a world that is to be filled with humans. It remains to be seen whether or not environmental memes are more than words. We can be optimistic only if these memes continue to replicate and function as benchmarks by which we judge and mediate our interactions with nature.

References

Bible, King James Version, Genesis 1.28. Available at the University of Michigan library online http://quod.lib.umich.edu/cgi/k/kjv/kjv-idx?type=DIV1&byte=1477, accessed February 19, 2014.

Blackmore, Susan. 2009. Dangerous Memes; or, What the Pandorans Left Loose. In *Cosmos & Culture: Cultural Evolution in a Cosmic Context*, edited by Steven J. Dick and Mark L. Lupisella, 297-318. Washington, DC: NASA.

Bloch, Maurice. 2005. *Essays on Cultural Transmission*. Oxford, London: Berg.

---

[34] Quinn 1992, 262-263.
[35] Jowit 2010.

Conca, Ken, Thomas Princen, and Michael F. Maniates. 2001. Confronting Consumption. *Global Environmental Politics* 1(3): 1-10.

Constitution of the Republic of Ecuador. 2008. English translation from Political Database of the Americas, Georgetown University. Available at http://pdba.georgetown.edu/Constitutions/Ecuador/english08.html, accessed February 18, 2014.

Dawkins, Richard. 2006. *The Selfish Gene: 30th Anniversary Edition*. London: Oxford University Press, Inc.

Dennett, Daniel C. 2009. The Evolution of Culture. In *Cosmos & Culture: Cultural Evolution in a Cosmic Context*, edited by Steven J. Dick and Mark L. Lupisella, 125-43. Washington, D.C.: NASA.

Global Footprint Network. 2014. Word Footprint. Available at http://www.footprintnetwork.org/en/index.php/GFN/page/world_footprint/, accessed February 14, 2014.

Hari, Johann. 2010. The Wrong Kind of Green: How Conservation Groups Are Bargaining Away Our Future. *The Nation*. Posted, March 22, 2010. Available at http://www.thenation.com/article/wrong-kind-green, accessed February 19, 2014.

Hayward, Tim. 2005. *Constitutional Environmental Rights*. Oxford, London: Oxford University Press, Inc.

Jowit, Juliette. 2010. Humans Driving Extinction Faster Than Species Can Evolve, Say Experts. *gaurdian.co.uk* Press Release, 7 March 2010. Available at http://www.guardian.co.uk/environment/2010/mar/07/extinction-species-evolve, accessed February 14, 2014.

King James Version of The Holy Bible (publication date) Genesis 1:28. Cleveland and New York: The World Publishing Company.

Letcher, Andy. 2003. Gaia Told Me to Do It: Resistance and the Idea of Nature Within Contemporary British Eco-Paganism. *Ecotheology* 8 (1): 61-84.

Lissack, Michael R. 2004. The Redefinition of Memes: Ascribing Meaning to an Empty Cliche. *Journal of Memetics - Evolutionary Models of Information Transmission* 8. Available at http://cfpm.org/jom-emit/2004/vol8/lissack_mr.html, accessed February 14, 2014.

Lovelock, James. 2003. The Living Earth. *Nature* 426 (18/25): 769-70.

Poulshock, Joseph. 2001. The Problem and Potential of Memetics. *Journal of Psychology and Theology* 30 (1): 68-80.

Quinn, Daniel. 1992. *Ishmael: An Adventure of the Mind and Spirit*. New York, NY: Bantam Books.

Rosales, Jon. 2008. Economic Growth, Climate Change, Biodiversity Loss: Distributive Justice for the Global North and South. *Conservation Biology* 22 (6): 1409-1417.

Rosenthal, Erika. 2010. Cancun Conference Results in Critical Steps Forward. *Earth Justice. News Release: Because the Earth Needs a Good Lawyer*, December 15, 2010. Available at http://unearthed.earthjustice.org/blog/2010-december/cancun-conference-results-small-critical-steps-forward, accessed February 14, 2014.

Scott, W. Richard. 2014. *Institutions and Organizations: Ideas and Interests*. 4th
    ed. Los Angeles, CA: Sage Publications, Inc.

Senge, Peter M., Bryan Smith, Nina Kruchwitz, Joe Laur, and Sara Schley. 2008.
    *The Necessary Revolution: How Individuals and Organizations Are
    Working Together to Create a Sustainable World*. New York, NY:
    Doubleday.

Stober, Spencer S. 2010. Ecuador: Mother Nature's Utopia. *The Journal of
    Environmental, Cultural, Economic & Social Sustainability* 6 (2): 229-
    39.

Stober, Spencer S. 2009. Mother Nature and Her Discontents: Gaia as a Metaphor
    for Environmental Sustainability. *The International Journal of
    Environmental, Cultural, Economic and Social Sustainability* 5 (1): 37-
    48.

The Greenlife. 2014. Simple, Healthy, Sustainable. Available at
    http://site.thegreenlifeonline.org/, accessed February 14, 2014.

The Pachamama Alliance. 2014. Available at https://www.pachamama.org/,
    accessed February, 2014.

ULSF. 2014. *The Talloires Declaration*. Association of University Leaders for a
    Sustainable Future (ULSF). Available at
    http://www.ulsf.org/programs_talloires.html, accessed February 14,
    2014.

US Green Building Council. 2014. Available at
    http://www.usgbc.org/DisplayPage.aspx?CMSPageID=222,   accessed
    February 14, 2014.

UNFCCC. 2010. *United Nations Framework Convention on Climate Change
    (UNFCC)*. Available at http://unfccc.int/2860.php, accessed February
    14, 2014.

van den Bergh, Jeroen. 2010. Environment versus growth: A Criticism of
    "Degrowth" and a Plea for "A-growth." *Ecological Economics* 70 (5).
    881-890.

World Bank. 2014. GDP growth. Available at
    http://data.worldbank.org/indicator/NY.GDP.MKTP.KD.ZG,   accessed
    February 14, 2014.

Part II: Systems

Editors' note: The study of systems is a vital part of the study of sustainability. We inhabit social systems, both face to face and electronic. If people are to contribute positively to sustainability and to make a difference for the better it is necessary for scholars to consider how social systems can be changed to rid them of those attributes that generate environmental degradation. The agent-structure debate is essential here: to what extent do social structures (established patterns of practice and relationships that provide order and predictability to social life) determine behavior; and to what extent may agents generate new, and environmentally more hopeful, structures? This conundrum is addressed in the following quote:

> One of the fundamental differences between social systems (like a business or supply chain) and natural systems (like a rainforest) is that social systems are created by human beings. There can be no "system" without the human actors who inhabit it and take actions that bring it to life. Put differently, how the system works arises form how we work; how people think and act shapes how the system as a whole operates...
>
> Seeing systems and understanding our role in shaping those systems are two sides of the same coin. It is easy to acknowledge this philosophically; it is far more difficult to see how this connection occurs in practice.
>
> Peter Senge, Bryan Smith, Nina Kruschwitz, Joe Laur and Sara Schley (2008)[1]

---

[1] Senge, Peter, Bryan Smith, Nina Kruschwitz, Joe Laur, and Sara Schley. 2008. *The Necessary Revolution: How Individuals and Organizations are Working Together to Create a Sustainable World*. New York: Doubleday, 169.

Chapter 6: The Perfect Storm: Catastrophic Collapse in the 21st Century

Glen Kuecker

Introduction: Framing 21st Century Challenges

Former US vice president Al Gore's documentary *An Inconvenient Truth* has awakened people to the problem of anthropogenic climate change. An April 2007 poll shows that even 60 percent of Republicans now think this issue is serious.[1] As follow-up to the documentary, a group known as the Focus the Nation organized discussions about Gore's dire warnings and possible solutions. Planners envision an energized civil society participating in grassroots forums at their churches, grade schools, offices, and universities. In April 2007 they held a national conference in Las Vegas, where people convened to share results of their meetings and plan the next step in saving the planet. Gore has stimulated a remarkable mobilization, one that has the potential to reach millions of people in consideration of what many think to be the most important issue of the 21st century.[2] It may influence the United States government to take more aggressive measures in combating global warming, especially because other grassroots campaigns, like Bill McKibben's Step It Up are underway.[3] If we depart from hydrocarbon civilization, the mobilization will have resulted in a revolutionary shift in economy and society comparable to the industrial revolution. The paradigm shift called for by Gore's followers would be one of the greatest in history, similar to the abolition of slavery movement.

The current global warming movement, however, has several critical flaws, which may cripple its mission. Foremost, it is a single-issue approach, one that neglects substantive consideration of the interconnections between climate change and several other major crises. Failure to see climate change as part of a larger systemic crisis results in misguided public policy, and a naive confidence in the ease of a paradigm shift. The core of the flaw is the emphasis on

---

[1] Pooley 2007.

[2] http://www.focusthenation.org/

[3] http://stepitup2007.org; McKibben 2007a.

sustainability, which falsely sees collapse not as a current reality but a future event. To avoid collapse the sustainability view takes a reformist position; with just enough timely tinkering the system can go on forever without any radical change. To use the coal miner's canary metaphor, this view sees the canary as alive, although gasping for air and needing better circumstances. Instead, we need to understand that the canary is dead, and we need to evacuate the coalmine. The time for making sustainability happen has long passed, and its belated pursuit will prevent us from addressing the present reality of catastrophic collapse.

Hurricane Katrina, which devastated New Orleans in 2005, may help people in the United States to see catastrophic events as immediate and long lasting challenges to humanity. Katrina woke people up to how inadequately prepared local, state, and federal governments are in the face of large-scale catastrophe, how sophisticated public policy crumbles in application, and how scarce public resources drained by colonial projects undermine vital infrastructure required for sustaining modernity's complex systems. Severe weather events such as Hurricane Katrina also teach how pre-existing race and class inequalities shape catastrophic outcomes by defining who is evacuated, rescued, and attended. Hurricane Katrina also re-mapped the public's geography of catastrophe by bringing an event associated with the global South to the global North.[4] Some wisely question the state of preparation for other inevitable catastrophes.[5] Despite such advances there exists a delusional quality to our thinking, as if we are unwilling to stare directly at the obvious reality that our modern life is structurally flawed on multiple fronts that have converged in starting the preliminary phases of catastrophic systemic collapse.

Optimism in the Face of Collapse

Even among those who approach global crises from a system perspective, there is an overwhelming propensity to construct a master narrative of optimism. Such a narrative maintains that while the system is in deep crisis we still have the opportunity to escape if only we act now. David Korten, clearly illustrates his conviction that the global system is not on a sustainable course, the "Siren Songs" have been played, and we are not listening to the warning because of delusion, ignorance and/or benign neglect.[6] Jared Diamond states:

> The risk of such collapses today is now a matter of increasing concern.... Many people fear that ecocide has now come to overshadow nuclear war and emerging diseases as a threat to global civilization.... Most of these threats.... it is claimed, will become globally critical within the next few decades: either we solve the problems by then, or the problems will undermine not just Somalia but also First World societies. Much more likely than a doomsday scenario involving human extinction or an apocalyptic collapse of industrial civilization would be 'just' a

[4] Walter 2006.
[5] Brinkley 2006; Copper and Block 2006; Horne 2006; and Dyson, 2005.
[6] Korten 1999; 2001.

future of significantly lower living standards, chronically higher risks, and the undermining of what we now consider some of our key values.[7]

He adds, "for the first time in history, we face the risk of a global decline."[8] The authors of *Limits to Growth* clearly argue that the world is in "overshoot," and we are set on an unsustainable course. Their multiple models demonstrate a short time period, within 50 years, before systemic collapse will replace overshoot. They state:

> The global challenge can simply be stated: To reach sustainability, humanity must increase the consumption levels of the world's poor, while at the same time reducing humanity's total ecological footprint. There must be technological advance, and personal change, and longer planning horizons. There must be greater respect, caring, and sharing across political boundaries. This will take decades to achieve even under the best circumstances. No modern political party had garnered broad support for such a program, certainly not among the rich and powerful, who could make room for growth among the poor by reducing their own footprints. Meanwhile, the global footprint gets larger day by day.[9]

The 2006 edition of *State of the World* explains, "Unless we find a couple of spare planets in the next few decades…. it is clear that the current western development model is not sustainable. We therefore face a choice: rethink almost everything, or risk a downward spiral of political competition and economic collapse."[10] Thomas Homer-Dixon also clearly demonstrates the likeliness of systemic collapse, but his *Ingenuity Gap* maintains an optimistic line of analysis.[11] In a subsequent book Homer-Dixon finds collapse to be almost certain, so much so he invites us to prepare for it by considering how catastrophe allows for the total regeneration of civilization.[12] These master narratives each have a firm conviction that the coal miner's canary is very much alive, and inhibit us from considering the canary's actual death. Clifford Geertz suggests the reasons for why we deny the canary's death. He states:

> The main problem, over and above their mind-bending dimensions, is that these various sorts of mega catastrophes seem to most people either so far off, so unlikely, or so thoroughly beyond what they have even vicariously experienced – psychologically off-scale, conceptually out-of-sight – as to be beyond the range of rational estimation or practical response. We are both emotionally disinclined and intellectually ill equipped to think systematically about extreme events. Absorbed as we are in the dailiness of ordinary life, and enfolded by its brevity, the

---

[7] Diamond 2005, 7.
[8] Diamond 2005, 23.
[9] Meadows, Randers and Meadows 2004, xv.
[10] Flavin 2006, xxi-xxii.
[11] Homer-Dixon 2000.
[12] Homer-Dixon 2006.

...calculation of remote possibilities and the comparison of transcendent cataclysms look pointless; comic, even.[13]

Denial invites us to frame catastrophes as unique, disconnected, and particular events. We see unexpected tragedy as the only common ground between September 11th, the Indian Ocean Tsunami of 2004 and Hurricane Katrina. Wrongly framed, our picture of what catastrophes are results in bad public policy that make disasters so much worse when they happen. The consequences are immense, as each catastrophic event carries a bigger punch, and our ability to survive is further compromised.[14]

To correct this problem we need to see catastrophic events as part of a larger process, as multiple points of structural crisis converging to form the Perfect Storm of the global system's catastrophic collapse. We need to understand catastrophic collapse is currently in process, and it is not something looming in the distant future. We need to understand that avoiding the collapse by altering course is not possible, because the time for changing course has long passed. Catastrophic collapse means that framing current reality with analysis defined by sustainability misunderstands reality. Sustainability analysis remains within an old paradigm defined by a conviction that applied reason can solve the problems that may cause collapse. It is premised upon the Enlightenment's project of perfecting the human condition. The catastrophic collapse paradigm takes a post-Enlightenment position. In collapse the perfectibility of the human condition yields to the basics of survival. Of course, surviving collapse will require implementing sustainability and wise application of the Enlightenment's paradigm of science and technology.

Complexity Thinking for the 21st Century

To appreciate this argument an explanation of complex systems is needed. Systems theory, while complex in application, has basic propositions. A system is constituted from multiple sub-systems, which in turn are the product of many interacting parts. Systems, especially as they evolve and become more complex in their structure and functioning, have three paths they can follow. They can reproduce, becoming ever more complex with each successive reproduction. Conversely, their interactions can set-off a catastrophic structural flaw that causes the interacting parts and sub-systems to destroy the system. In this case, the system stops functioning and enters a state of collapse. Between these two paths, rests the third possibility, oscillation. It occurs when systems wobble between the states of reproduction and collapse. Oscillation is an unstable state, sometimes called the "edge of chaos," in which a system can have periods of reproduction before yielding to periods of crisis. Meadows, Randers and Meadows argue we are currently experiencing either oscillation or collapse because the system is in a state of overshoot, in which it has surpassed any capacity for reproduction.[16]

---

[13] Geertz 2005, 5.
[14] Diamond 2005; Geertz 2005;Homer-Dixon 2006, 29-30.
[15] Clark 2002; Perrow 1984; Waldrop 1992; Homer-Dixon 2000, 2006.
[16] Meadows, Randers and Meadows 2004.

Complex systems, especially those that have humans as key components, have remarkable survival mechanisms focused around the capacity to adapt to the unpredictable ways a system can evolve. An adaptive system finds solutions to potentially catastrophic interactions within the system. These solutions are technological fixes, and they are a driving force in systemic reproduction. Such adaptation, however, has important limitations. Often solutions only treat the symptom of the original problem, making the evolving system inherently unstable. The system eventually becomes so complex that it is not possible to predict possible outcomes of systemic interactions.[17] Unpredictability can throw a healthy system into oscillation, and an oscillating system into collapse. Likewise, and more dangerous, adaptation produces new and more complicated interactions in the system – what Homer-Dixon calls "unknown unknowns." They inevitably produce a catastrophic interaction for which innovative technology cannot fix.[18]

The current global system – the totality of systems, subsystems, and interacting parts – is producing a phenomenal level of "unknown unknowns." It is built upon centuries of technological innovations that create stopgap, band-aid solutions to underlying structural problems. We often are unable to anticipate the next systemic crisis, while we continue to compound the problem by adding temporary and misguided solutions. Our ignorance about complexity and unwillingness to take what we do know seriously means catastrophic events increasingly define the human condition. We continue to select public policies that are disastrous in their consequences, and we frequently mask them with professional, authoritative, or expert sounding language. The implications for wrong thinking are immense.

We confront multiple points of structural crisis within the global system. Even in isolation, any one of these points could trigger a tipping point in the system, pushing it away from oscillation and into collapse. Each is capable of driving collapse deeper and faster, overwhelming our capacity to cope. A short list of the ongoing structural crises includes energy, environment, climate, disease, population, economics, and conflict. As the current global system is tightly coupled with a historically unprecedented degree of intensity, we need to recognize that each of the points of crisis is related to the others.[19] Tight coupling can generate positive feedback loops within the system, causing a convergence of structural crises resulting in the Perfect Storm. Our current historical moment is defined by the initial phases of the storm.

The Challenge of Energy Transition

With the price of gasoline floating at $3 a gallon, people in the United States are slowly waking-up from their SUV slumber to discover oil is not a renewable resource. They are also learning that the historical moment when oil is no longer available is rapidly approaching. In fact, most oil companies, economists, and academics agree that peak oil production is likely to occur within the next twenty years, possibly ten. When Hurricanes Katrina and Rita disrupted the flow of

---

[17] Perrow 1984.
[18] Homer-Dixon 2000, 171-187.
[19] Homer-Dixon 2000, 112.

petroleum into and out of the southern refinery sector in 2005, US citizens became keenly aware of our dependence on petroleum as well as the extreme vulnerability of the petroleum infrastructure, the system of production and delivery needed for our energy system to survive. As Daniel Yergin explains, a structural crisis in any part of the infrastructure can cause significant ripples or waves through the entire system – what is called a positive feedback loop – potentially causing regional, national, or global economic crisis.[20] As our pocket books suffer, awareness increases. But, this awareness is profoundly misguided about the problem.

As people pay more at the gas pump, they are seldom aware that peak oil production has profound consequences for the meaning of life on Earth. Our civilization, as Yergin illustrates, is defined by hydrocarbons.[21] They are omnipresent. Petroleum is not only essential for powering transportation systems, but also provides petrochemicals needed for synthetics in manufacturing, those plastics that are in just about everything we consume.[22] The end of oil means we need to substitute petroleum with something new if we are to sustain the current levels and forms of production.[23] Furthermore, hydrocarbons are the key component in manufacturing the fertilizers that drove the Green Revolution's 250 percent increase in grain production and which sustains the current system of intensive agriculture that nearly seven billion people depend upon.[24] While we already experience the preliminary phases of collapse within the fuel sector, peak oil production means we will soon have structural crises in the industrial production and food sectors of the global system.

Transitioning from petroleum to an alternative energy, food, and production source is an extreme challenge. We know that future oil consumption, especially with the needs of huge consumers like China and India, will continue to rapidly increase. Current demand is nearly two million barrels per day, and has an annual rate of increase at 2.25 percent. Global energy consumption will rise by 71 percent, between 2003 and 2030.[26] This increased demand is occurring concurrently with the era of peak production, consequently decline in available oil will be rapid not gradual. Rapid depletion will cause major shocks within the system and make gradual and smooth transition exceptionally difficult to accomplish, especially on a coordinated and cooperative global level. Second, we know that finding alternatives to hydrocarbons is an exceptionally difficult if not impossible proposition.

Those seeking hydrocarbon substitutes focus mainly on energy. There is little discussion about how to supplant petrochemicals that underpin food and manufacturing systems. But even in the area of energy, the prospects are not good. Jeremy Rifkin illustrates that fuel substitutes range from solar, wind, hydro, and nuclear power, but they are not practical replacements. Many have invested great faith in the natural gas "bridge" to carry us over to hydrogen. Our future

---

[20] Yergin 1991.

[21] Yergin 1991, 541-560.

[22] Kunstler 2005, 23.

[23] Roberts 2005, 1-17.

[24] Kunstler 2005, 159-160.

[25] McKibben 2004, 34.

[26] Runge and Senauer 2007, 44.

depends on making that bridge function while transitioning to a new fuel source. If we make it, hydrogen is a safe, renewable, clean, and long-term solution. But, getting to hydrogen is a dubious proposition, because the technology needed is closer to science fiction than application, costs are immense, and public policy is decades late in pushing the change. Oil interests also prevent rapid change, as they are committed to the dead energy system until the market tips in favour of hydrogen. Furthermore, the market's lag time is too long, and, most likely, we have passed the point of seamless transition to hydrogen. With a $600 billion price tag for its infrastructure just in the United States, the costs of a global hydrogen transition may be prohibitive.[27]

Many people think biofuel provides a good hydrocarbon substitute. There are, however, several limitations with biofuel. First, producing biofuel still consumes high levels of energy, which often comes from coal or natural gas. Second, biofuel requires converting large tracts of land for sugar, soybeans, or corn. In many places, tropical forests are being cleared for biofuel production. With continued consumption of hydrocarbons and forest depletion, biofuel actually has the potential to increase greenhouse gases.[28] Third, biofuel deepens food insecurity by putting people and cars in direct competition for fuel.[29] For example, Runge and Senauer observe that "filling the 25-gallon tank of an SUV with pure ethanol requires 450 pounds of corn – which contains enough calories to feed one person for a year." Increasing ethanol production in the United States thus has consequences in countries like Mexico. Ethanol production has driven the price of corn from $2 per bushel to $4.35. This jump resulted in a doubling of Mexico's corn tortilla prices during the last months of 2006, which caused a subsistence crisis for nearly 50 million Mexicans living in poverty. Runge and Senauer warn, "resorting to biofuels is likely to exacerbate world hunger." They state, "the number of food-insecure people in the world [may] rise by over 16 million for every percentage increase in the real prices of staple foods. That means that 1.2 billion people could be chronically hungry by 2025."[30] Biofuel illustrates how the promise of a quick and cheap fix to the energy problem, one that will allow the current system to continue without dramatic change, causes positive feedback loops within the larger system, resulting in grave, unintended consequences. The quick fix is part of the fallacy of the sustainability approach, because it glosses over the problems that emerge when radical change is avoided by shallow thinking.

Climate Change and the Degrading Ecosystem

If we navigate the transportation, production, and food crises, we still have to contend with the climatic consequences of hydrocarbon civilization. Climate is perhaps the most complex of complex systems. A multitude of inputs feed the system generating a "butterfly effect" of positive feedback loops resulting in potentially catastrophic climate change. Climate change illustrates how complex

---

[27] Rifkin 2003.

[28] Moberg 2007, 24-26; Farrell et al 2006, 506-508.

[29] Azar 2005.

[30] Runge and Senauer 2007, 42, 50, and 51.

systems produce "unknown unknowns." While we do, in fact, know global warming is happening, its certain consequences are less known. The 2007 round of reports from the Intergovernmental Panel on Climate Change (IPCC) makes it blatantly clear that there will be negative consequences that require immediate and deep action to mitigate.[31] Kerry Emanuel's research shows that the power and duration of catastrophic storms is directly related to such temperature increases.[32] The IPCC supports this scenario by predicting more frequent and severe storms.[33] Melting of the polar ice caps also influences water temperature. Such change can alter the flow and temperature of the Gulf Stream, the engine that drives weather from North America to Europe. Such change may trigger droughts where predictable rains are known, and change the geography of climate that influences global economic systems. Climate change might do what "Ché" never could by flipping Eduardo Galeano's "upside down world" right side up.[34]

It is possible that the current increase in catastrophic weather events can tip toward conditions unfavorable to life on Earth. We simply do not know. The Pentagon is concerned enough about climate change that it has ordered studies of worse case scenarios and how they impact national security. One study describes how climate change causes the breakdown of state systems resulting in civil wars, social upheaval, and mass migration. This scenario is beyond the capacity of the United States to manage.[35] The United Kingdom's Ministry of Defence is likewise concerned, listing climate change as one of three global security threats along with globalization and global inequality.[36] We only need visit post-Katrina New Orleans, a comparatively tiny example of response failure, to appreciate this concern.[37]

Climate change directly interacts with the global ecological system, one that is under immense stress even without climate change. As with climate change, the impact of altering ecological systems is very hard to predict. There are some basic guidelines, however. We know that life depends upon diversity for evolution.[38] Extinction rates are far greater than at any point in human history, which means we are pulling key pieces from nature's design for a healthy complex adaptive system.[39] Edward O. Wilson estimates 27,000 species are lost per year, and by 2022 an amazing 22 percent of all species will be extinct if we do not change course.[40] The Living Planet Index, which tracks population trends for over 1,100 species, finds that biodiversity has declined by 40 percent between 1970 and 2000.[41] Each forest cut-down, each mega-dam constructed, each car that pollutes adds to the destruction of diversity. The Earth has 24 major ecosystem "services" that sustain life. Scientists estimate 15 of them are degraded or have already

---

[31] McKibben 2007b; Intergovernmental Panel on Climate Change 2007.
[32] Emanuel 2005.
[33] Intergovernmental Panel on Climate Change 2007.
[34] Galeano 2000.
[35] Schwartz and Randall 2003.
[36] Development, Concepts, and Doctrine Centre 2007, xii.
[37] Brinkley 2006; Horne 2006.
[38] Berry 1988, 45; Wilson 1992.
[39] Meadows, Randers and Meadows 2004, 85-86.
[40] Wilson 1992.
[41] Assadourian 2006, 92.

tipped past sustainable limits. This destruction is a defining feature of catastrophic collapse, and it generates the positive feedback loops that prevent soft landings and post-collapse survival.[42]

Adaptive responses to problems in one part of the global system can cause unanticipated, negative reactions in other parts of the system.[43] The World Bank, for example, promotes shrimp farming throughout Latin America's Pacific coast in order to diversify economies. The farms, however, destroy coastal ecologies that ripple through interior ecological systems.[44] Such destruction makes poverty worse, and can stimulate forces like migration and urbanization, causing problems in other parts of the system.[45] Resource extraction, especially to meet high consumer demand in the first world, harms important ecological regions such as the Amazon and Indonesian forests. The "sinks" necessary for ecological systems to reproduce are increasingly clogged if not eliminated.[46]

Pandemics and Demographic Transitions

Some parts of ecology's complex system are deadly. An important one is disease. We already see the reality of catastrophe with the AIDS epidemic in Africa. Our inability to stop positive feedback loops that reproduce the epidemic is a grave cause of concern for how we respond to the next great pandemic, which many think will be the avian flu, a lethal strain known as H5N1. While a catastrophic outbreak has not yet happened, public health officials are in agreement that the question is not if it will happen but when.[47] This warning means that all the systemic factors for a major pandemic are now in place, with exception of the virus's ability to spread from human to human, which could happen at any unpredictable point in the future. H5N1 may kill more people than the great influenza pandemic of 1918-1919, when upwards of 40 million people died globally. Illustrating the centrality of complex systems for understanding pandemics, H5N1 is geographically located in Asia, and is spreading there because of the region's dramatic economic boom, which has caused mass movements of people, and stimulated the production of poultry. As more chickens are produced to feed exploding urban populations – China produces 13 billion chickens today, but only 12.3 million in 1968 – the greater the possibility the virus will jump from birds to humans.[48] When that happens, the consequences for humanity will be catastrophic. Human cases so far have a 30-70 percent mortality rate.[49] With no immunity humans are vulnerable to a pandemic that could rapidly kill millions of people. Maximum global vaccine production is estimated at 300 million, a fact that raises the disturbing question of whom among nearly 7 billion people would get the vaccine, assuming we can develop one that

---

[42] Millennium Ecosystem Assessment 2005.

[43] Homer-Dixon 2000.

[44] Public Citizen 2005; Chafe 2006, 100-101.

[45] Rich 1994.

[46] Meadows, Randers and Meadows 2004, 51-127.

[47] Osterholm 2005; Greger 2006.

[48] Osterholm 2005, 37; Garrett 2005, 10-11.

[49] Garrett 2005, 14.

responds to the complexities of H5NI. [50] Our experience with severe acute respiratory syndrome (SARS), which only lasted six months in 2002-03, teaches that the economic costs of a much larger event could be staggering. SARS cost the Asian-Pacific region $40 billion.[51] It is estimated that a H5N1 pandemic would kill 142.2 million people and cost $4.4 trillion in lost GDP.[52] A H5N1 outbreak would shut down trade, causing a collapse in the tightly coupled system of "just-in-time" globalized production. The influenza would strike labor and management, causing organizational hierarchies to falter. As quarantines are established, schools and factories would close. Health care systems, already stressed even in the most developed countries, would collapse.[53]

The fuel, productive, climate, environmental, and disease challenges facing humanity are compounded by demographic factors. In a few generations global population will be 9.5 billion people. The two billion additional people will live in a global South already struggling to survive. A dramatic jump in the megalopolis phenomena is happening. In 2007, demographers predict the majority of humanity will live in cities, a world historical milestone in the human experience. It is estimated that 70 million people a year migrate from rural to urban. That's 1.4 million individuals a week, 200,000 a day, 8,000 an hour, or 130 per minute. At this pace, by 2030 there will be three billion urban squatters. If we are to house them, 35 million new houses need to be constructed each year, about one every second.[54] Analysts like Mike Davis do not hesitate in calling our cities the dumping grounds for surplus population, as urbanization is defined by the slum phenomena.[55] Today's Mexico Cities are more common, taxing the limits of weak nation-states to cope. Global South demographic growth aggravates ecological problems, and accelerates the depletion of scarce resources. Provisioning clean water and sanitation in the megacities already stresses urban systems and their populations.[56] Lack of potable water is becoming more extreme, and some cities, such as El Paso-Ciudad Juárez are predicted to meet unsustainable scarcity by 2025.[57] Suketu Mehta estimates half of Mumbai's 13 plus million people do not have toilets or sewage. They produce 2.5 million kilos of excrement each day. As it evaporates, suspended fecal dust enters the air and is breathed by all of Mumbai's inhabitants.[58] Lack of basic infrastructure will increase in the twenty-first century as more global South cities become megalopolises. In China, 114 million people have migrated to cities; another 250 to 300 million will follow in the decades ahead.[59] In the last eleven years the number of Chinese cities with one or more million inhabitants has grown to 41. China currently has 16 of the world's 20 most populated cities.[60] Shanghai, in

---

[50] Garrett 2005, 17-18.
[51] Osterholm 2005, 28.
[52] Osterholm 2007, 48.
[53] Osterholm 2005; Greger 2006.
[54] Neuwirth 2006, xiii.
[55] Davis 2006, 174-198.
[56] McGranahan and Satterthwaite 2007.
[57] U.S. Water News Online 1998.
[58] Mehta 2004.
[59] Pei 2005, 57.
[60] Lee 2007, 9.

2005, built more building space than exists in all of New York City's offices, and every month China adds urban infrastructure equal to Houston, Texas, the fourth largest city in the United States.[61] Consequent increases in resource consumption and epidemic factors come with such dramatic transitions, as well as the problem of reproducing complex urban systems in a stressed world.

Those who are not absorbed by cities continue their migrations, most often to the global North. Colin Powell informs that at least 180 million people do not reside in their countries of birth. The income they send home, totalling about $93 billion, outpaces "official" development funds by $16 billion.[62] Such incomes often constitute the second and third most important items to the gross domestic product in the global South. Future survival in the global South will depend more on migration, while the global North will continue apartheid approaches, such as those not so subtly advocated by nativists like Samuel Huntington, to keep migrants out.[63] The collision between increasing levels of migration and deeper host country restrictions means the current humanitarian crisis facing migrants will most likely worsen.[64]

Plague and famine, as we know from the human made tragedy of Africa, are additional components to the demographics of a complex system in collapse. Demographic stress manifests itself in genocides and resource wars. As Michael Klare shows, millions have perished since the end of World War 2 in vicious civilian conflict.[65] Competition for scarce, non-renewable resources leads agents of developed world consumers to interact with local dynamics of failed states, ecological stress, and extreme poverty in causing these deadly wars.[66] Over five million people died in these wars during the 1990s, while six million fled to other countries, and roughly 15 million were internally displaced. The proliferation of weapons of mass destruction makes these conflicts more dangerous and harder to contain within the geopolitical periphery.

Geopolitics of Systemic Collapse

Paul Kennedy clearly illustrates how global hegemons succumb to decline and collapse because of imperial overreach.[67] History teaches that the increased military expenditures required for managing complex global systems undermine the actual foundations making the great powers great. History also teaches that during moments of imperial overreach major shocks to the global system result. Less understood is the connection between overreach and the inability to manage the vexations of systemic collapse, and there is very little doubt that it is already very expensive. The United States has spent nearly a trillion dollars on military operations since September 11, 2001, and there is no foreseeable end to that money drain. Compare these costs of empire to the $15 billion, five-year program pledged by the Bush administration to fight AIDS, tuberculosis, and malaria, and

---

[61] Hughes and Sawin 2007, 93.
[62] Powell 2005, 32.
[63] Huntington 2004a; 2004b.
[64] Walter 2006, 117-139.
[65] Klare 2001.
[66] Tabb 2007; Ross 2003.
[67] Kennedy 1987.

we begin to see how overreach prevents us from attacking the real mass killers.[68] With 26 million deaths and 40 million current afflictions, AIDS has killed far more people than Saddam Hussein. The 2006 discretionary Department of Defense budget is a staggering $419.3 billion, which constitutes nearly half of the entire discretionary budget of $840.3 billion. The military's budget is roughly equivalent to the current United States federal deficit, and represents an unhealthy portion of its GDP. The cost of empire means less money is spent on crucial domestic infrastructure, a truth painfully revealed by Hurricane Katrina and the Interstate 35 bridge collapse in Minneapolis in 2007. Such investments are necessary for competing in the globalized economy, especially with countries and regions that are not making comparable military expenditures.

Transitions in the global system's hegemonic power are moments of great instability and danger.[69] The last transition from Great Britain to the United States came with two global wars that killed millions of people. As the center of the global capitalist system increasingly moves away from the United States; uncertainty and stress in the complex system of international finance, trade, production, and wealth increase. Reasonable people think Asia is becoming the next geographic center of global capitalism. Jeffrey Sachs, for example, estimates that by 2050, China's economy might be 75 percent bigger than the US economy.[70] By 2003, $450 billion of foreign money entered the Chinese economy. Much of it flows from the United States as we consume vast quantities of Chinese production. China, in 2004, had a $480 billion stake in US securities markets.[71] China is the second largest importer of oil, constituting 31 percent of the world's oil demand growth. Its 9 percent economic growth rate equates to a need for 1 million barrels of oil every day.[72] This oil demand finds China aggressively forming close partnerships with oil producers in the Middle East, Africa, and Latin America, often with enemies of the United States. China negotiates trade and energy deals throughout the world, quietly positioning itself as a global economic power. Quality statesmanship is required to navigate the potential for serious military conflict in the years ahead, especially as the United States and China compete for scarce resources, investment opportunities, and markets.[73]

Economic Stress

The current sub-prime interest rate global financial crisis demonstrates several key points about the structural vulnerability of the global economy. First, keeping the global system in growth phase is accruing huge costs to the system. The economy is in overshoot, and we are extending more and more inputs to keep it alive. We are in a classic positive feedback loop, one defined by the "growth imperative" generating steeply negative returns that push the system deeper into collapse. The costs for reproducing the global economy are so immense they

---

[68] Garrett 2007, 19.
[69] Mahbubani 2005, 50.
[70] Fishman 2006, 17.
[71] Fishman 2006, 265.
[72] Fishman 2006, 117.
[73] Jianhai and Zweig 2005.

make the system unsustainable. The United States currently serves as the consumer of last resort, a role that sustains global economic growth, especially in China and India. This system rests upon an unending capacity of the US consumer to increase debt, an unsustainable proposition. Average credit card debt balance per cardholder is $4,956 at end of 2005, and unpaid credit card balances at end of 2005 reached $838 billion.[74] Other significant ticking time bombs lurk within the system. The aging population in the global North combined with its population decline means a major crisis in retirement is right around the corner.[75] Matching the pension problem is the growing crisis in health care and insurance in the global North.

These structural problems can merge with larger systemic problems in overwhelming the economy. The end of oil's transformation in energy, production, and food systems carries heavy price tag. Even if the global North can afford the transition, billions upon billions live in areas of the world that have no ability to pay for the new technologies, production, and distribution systems. Increased natural disasters are taxing the system's ability to pay for them. Caring for vulnerable populations might be well beyond the capacity of humanity to fund. The cost of war keeps increasing, and pulls crucial funds from other systemic needs, such as the transition away from oil and contending with natural disasters. There is a wide range of costs not calculated in how capitalism balances its accounts, especially when measured by GDP. These costs include ecological destruction, poor health conditions, resource depletion, and pollution. High levels of systemic stress combined with high costs, especially in a tightly coupled globalized economy, mean the system's resilience is compromised.

Conclusion: The Coming Global Apartheid

When confronted with the Perfect Storm, optimists may think that we need to focus on changing course. This view, however, is part of the problem. We have long ago passed the time when we could have changed course. Our complex system carries immense momentum generated by the positive feedback loops driving it.[76] We are a freight train attempting to stop quickly when its speed and weight will carry it into doom. It is foolish to only speak of alternatives. Instead, what we need is sober and sophisticated discussion about preparing as best we can for catastrophic collapse.

A core problem exists in our ability to understand the meaning of the Perfect Storm for the human experience. The systems we have designed are so complex we have trouble understanding how they work in totality. As Homer-Dixon explains in *Ingenuity Gap*, we are good at seeing the parts of the system, but falter with the bigger picture.[77] In part, this problem reflects the consequences of the fragmentation of knowledge within academia, as well as the specialization needs of capitalist society. Consequently, we conflate symptoms for causes, and tend to see problems as isolated, unfortunate occurrences within an otherwise

---

[74] Foster 2006.

[75] Barnett 2004, 206-214; Ghilarducci 2006.

[76] Meadows, Randers and Meadows 2004, 141-145.

[77] Homer-Dixon 2000, 171-187.

healthy system. Few people frame discussion of Hurricane Katrina, the 2004 Tsunami, the Iraq War, China's emergence, AIDS, mass migration, genocides, or September 11th as interrelated outputs of a complex system that is in the process of collapse. Until the bigger picture starts to shape our discussions, we will have immense troubles surviving the Perfect Storm.

More troublesome, the failure to frame the bigger picture distorts our ability to devise public policy that can reduce the impact of the Perfect Storm. Very powerful analytical tools, such as data mining, can do wonders in navigating catastrophe, but they are only as good as the information we put in, and that information is flawed if we fail to understand what the complex system is doing. Likewise, misuse of tools from the social and hard sciences can exacerbate an immense problem already generated by the early salvos of the Perfect Storm, that being global apartheid.

As the complex system continues to become unglued, the historical distinctions between rich and poor, racial minority and majority, men and women, the global North and South will intensify. Social and economic injustice will define how we respond to particular crises and the overall condition of humanity during collapse. We will have to decide among the 7 to 9 billion people who gets medicine, water, heat, shelter, and food. We will decide who lives and who must die. As environmental justice scholars demonstrate, we already make these decisions, and the results are brutal for much of humanity.[78] An estimated 2.7 billion people survive on $2 a day, while one billion children face severe nutritional deprivation.[79] Yet, we pretend these decisions are not made, and we ignore the brutal mechanisms of repression and exclusion required to enforce these decisions. As catastrophic collapse deepens, our global apartheid will become more severe. The walls we have already built and militarized in places like Palestine and the United States-Mexican border will become ever more present. Sophisticated technologies will be horded by the haves and deprived for the have-nots. Ever increasing segments of the global population will be excluded from health care, food, water, shelter, and work. Exactly what this reality will mean to the human condition is hard for us to imagine. It means the Enlightenment's faith in the perfectibility of the human condition will give way to a new vision for humans, one rooted in the basic need to survive as a species. The conservative view of the state of nature is well positioned to further manipulate our darker angels.

References

Agyeman, Julian. 2005. *Sustainable Communities and the Challenge of Environmental Justice*. New York: New York University Press.

Assadourian, Erik. 2006. Global Ecosystems Under More Stress. In *Vital Signs: 2006-2007. The Trends that are Shaping our Future*, edited by Linda Starke, 92-93. New York: W.W. Norton/Worldwatch Institute.

Azar, Christian. 2005. Emerging Scarcities: Bioenergy—Food Competition in a Carbon Constrained World. In *Scarcity and Growth Revisited: Natural*

---

[78] Agyeman 2005.
[79] Homer-Dixon 2006, 186-187.

Resources and the Environment in the New Millennium, edited by R. David Simpson, Michael A. Toman, and Robert U. Ayres, 98-120. Washington, DC: Resources for the Future.

Barnett, Thomas. 2004. *The Pentagon's New Map: War and Peace in the Twenty-First Century*. New York: Berkley Books.

Berry, Thomas. 1988. *The Dream of the Earth*. San Francisco: Sierra Club Books.

Brinkley, Douglas. 2006. *The Great Deluge: Hurricane Katrina, New Orleans, and the Mississippi Gulf Coast*. New York: Harper Collins.

Chafe, Zoë. 2006. Disappearing Mangroves Leaves Coasts at Risk. In *Vital Signs: 2006-2007. The Trends that are Shaping our Future*, edited by Linda Starke, 100-101. New York: W.W. Norton/Worldwatch Institute.

Clark, Robert. 2002. *Global Awareness: Thinking Systematically About the World*. New York: Rowman and Littlefield Publishers.

Cooper, Christopher and Robert Block. 2006. *Disaster: Hurricane Katrina and the Failure of Homeland Security*. New York: Times Books.

Davis, Mike. 2006. *Planet of Slums*. New York: Verso.

Development, Concepts, and Doctrine Centre. 2007. *Global Strategic Trends*. United Kingdom: Ministry of Defence.

Diamond, Jared. 2005. *Collapse: How Societies Choose to Fail or Succeed*. New York: Viking.

Dyson, Michael Eric. 2005. *Come Hell or High Water: Hurricane Katrina and Natural, Racial and Economic Disasters*. Cambridge, MA: Basic Civitas.

Emanuel, Kerry. 2005. Increasing Destructiveness of Tropical Cyclones Over the Past Thirty Years. *Nature* 436: 686-688.

Farrell, Alexander, Richard J. Plevin, Brian T. Turner, Andrew D. Jones, Michael O'Hare, Daniel M. Kammen. 2006. Ethanol Can Contribute to Energy and Environmental Goals. *Science* 311: 506-508.

Fishman, Ted. 2006. *China Inc.: How the Rise of the Next Superpower Challenges America and the World*. New York: Scribner.

Flavin, Christopher. 2006. Preface. In *State of the World: China and India*, edited by Linda Starke, xxi-xxii. New York: W. W. Norton/Worldwatch Institute.

Foster, John Bellamy. 2006. The Household Debt Bubble. *Monthly Review* 58 (1): 7-11.

Galeano, Eduardo. 2000. *Upside Down World: A Primer for the Looking-Glass World*. Translated by Mark Fried. New York: Picador.

Garrett, Laurie. 2005. The Next Pandemic. *Foreign Affairs* 84 (4): 3-23.

Garrett, Laurie. 2007. The Challenge of Global Health. *Foreign Affairs* 86 (1): 14-38

Geertz, Clifford. 2005. Very Bad News. *The New York Review of Books*, March 24, 4-6.

Ghilarducci, Teresa. 2006. The End of Retirement. *Monthly Review* 58 (1): 12-27.

Greger, Michael. 2006. *Bird Flu: A Virus of Our Own Hatching*. New York: Lantern Books.

Homer-Dixon, Thomas. 2000. *The Ingenuity Gap*. New York: Alfred A. Knopf.

Homer-Dixon, Thomas. 2006. *The Upside of Down: Catastrophe, Creativity, and the Renewal of Civilization*. Washington, DC: Island Press.

Horne, Jed. 2006. *Breach of Faith: Hurricane Katrina and the Near Death of a Great American City*. New York: Random House.

Hughes, Kristen and Janet Sawin. 2007. Energizing Cities. In *State of the World: Our Urban Future*, edited by Linda Starke, 90-107. New York: W.W. Norton/Worldwatch Institute.

Huntington, Samuel. 2004a. The Hispanic Challenge. *Foreign Policy* 141: 30-45.

Huntington, Samuel. 2004b. *Who Are We? The Challenges to America's National Identity*. New York: Simon and Schuster.

Intergovernmental Panel on Climate Change. 2007. *Climate Change 2007: The Physical Science Basis: Summary for Policymakers. Contribution of Working Group I to the Fourth Assessment Report of the Intergovernmental Panel on Climate Change*. Geneva: IPCC.

Jianhai, Bi and David Zweig. 2005. China's Global Hunt for Energy. *Foreign Affairs* 84 (5): 25-38.

Kennedy, Paul. 1987. *The Rise and Fall of the Great Powers*. New York: Vintage Books.

Klare, Michael. 2001. *Resource Wars: The New Landscape of Global Conflict*. New York: Henry Holt and Company.

Korten, David. 1999. *The Post Corporate World: Life After Capitalism*. West Hartford, CT: Kumarian Press.

Korten, David. 2001. *When Corporations Rule the World*. Second edition. West Hartford, CT: Kumarian Press.

Kunstler, James Howard. 2005. *The Long Emergency: Surviving the Converging Catastrophes of the Twenty-First Century*. New York: Atlantic Monthly Press.

Lee, Kai. 2007. An Urbanizing World. In *State of the World: Our Urban Future*, edited by Linda Starke, 3-21. New York: W.W. Norton/Worldwatch Institute.

Mahbubani, Kishore. 2005. Understanding China. *Foreign Affairs* 84 (5): 49-60.

McGranahan, Gordon and David Satterthwaite. 2007. Providing Clean Water and Sanitation. In *State of the World: Our Urban Future*, edited by Linda Starke, 26-45. New York: W. W. Norton/Worldwatch Institute.

McKibben, Bill. 2004. Crossing the Red Line. *The New York Review of Books*, June 10, 32-36.

McKibben, Bill. 2007a. Warning on Warming. *The New York Review of Books*, March 15, 44-45.

McKibben, Bill. 2007b. This April... Red + Blue Go Green. *In These Times*, April, 20-23.

Meadows, Dennis, Jorgen Randers and Donella Meadows. 2004. *Limits to Growth: The 30-Year Update*. White River Junction, VT: Chelsea Green Publishing Company.

Mehta, Suketu. 2004. *Maximum City: Bombay Lost and Found*. New York: Alfred A. Knopf.

Millennium Ecosystem Assessment. 2005. *Living Beyond our Means: Natural Assets and Human Well-Being*. Washington, DC: United Nations Environment Program.

Moberg, David. 2007. Biofuels: Promise or Peril? *In These Times*, April, 24-26.

Neuwirth, Robert. 2006. *Shadow Cities: A Billion Squatters, a New Urban World*. New York: Routledge.

Osterholm, Michael. 2005. Preparing for the Next Pandemic. *Foreign Affairs* 84 (4): 24-37.

Osterholm, Michael. 2007. Unprepared for a Pandemic. *Foreign Affairs* 86 (2): 47-57.

Perrow, Charles. 1984. *Normal Accidents: Living with High-Risk Technologies*. New York: Basic Books.

Pei, Minxin. 2005. Dangerous Denials. *Foreign Policy* 146: 56-58.

Pooley, Eric 2007. The Last Temptation of Al Gore. *Time*, May 28, 30-39.

Powell, Colin. 2005. No Country Left Behind. *Foreign Policy* 146: 28-35.

Public Citizen. 2005. Fishy Currency: How International Finance Institutions Fund Shrimp Farms. Available at http://www.foodandwaterwatch.org/doc/FishyCurrency-WEB.pdf, accessed July 4, 2014.

Rich, Bruce. 1994. *Mortgaging the Earth: The World Bank, Environmental Impoverishment, and the Crisis of Development*. Boston: Beacon Press.

Rifkin, Jeremy. 2003. *The Hydrogen Economy*. New York: Tracher/Penguin.

Roberts, Paul. 2005. *The End of Oil: On the Edge of a Perilous New World*. New York: Houghton Mifflin Company.

Ross, Michael. 2003. The Natural Resource Curse: How Wealth Can Make You Poor. In *Natural Resources and Violent Conflict: Options and Actions*, edited by Ian Bannon and Paul Collier, 17-42. Washington, DC: The World Bank.

Runge, C. Ford, and Benjamin Senauer. 2007. How Biofuels Could Starve the Poor. *Foreign Affairs* 86 (3): 41-53.

Schwartz, Peter and Doug Randall. 2003. *Imagining the Unthinkable: An Abrupt Climate Change Scenario and Its Implications for United States National Security Policy*. Washington, DC: United States Department of Defense.

Tabb, William. 2007. Resource Wars. *Monthly Review* 58 (8): 32-42.

U.S. Water News Online. 1988. Southwest Cities will Face Future Water Shortages. Available at http://www.uswaternews.com/archives/arcconserv/8soucit11.html, accessed March 31 2014.

Waldrop, M. Mitchell. 1992. *Complexity: The Emerging Science at the Edge of Order and Chaos*. New York: Simon and Schuster.

Walter, Jonathan. 2006. Death at Sea: Boat Migrants Desperate to Reach Europe. In *World Disasters Report: Focus on Neglected Crises*, edited by Jonathan Walter, 117-140. Geneva: International Federation of Red Cross and Red Crescent Societies.

Wilson, Edward O. 1992. *The Diversity of Life*. Cambridge: Harvard University Press.

Yergin, Daniel. 1991. *The Prize: The Epic Quest for Oil, Money, and Power*. New York: Simon and Schuster.

Chapter 7: Creating Sustainable Organizations in a Globalizing World: Integrating Anthropological Knowledge and Organizational Systems Theory

Kimberly Porter Martin

Western cultural views of how best to organize and lead (the majority paradigm in use in the world) are contrary to what life teaches. Western practices attempt to dominate life; we want life to comply with human needs rather than working as partners. This disregard for life's dynamics is alarmingly evident in today's organizations. Leaders use control and imposition rather than self-organizing processes. They react to uncertainty and chaos by tightening already feeble controls, rather than engaging our best capacities in the dance. Leaders use primitive emotions of fear, scarcity, and self-interest to get people to do their work, rather than the more noble human traits of cooperation, caring, and generosity. This has led us to this difficult time, when nothing seems to work as we want it to, when too many of us feel frustrated, disengaged, and anxious.[1]

One of the contemporary buzzwords in today's world is "sustainability." Unlike most fashionable terms, sustainability is not likely to become obsolete. In fact, sustainability will surely become more central to how human beings cope with their world as we move into the future. The ultimate challenge for the human species is how to survive without exhausting Earth's resources and significantly impacting the environment. In the past, human cultures operated with a set of arrogant, largely unconscious assumptions as they evolved into the densely populated, urban, industrialized societies that dominate the world today. It was assumed that the resources needed to support human societies would continue to be available, or that humans would be able to manipulate resources (through domestication, industrialization, etc.) in increasingly efficiently ways to insure

---

[1] Wheatley 2005, 1.

plenty. If these strategies didn't work, humans could always migrate to new places where resources were plentiful. Alternatively, they could use force or global corporations to get what they wanted from other parts of the world.

Today we have international transportation and communications systems that allow us to see what is happening to the world's resources. And we have found that our old assumptions don't work anymore. As a species we long ago outstripped natural resources in most parts of the world. We have domesticated animals and plants with ever increasing efficiency to the point where we can now produce them artificially through cloning. We have replaced biological diversity with monoculture crops and domesticated animals, and converted natural habitats for agriculture. Industrialization has further increased our efficiency, but also creates pollution that has had significant environmental impact. Even many of the habitats and life forms in the vast oceans are vanishing victims of our actions. If we go on as we have begun, we will dramatically change environmental conditions and exhaust the resources that have allowed the human species to thrive. We must change the ways we think about and use resources. The sustainability movement is the result of that fact.

Until recently, the concept of sustainability has been associated only with the physical and biological environments upon which we depend. Beginning in the 1960's, the writings of people like Rachael Carson and Paul Ehrlich have urged us to better understand our relationship to the Earth and its resources.[2] But the sustainability movement has been naïve in asking for change without adequately addressing culture and the cultural systems and core values that underlie our relationship with the environment. It has assumed that if people are informed about the future consequences of their actions, they will act rationally and change. However information is not enough. Real change in sustainability must be built on real systemic cultural change at the individual, local, regional and global levels. Our most urgent challenge is how to create cultures that value sustainability and are organized to accomplish it. Modern Western culture does not have experience tracking the rapid rates of change we face, much less responding to it. Our dominant organizational paradigms are based on traditions that have worked in the past, and are characterized by high levels of competition and conflict. These paradigms will not work in times of increasingly rapid change. We need different kinds of complex cultural systems if we are to cope effectively with the future.

Anthropology, Culture, and Sustainability

Once we recognize that understanding and managing culture change is essential to the success of any and all sustainability efforts, the anthropological knowledge base becomes critical. Anthropology is the only discipline that formed around culture as its central concept. For the past 100 years, anthropologists have examined culture from every possible perspective including its relationship to biology, psychology, language, material culture, belief systems and lifestyles. Documenting and understanding culture change has always been central to work in anthropology. The comparative method has provided insights about why

---

[2] Carson 1962; Ehrlich 1968.

individual societies have followed different cultural paths, while shedding light on what social and psychological qualities are inherently human and what qualities are acquired or learned as a part of membership in a social group.

Theory in anthropology has long recognized the complex integration of customs, beliefs and behaviors that make up all cultural systems. One small change may ripple throughout the society's culture, affecting economic, political, religious and social systems. A classic example is the introduction in the 1930s of steel axes by missionaries to the culture of the Yir Yiront, an Australian Aboriginal group. Before the missionaries arrived, the Yir Yiront were hunter-gatherers. They had a relatively simple material culture in which stone axes were important both practically and symbolically. The stone from which the axes were made was largely acquired through trade and ritual relationships with other aboriginal groups. Axes were important tools that were used to build huts, and to acquire food and the fuel for cooking it. In addition, axes had important religious and ritual significance for the Yir Yiront. They belonged only to high status adult men. Women, children and men who did not own an axe depended on axe-owning clansmen to complete their daily work, and no one outside a man's clan was allowed to use his axe. An axe owner had status and prestige, and was the keeper of an important religious and survival tool for members of his clan.

The missionaries saw that axes were an important practical necessity for Yir Yiront daily life, and, wanting to make people's lives easier, they imported steel axes. They gave them to anyone who wanted an axe, including women and children. Suddenly trade relationships essential to maintaining social networks between Aboriginal groups were rendered useless. Age and gender social hierarchies were undermined. Eventually the religious and ritual system collapsed. The well-meaning missionaries had not understood the multiple functions that the stone axes played in the society. They had wanted to help; instead, they destroyed a way of life. The anthropology literature is full of stories like this one, stories that demonstrate the complex interaction and interdependence of culture traits and customs.

For many decades, anthropologists have used their knowledge of the nature and consequences of culture change to act as cultural brokers for traditional societies trying to figure out to what degree and in what ways they want to enter the modern world. Applied anthropologists work in companies like General Motors, helping corporate executives understand the complexities of the work cultures that form the basis of their production lines. They consult for marketing, health care and educational organizations, helping to build sustainable change in the ways that modern cultural systems interact with one another. Awareness of the tremendous complexity of engineering culture change is essential to creating a sustainable world, and anthropologists have a great deal to offer in understanding how such change might be accomplished

So far I have used the word "culture" a great deal without defining it. There are many working definitions for this term, each of which is appropriate to certain contexts of analysis or action. Here I will provide a definition that is very broad, and can be used in many contexts. Culture is the abstract, learned, shared rules/standards/patterns used to interpret experience and to shape behavior. This

---

[3] Sharp 1952.

definition differentiates between behaviors and culture. Here behavior and material goods are a product of the culture, the abstract ways of thinking that characterize people in a particular group. The capacity for culture may be genetically programmed in humans, but the culture they use is learned. The shared quality of culture is what makes the behaviors and beliefs of one individual intelligible to others in his or her group. Because of shared culture, you know what to expect from others, and they from you.

The entirety of any given culture is not recorded anywhere, but exists in the minds of each individual. According to this definition, culture is an emergent property resulting from the interaction of individuals living in a group. No two individuals will have identical cultures, but the abstract patterns they use will have enough in common so that they can relate to one another. Individuals experiment with new rules or patterns, and these innovations either spread because others find them useful and superior to the old ways, or they are rejected as deviant and disruptive to the system and soon die out. In this way, culture is a fluid phenomenon, created every day by the behaviors of the individual members of the group for which it organizes the world. Each individual in the group is constantly balancing conservation of patterns that have served well in the past with trials of innovative new strategies that may or may not work better in new circumstances.

Culture is an essential component of sustainability because all social and organizational systems are inherently cultural. Culture is, after all, the primary human adaptation that has allowed us to adjust to and manipulate our environment and resources so that humans are able to live in an astoundingly wide range of environments around the world. Overall, culture has evolved generally in the direction of increasing social and economic complexity that has allowed dense populations to flourish in urban environments. Increased population size and density have occurred in tandem with increasing social hierarchy, intensification of competition over resources, and "command and control" organizational principles that are based on enforcement by those at the top of the hierarchy. The cultural patterns associated with population density and cultural complexity are precisely those that have been based on the assumption that Earth's resources are limitless, and will always be able to provide for human life on Earth. These cultural assumptions have not allowed us to recognize in any realistic way that resources are truly limited. Our challenge now is to engineer culture change that makes these assumptions explicit and acknowledges that they are no longer tenable. This means massive culture change, locally, regionally and globally, a daunting task at best.

Organizational Culture from Systems Theory

Systems theory was first introduced during the mid-twentieth century by scientists like von Bertalanffy and Wiener, and was transformed as it spread from the world of mathematics through the biological, physical and social sciences.[4] Anthropologists began to use systems theory to model human cultures and their

---

[4] Von Bertalanffy 1950; Wiener 1967.

relationship to their environments in the 1960s.[5] This paper focuses on one branch of systems theory developed in the 1990s that addresses how complex systems self-organize, learn and adapt to changing environments.

Peter Senge, Margaret Wheatley and Dee Hock are three theorists who represent the kinds of ideas currently being used to understand self-organizing, complex systems as represented in human organizations.[6] Their work has primarily been done in corporate, government and educational contexts within modern Western cultures. Each of these three theorists has a slightly different focus, but all believe that organizations, whether local, regional or global, must be living systems that are highly sensitive and continuously responsive to the unprecedented rates of change that characterize our world today. A full overview of the ideas of these three theorists is beyond the scope of this paper. I will only provide a brief glimpse of the kinds of insights that permeate their writings.

Peter Senge writes about "learning communities," a term that addresses the need for organizations to be continuously monitoring the social and physical environment for new information and incorporating that new information as the group decides upon action.[7] He describes a learning community as "an organization that is continually expanding its capacity to create its future."[8] In the past, when culture change has occurred relatively slowly over many generations, there was time to evaluate the changes and to engineer responses. The Industrial Revolution, for example, took 200 years and many generations to produce modern industrialized society. Traditions could be suspended or altered one by one, with time to evaluate the impact of one change before more were needed. We are now in a period where children born five years apart will have significantly different experiences growing up, and where we barely have time to acknowledge that one change has occurred before we are dealing with dozens more. The concept of learning communities is essential to understanding and dealing with change on this scale.

Senge is famous for his "five discipline" model for achieving a learning organization. The first discipline is personal mastery, which challenges individuals to explicitly examine their personal perspectives, assumptions, vision for the future, and thought processes. The second discipline is labeled "mental models," and corresponds to what anthropologists would call culture. Here the focus is on the sources of thought patterns and how inferences are derived from experience. The third discipline is the development of a shared vision to which each member of an organization is willing to commit. Team learning that involves collaborative thinking and alignment rather than agreement is the fourth discipline. Senge asserts that learning organizations do not require everyone to always agree on a single outcome in decision making. He describes instead a process of alignment that allows people to think differently even as they move toward a common goal. The final discipline is systems thinking, which focuses not of the parts, but on the big picture and how the parts relate to one another to create the big picture.

---

[5] Geertz 1963; Rappaport 1967.
[6] Senge 1990; Wheatley 2005; Hock 2005.
[7] Senge 1990.
[8] Senge 1990, 14.

Margaret Wheatley's ideas about change as the essential challenge for organizations today and in the future can be summarized by four principles she attributes to the process of change.[9] She points out that participation in change is not a choice. All individuals and organizations must deal with the accelerating rates and myriad types of change that we face today. She reminds us that each of us creates a unique, individual perspective about what is real, all of which are valid and worthwhile; it is her position that all of those perspectives have a place in how we adjust to change in our own lives and in the groups in which we are members. She points out that the best-laid plans never unfold exactly as we envisioned them, but take on a life of their own. Finally, she asserts that the healthiest organizations can accomplish change gracefully because they have open internal networks of connection that are in continuous dialog about how to improve.

Wheatley uses the term "self-organizing groups" to highlight two aspects of healthy organizations. First, living systems in nature have emerged from individual responses to environmental conditions that collectively form a systemic pattern of interconnectedness between individual elements. This is in stark contrast to "top down," human organizations whose systems are dictated by leaders who assume that the future will look a lot like the present. Second, a successful organization in today's climate of accelerating change must re-form itself continuously to adjust to new information and circumstances; the more rapid the change, the more fluid the organization has to be. Tradition is, by definition, the enemy of flexibility. Tradition is based on accumulated wisdom about what has consistently worked in the past. One form of tradition that is currently a fad in a variety of types of organizations is "best practices." The new and updated version of tradition called "best practices" is only valid for those organizations that fall in the center of the normal curve, and may not be relevant for the majority of problems and organizations. Best practices focus on what has worked in the past and this makes them questionable as panaceas for problems that will emerge in the future. Wheatley argues that instead of building organizations based on known content, we must build organizations on what we know about the processes of change and responsiveness because we cannot know the content of the future. She points out that this kind of adaptability is an essential part of the natural world, and all we need to know we can learn from a better understanding of self-organization in nature.

Dee Hock coined the term "chaordic organizations" based on his work as the founder and CEO of Visa International.[10] He used the term chaordic to refer to a kind of organization that blends chaos and order in equal parts. His model of what will make organizations successful in the future also incorporates ideas about the necessity of "learning" and "self-organizing" in healthy organizations. Hock provides practical guidelines that he used to insure sensitivity to change and fluidity of structure in Visa International based on five principles. First, all the members of the group must share and be committed to a vision of what is to be accomplished. Second, no decision will be made above the level at which it will be implemented; those whose contribution will be affected are best equipped to

---

[9] Wheatley 2009.
[10] Hock 2005.

evaluate decisions that impact their work. Third, there must be mutual respect that extends to every individual in the organization, regardless of their position in any organizational hierarchy. The fourth requirement is trust; the assumption must be that each individual is competent, and that they will do their best to make the organization's vision come true. Finally, a chaordic organizational structure must be based on cooperation and interdependence, not on competition. This is contrary to the Western culture assumption that people will perform better if they are pitted against others in a win-lose battle. Hock's argument is that the "win-lose" model stifles creativity and intimidates innovators, who are afraid to try new things for fear of losing.

Senge, Wheatley and Hock do not agree on everything; however their theories embrace a number of common elements. If we put these three theorists' ideas together, the picture they present of sustainable organizations in the future is very different from that of existing complex organizations based on Western cultural core values and traditions. Today's Western and global organizational models are based on assumptions of environmental plenty and relatively stable social and physical environments. These factors have combined with complexity to favor command and control organizations that are characterized by the following:

1. Vision comes from those who have power/status
2. Deference to those of higher status is expected.
3. Trust is expected on the basis of status.
4. There is pressure to conform to tradition and "best practices."
5. Competition is believed to result in excellence.
6. Only a few perspectives dominate through power and control.
7. Independence is equated with self-interest.
8. Change comes from the top, from power/status.
9. Agreement on and commitment to a single strategy is the goal.

In contrast, organizational systems theorists such as Senge, Wheatley and Hock argue that environmental degradation and rapidly changing social and physical environments will require a different organizational strategy based on:

1. A shared vision of what is to be accomplished.
2. Mutual respect, leading to mutual trust.
3. Interdependence rather than competition.
4. Decisions that are made at the level at which they will be implemented.
5. Respect for many perspectives.
6. Use of collaborative thinking.
7. Alignment of differing perspectives rather than agreement on a single perspective.
8. Flexibility to respond to environmental change.
9. Organizations that are "intelligent" and "learning," consciously aware of the environment and continuously trying on new perspectives for change.

Many would look at the systems theorist's list of organizational qualities and dismiss them as unrealistic. I have many times heard people say that human nature will not allow organizations to operate without the four C's of competition, conflict, command and control that are so central to Western organizations. This is where anthropology can provide insight about whether an alternative kind of organization is within our reach, or whether it is a naïve fantasy based on unrealistic expectations.

Insights from Anthropology

As we have seen, the formal structures of today's dominant Western industrial societies are based on "command and control" traditions that are in stark contrast to the chaordic model. However, Western cultural systems have not entirely discarded these chaordic qualities; they have relegated them to the private domain and frequently to the pervue of women. There are also examples of non-western cultures that base their social, economic, political and religious systems on chaordic principles, including hunting and gathering societies (e.g. Lee 1984), "big-man" systems of redistribution (e.g. Sillitoe 1998), women-centered societies (Sanday 2004), and communities living in poverty (e.g. Stack 1974).[11] The remainder of this paper will address only one example, that of the original human culture, hunting, and gathering.

Hunting and gathering societies are typically broken up into small groups of 30 to 50 people that anthropologists call "bands." They are, of necessity, mobile, moving frequently either to follow resources such as herd animals, or to collect plant foods as they become available. This mobility requires flexibility to respond to changes in the natural world, and makes minimal material possessions a necessity. Small groups cannot defend territories. Thus resources, like water holes or stands of fruit-bearing trees, may be used by a number of groups in turn in what anthropologists call home ranges, a term that contrasts with territories that are "owned" and defended against use by others.

Reliance on naturally occurring plant and animal populations is combined with a reverence for nature that frequently results in belief systems like animism, in which all natural life and physical forms (e.g. landmarks, rocks, streams) have spirits with whom humans must live in harmony. This often results in religious rituals such as that traditionally performed by the !Kung San in South Africa after hunting and killing a giraffe. The ritual is a formal apology to the giraffe for taking its life, with assurances that the hunters will not kill more than they need to feed their families.[12]

Hunting and gathering societies tend to be egalitarian; there are no classes or ranks within the group based on access to resources. Instead, status differences are based on personal qualities and abilities that are valued by the group. In such a group, the qualities and abilities of each are well known to all, and people are judged by whether they contribute to the group according to these qualities and abilities. Thus leadership is frequently situational, with the group turning to the best-qualified person depending on what problem or decision they face. The most skilled hunter will be deferred to in the hunt, while people will turn to the most skilled storyteller to describe the hunt. When a pregnant woman goes into labor,

---

[11] Lee 1984; Sillitoe 1998; Sanday 2004; Stack 1974.
[12] Lee 1984.

she will turn to one or more older women who have experience with the process and have proven their worth helping others to deliver their infants. Thus authority is diffused through the group, with no one person in charge. People are able to make their own decisions about most things, and each one decides independently whose advice to take when they are uncertain. Group decisions are made by consensus rather than by delegated authority.

Finally, and perhaps most importantly, these groups are characterized by their use of generalized reciprocity as the means of distributing goods and services within the group. Generalized reciprocity means sharing what you have with others, with an underlying expectation that they will share what they have with you. !Kung San women who happen upon a stand of mongongo nut trees that are in full fruit and collect more than they or their families can eat in one day will share with others what they bring back to camp. Hunters who kill an animal that provides more meat than their families can eat will distribute the kill throughout the group. Shaman will perform curing rituals when someone is sick, regardless of whether the person is a member of their family. This is not barter, as there is no expectation of immediate equivalent exchange of goods or services. It is the obligation to give when others need, combined with the obligation to receive when others can give. It is based on trust earned through personal relationships and first-hand knowledge of whether others can be trusted. Reciprocity is the social glue that underlies every aspect of band society. When a member of the group does not share when he or she can, trust is eroded and ultimately that individual may be ostracized from the group. Fear of being ostracized is one of the most powerful sanctions that humans face, and provides strong motivation to participate as a productive, cooperative member of the group.

Hunting and gathering bands have all of the qualities that systems theorists predict will facilitate the most sustainable kinds of human organization for the future. They have a shared vision that includes knowledge of their environment complemented with a worldview and religion that honors the natural world as a supernatural force that must be respected. They have a common interest in surviving within the parameters of that shared vision. The size of their groups facilitates social glue based on reciprocal personal relationships, earned trust, tolerance for individual perspectives and opinions, and a sense of interdependence rather than competition. Flexibility is built into the way they are organized in their willingness to move, the fact that they are unencumbered with material possessions, and the variety of individual perspectives that can be used to create solutions to daily problems. They are "intelligent" and "learning," sensitive to their social and physical environments, and able to respond immediately to changes that may affect their well-being. And they have been around in many types of environments for more than two million years.

Toward a New Organizational Structure

Hunting and gathering bands provide a glimpse of the kinds of "cultural infrastructure" that will be necessary to allow sustainable chaordic organizations to emerge as we move into the future. We need to think in terms of networks of small groups in which relationships are personal and based on reciprocity. Trust built on personal experience facilitates situational leadership that allows people to solve problems creatively and at the level at which the problems occur. Leadership will be about motivation, encouragement and permission to do one's best, rather than fear and control. This kind of organization will use praise, status based on performance and peer pressure as major sanctions to motivate members of the group. Finally, these networks of small groups will be connected through reciprocity that includes both the obligation to give and the obligation to receive, acknowledging the contributions of each group and individual as worthy and desirable. As Margaret Wheatley has said, we need to reach out to the best of human nature that include cooperation, caring and generosity.[13] These traits flourish in a context of encouragement and respect.

I have often been told that the qualities of organization that characterize hunting and gathering bands cannot succeed realistically in modern society because they are contrary to human nature as we know it in modern industrialized societies. We modern Western folks, especially in the United States, view humans as inherently competitive, selfish, materialistic, and lazy, characteristics that are antithetical to the hunting and gathering or chaordic cultural model. But the anthropological record clearly shows that both in the past and in the present, there are examples of human societies that are not based on modern Western ideas of "human nature." Human organizations can operate with values of cooperation, reciprocity and flexibility, even within the constraints of modern Western culture. Sustainable, flexible, cooperative and adaptive organizations and societies that are built on principles like those of chaordic organizations exist and thrive, even if they do not dominate. Dee Hock's Visa International is an example of a chaordic organization that functions beautifully in the world of global finance as the most efficient and successful model for the exchange of monetary value that has ever been developed. The Visa International model is at the heart of developing globalization today. And it is there because it is a self-organizing, adaptive, learning organizational model that facilitates responsiveness to change. I will present one additional example here.

The Washington State Family Policy Council (WSFPC) has been one of the most successful programs for improving children's lives in the United States. It was structured as a chaordic organization that operates out of a government system that is traditionally command and control oriented.[14] The WSFPC is made up of networks of volunteers, one network for each county in the state. Each network is only allowed one staff member, and the network budget is tiny in comparison with normal government human services programs. The networks' mandate is to improve children's lives. They work to understand the problems of children in their county, and are encouraged to think outside the box to create solutions custom-designed to fit their children and their community. Flexibility and creativity are the hallmark of this program, and it has proved extraordinarily

---

[13] Wheatley 2005.

[14] Laura Porter, director of the Washington State Family Policy Council (WSPFC), personal communications. The State of Washington closed this program in the fall of 2013. Many of the WSPFC networks are still in operation, continuing to change children's lives. Laura Porter, who coordinated the networks for over 15 years, is currently working with individuals and organizations in more than seven states to set up networks based on the WSPFC model

successful in reducing truancy, teen pregnancies, underage drinking, and juvenile delinquency and recidivism. Along with these significant changes, networks have been effective in strengthening community capacity in ways that create a new sense of community in the counties as a whole, which, in turn, improves the quality of everyone's lives.[15]

Both Visa International and WSFPC are the kinds of organizations that eminent anthropologist Mary Catherine Bateson envisions in her writings about sustainable social organizations.[16] She focuses on the concept of ambiguity as one of the central issues that must be addressed as we move into the future. Ambiguity is the opposite of certainty. Modern Western culture is wedded to certainty and an arrogant assumption that the way we do things now will continue to work in the future. Bateson argues that we will have to embrace ambiguity to succeed in the future, because we cannot know for certain what the future will bring. She believes that global organization has begun to move gradually in the direction of decentralizing power, embracing rich cultural and political diversity, responsiveness to ongoing change, and viewing ambiguity as a virtue. New forms of organization will emerge, much as new life forms do. We cannot plan, but we can be on the lookout for and support emergent patterns that embrace the more noble human traits of cooperation, caring and generosity. We need the best of human nature and human organization to ensure sustainability for the future.

References

Bateson, Mary Catherine. 2004. *Willing to Learn: Passages of Personal Discovery*. Hanover, NH: Steerforth Press.

Carson, Rachel. 1962. *Silent Spring*. Boston, MA: Houghton Mifflin.

Ehrlich, Paul. 1968. *The Population Bomb*. New York: Sierra Club-Ballantine Books.

Geertz, Clifford. 1963. *Agricultural involution: the process of ecological change in Indonesia*. Berkeley, CA: University of California Press.

Hall, Judy, Laura Porter, Dario Longhi, Jody Becker-Green and Susan Dreyfus. 2012. Reducing Adverse Childhood Experiences (ACE) by Building Community Capacity: A Summary of Washington Family Policy Council Research Findings. *Journal of Prevention and Intervention in the Community* 40: 325-334.

Hock, Dee. 2005. *One from Many: Visa and the Rise of Chaordic Organization*. San Francisco: Berrett-Koehler Publishers.

Lee, Richard. 1984. *The Dobe !Kung*. New York: Hold Rinehart & Winston.

Rappaport, Roy. 1967. *Pigs for the Ancestors*. New Haven CT: Yale University Press.

Sanday, Peggy. 2004. *Women at the Center: Life in a Modern Matriarchy*. Ithaca, NY: Cornell University Press.

Senge, Peter. 1990. *The Fifth Discipline: The Art and Practice of the Learning Organization*. New York: Doubleday Dell Publishing Group.

---

[15] Hall et al. 2012.

[16] Bateson 2004.

Sharp, Lauristan. 1952. Steel Axes for Stone-Age Australians. *Human Organization* 11(1): 17-22.

Sillitoe, Paul. 1998. *An Introduction to the Anthropology of Melanesia: Culture and Tradition.* Cambridge: Cambridge University Press.

Stack, Carol. 1974. *All Our Kin: Strategies for Survival in a Black Community.* New York: Harper and Roe.

Von Bertalanffy, Ludwig. 1950. An Outline of General Systems Theory. *British Journal for the Philosophy of Science* I(2): 134-165.

Wheatley, Margaret. 2005 *Finding Our Way: Leadership for an Uncertain Time.* San Francisco: Berrett-Koehler Publishers

Wiener, Norbert. 1967. *The Human Use of Human Beings. Cybernetics and Society.* New York: Avon.

Chapter 8: Technology and Ecology

David Grierson

Anticipating a shift in our collective worldview the holistic philosopher Lewis Mumford, in *Technics and Civilization,* advocated a new culture in which, rather than simply shaping our lives, a new form of humanistic technology immersed in the social milieu would become an evolutionary instrument enabling a better quality of life by actively enhancing our environmental setting. Mumford believed that what distinguished humanity from other species, was not primarily our use of tools, but our use of language. Widespread electrification and mass communication, in better connecting us, would allow us to share our, "wishes, habits, ideas, and goals," and so build "a better world for all."[1] Technology was one part of technics but for Mumford for it to be durable, effective, and efficient it had also to be fused with human spirit and creativity. Later Thomas Kuhn in *The Structure of Scientific Revolutions* challenged the prevailing view of progress in science as a continuous accumulation of accepted facts and theories, arguing that often advances in science have been more sporadic in nature. Asking new questions of old realities, he said, had led to game-changing periods of ascendant transformation. Rather than accepting the unwavering logic of constant growth, Kuhn's concept of paradigm shift[2] offers the tantalizing prospect of advancement through discontinuous revolutionary breaks with our earlier thoughts, beliefs, values, and experiences. Both concepts sanction a transition towards an ecological view[3] of the world that is achievable, and provoke the prospect of realizing a different world through a radical redefinition of the faltering relationship between society, technology, and nature.

The mechanistic worldview that has determined nature as a machine composed of related but discrete components helps to support the commonly held

---

[1] Mumford 1934.

[2] Kuhn 1962.

[3] The word ecology, coined by the German biologist and philosopher Ernst Haeckel (initially as oecology) in 1866 derives from the Greek oikos, referring originally to the family household and its daily operations and maintenance. Haeckel described ecology as the study of the relationships between living things and the environment in which they live. Today this definition has been expanded to refer to the larger cosmic household here upon Earth. Cited in Worster 1985. See also Dobson 1995.

idea that humans are at the pinnacle of creation, the source of all value, the measure of all things. In offering resistance to this way of thinking and rejecting the assumption of human self-importance in the larger scheme of things, physicist Fritjof Capra has argued that our society is now embarking on a fundamental shift towards a more ecological, holistic, organic, or systemic view of the world.[4] This chapter identifies threads of the mechanical and ecological paradigms and describes some characteristics that seem to signal a shift from one to the other giving emphasis to the importance of aligning future technological developments with ecological values and the practice of sustainability.

Constant Craving and Ecological Limits

Developments in science and technology have without question driven our evolutionary progress and helped shape our modern lives. Advancements in scientific knowledge and technological skills and expertise have been beneficial in helping improve health, raise standards of living, and enhance global communications. Nevertheless, when viewed through a critical lens, such progress has been achieved at considerable environmental and human cost.[5]

Contemporary discourse within the sustainability agenda upholds the belief that social and environmental problems arise largely from seeing ourselves as separate from nature. The mechanistic (or reductionist) paradigm has roots within the European scientific revolution influenced by Copernicus, Kepler, Newton, and Galileo, and has dominated our culture for several hundreds of years, having shaped Western society and significantly influenced the rest of the world. While a mechanistic view of the world has underpinned enormous changes impacting on all aspects of modern life and brought many benefits, relatively recent discourse around the sustainability agenda has brought into question long held assumptions with respect to aspects of economic growth, along with previously entrenched ideas and values (e.g. nature exists to serve humanity). Sustainability involves a move from a current condition of unsustainable activity towards a process of improvement and increased quality. Essentially the term is used to indicate a change of attitude prioritizing ways of life that are in balance with the current renewable resources of the ecosystem and the biosphere. Although we are unclear about how much damage has already been inflicted on the biosphere the thesis proposes a precautionary approach as a practical way forward. In the face of inherent uncertainty, risk is deemed inappropriate, since failure to maintain a viable biosphere will be catastrophic and irreversible. The widespread interest in theories, ethics, and practice concerning sustainability indicates an increasing concern around the adverse impacts that conventional models of development have had on the environment, in both the developed and undeveloped parts of the world.[6]

Capra argues that the major problems of our time require a radical shift in our perceptions, our thinking and our values as radical as the Copernican revolution; one which is underpinned by a new perspective pre-figuring an integrated

---

[4] Capra 1986.
[5] Barbour 1991.
[6] Grierson 2003.

network of all living and nonliving parts. With echoes of the medieval cosmology depicted in the Great Chain of Being, actions in any one part this network affect the equilibrium of the whole. In support of his assessment that a mechanistic paradigm is now giving way to an ecological one, Capra sees the convergence of a number of theoretical/conceptual positions (ecology, feminism, community politics, environmental economics, consciousness raising) which to varying degrees augment an ecological view and support different groups working on a variety of causes with common elements, increasingly being brought together in collective action. James Lovelock's Gaia hypothesis, for example, presents the Earth as a self-regulating system within which conditions suitable for life are maintained by feedback processes involving both living things and the non-living part of the planet.[7] But rapid growth in populations, economies, and cities since the industrial revolution has placed the system under huge stress. Richard Douthwaite in *The Growth Illusion* (1992) asked how it is that we could have progressed along the path of economic growth, technical innovation and increasing efficiency for so long and yet end up with massive unemployment, widespread poverty and the fear of economic and ecological collapse. His answer is that "economic growth has enriched the few, impoverished the many, and endangered the planet."[8] He argues that as economic growth continues it takes more and more resources to achieve additional increments of growth. The whole process, in effect, becomes progressively more inefficient.

The last fifty years has been marked by an intensification of concern about pollution and an awareness that environmental problems arise within the context of a complex interrelationship between humans, their resource base, and the social and physical environments.[9] Consequentially questions about the objectives and strategies of conventional growth policies have been brought to the forefront of public debate. In 1972 The *Limits to Growth* report by the Club of Rome (the name given to an "invisible college" of scientists, researchers, industrialists who conducted the research), described the results of a complex computer model, which they argued outlined the "predicament of mankind." In neo-Malthusian fashion the report indicated that if the then current growth trends in world population, industrialization, pollution, food production, and resource use continued, the planet's carrying capacity would be exceeded within a hundred years, bringing about a disastrous "overshoot and collapse," ultimately leading to "eco-catastrophe," famines, and wars. The fundamental problem according to the Limits thesis is that global growth in resource use, industrial output, population and pollution is exponential. The report explained that, "A quantity exhibits exponential growth when it increases by a constant percentage of the whole in a constant time period."[10] This kind of growth displays a gentle and gradual curve for a long time but then rapidly shoots up in a very short period. Translated to the arena of industrial production, resource depletion, and pollution, what seems an innocuous rate of use and waste disposal can quickly result in dangerously low levels of available resources and dangerously high levels of pollution. The theory, in applying thermodynamic laws to economics, argues that all production that

---

[7] Lovelock 1979.
[8] Douthwaite 1992.
[9] Turner 1988.
[10] Meadows et al 1972.

uses material and energy eventually transforms them into a more random, chaotic, or disordered, state.

Inspired by nineteenth-century liberal philosopher John Stuart Mill, who proposed the idea of the "stationary state" as a counterpoint to the relentless selfish and competitive drives at the heart of capitalism,[11] Herman Daly suggests that there is a limit to the use we can make of scarce resources when exponential extraction leads to sudden exhaustion. Waste, he says, is an inevitable by-product of the extraction and use of resources. For Daly, "living in intimate contact with garbage and noxious wastes is a by-product of growth."[12] He sees environmental degradation as a disease induced by economic "doctors" who have tried to treat the basic sickness of unlimited wants by prescribing unlimited production.

Economic growth in the conventional sense is ultimately more of a problem than a solution because it damages the environment and leads to social injustice. While it may be anathema to many to abandon our constant craving for material wealth and redefine our notion of growth, modern environmentalists recognize the necessity of an ecological society based on a comprehensive set of sustainable policy objectives that cover all aspects of our lives; economic, social, cultural, political, technological and environmental. Such a society, if it can be constructed at all, will acknowledge that there are ecological limits to material growth.

Imperceptible Changes and the Illusion of Free Will

> Changes from one kind of civilization to another do not happen often in history: the invention of agriculture, the rise and fall of conquest states...and the coming of industrialism. An earlier generation may have been justified in discounting any further such radical changes. We cannot. Most trends of the past are simply not sustainable.[13]

The dramatic changes of thinking that took place in the field of atomic and subatomic physics at the beginning of the twentieth century led Kuhn, in 1962, to define the idea of a scientific *paradigm* as, "a constellation of achievements – concepts, values, techniques, etc. – shared by a scientific community and used by that community to define legitimate problems and solutions."[14] Kuhn argued that a paradigm gains its status because it is more successful than a competitor at solving some problems that have been recognized as acute, and that changes in world views occur in sporadic, progressive spaces, which he called *paradigm shifts*. The nature of these transitional periods is such that it is not always possible to accurately trace the rise and fall of new ideas as their beginnings may be barely perceptible and they might end unnoticed. Kuhn also displays a sense in which paradigms not only belong to a scientific community but also apply to wider society and its relationship to nature. Although some question his analysis, suggesting that the shift from classical to quantum theories in the twentieth century failed to display all the characteristics he suggests, the notion of

---

[11] Daly 1977.
[12] Daly 1991.
[13] Goerner 1999, 32.
[14] Kuhn 1962, cited in Capra 1986, 3.

paradigms is useful in allowing us to theorize on how societal and environmental change may occur.

The underlying causes of the modern environmental crisis in its widest sense lie in the revolutions of science, religion and economics in the early modern age, which helped to lay down the foundations of the dominant Western worldview, and shaped institutions such as the systems of capitalism and state socialism. From the middle of the sixteenth century to the end of the seventeenth, early modernism and the principles of classical science established ways of thinking about the world and our position in it, which were vastly different from the medieval cosmologies and pre-modern notions that had preceded them. Seeing ourselves as separate from nature follows ideas of Rene Descartes, who saw science as rendering us the "masters and possessors of nature,"[15] and in particular, Francis Bacon who saw it's potential in "enlarging the bounds of Human Empire."[16] Although now refuted, classical science held that the machine of nature is composed of discrete components. Its fundamental particles, like atoms, electrons, and quarks are solid bodies in empty space. We, as observers of nature (subjects) were separate from it (the object) so we could be "objective," impersonal, or detached about it. The widespread acceptance of this view, which led to the belief that we humans are at the highpoint of creation, became deeply embedded in our culture and consciousness. The historical roots of this perspective coincide with the beginnings of industrial capitalism. The science associated with this period is characterized as being primarily concerned with achieving material progress and was imbued with values identified with liberalism and the French Revolution. From ancient times the main goal of science has been gaining wisdom and understanding while remaining in harmony with nature. However in the Western world, since Bacon, the goal of science has tended to be patriarchal and has largely involved the pursuit of knowledge in order to control and exploit nature. Dualism, as between mind and matter, championed by Descartes, sets the paradigm for understanding most of Western culture. Descartes doubted everything until he reached a definite conclusion in his famous dictum *Cogito ergo sum*.[17] He deduced from this that since thought was the essence of nature, mind and matter were separate and distinct entities. The material world was a machine without life or spirit. The natural world functioned in accordance with mechanical laws and nature could be explained in terms of the mechanistic movement of the parts. Even human beings belonged to a category of machine in which the human body was seen as a container activated by a soul that was connected to the body via the pineal gland in the brain. Thanks to Cartesian dualism the mechanical view of nature became the dominant view of "classical science." Isaac Newton derived a mathematical formulation that undertook Descartes' work and completed the mechanistic world-view. For Newton, God had set the whole universe in motion and it has continued to run ever since like a machine governed by immutable laws. Such a view is essentially deterministic and fatalistic. It says that given sufficient knowledge of nature's laws we could have predicted the present. The future is already cast, and free will is an illusion.

---

[15] Descartes 1637, Part VI.
[16] Bacon 1626.
[17] Descartes 1637, Part I.

Capra suggests that the mechanistic paradigm is now receding because, as a model, it has a number of entrenched ideas and values that have recently been brought into question, namely:

- the view of the Universe as a mechanical system composed of elementary building blocks;
- the view of the human body as a machine;
- the view of life in society as a competitive struggle for existence;
- the belief in unlimited material progress to be achieved through economic and technological growth;
- the belief that a society in which the female is subsumed under the male follows a basic law of nature.[18]

Against the Plague Wind

In *A Green History of the World*, Clive Ponting relates many examples where human societies have failed to achieve a sustainable balance between their own material demands and the environment's well-being. He describes the Sumerian empire as the first literate society on Earth to succumb to self-inflicted ecological collapse. The technical innovation of irrigation, which had been invented around 5500 B.C., eventually brought Sumeria to its nemesis. Irrigation increased crop yields substantially but it also increased the salt content and the ability of the soil to retain water. The rapid population growth, which resulted from increased crop production, meant that the land could not be left to lie fallow in order to recover. Crop yields remained high for a time, but collapsed abruptly in 2400 B.C. The food shortfall made it difficult for the empire to support its army and Sumeria was conquered within a few decades. The rise and fall of Sumeria illustrates a tendency that has shown itself time and again in the history of human society: a given technological development increases humanity's ability to extract a higher level of comfort from the natural world, but it does so at the cost of greater environmental damage. Ponting points out that damage to the environment was usually one among a number of factors, which caused these societies to come apart, and in such cases, "the decline and eventual collapse were usually prolonged...and generations living through this process would probably not have been aware that their society was facing long term decline."[19]

Classical science asserted that the world operated according to consistent physical laws, which could be discovered through reason and experiment and applied to practical effect. To a rising class of European modern capitalists during the sixteenth to eighteenth centuries, better knowledge resulted in better machines, which lowered production costs, attracting more and more people into the system, and accumulating the capital needed to develop better production methods and machinery. By the late eighteenth century the pursuit of technological advantage in Britain had culminated in the industrial steam engine, a machine, which would power the Industrial Revolution through the next hundred years. During the nineteenth century advances in technology raised the

---

[18] Capra 1996.
[19] Ponting 1991, 401.

standard of human welfare across Europe but inflicted greater environmental damage as forests were systematically destroyed. Eventually, because wood was becoming a scarce resource, it was replaced by coal as the primary source of fuel. For centuries humans had limited the burning of coal, because as a source of fuel it was inefficient, messy and difficult to extract from the ground. Dwindling wood stocks and technological breakthroughs in the 1840s enabled coal to be converted into heat much more efficiently and the industry grew rapidly. But the environmental trade-off for coal was worse than it was for the steam engine. As "progress" became the key word of nineteenth century philosophy and politics, massive increases in production and efficiency were accompanied by blackened skies, putrid rivers, and other side effects leading to William Blake's passionate assault on the "dark Satanic Mills" of industrial England.[20]

The ideas, beliefs and values held within sustainability (as the pursuit of an ecologically benign culture) are historically derived from a diverse range of philosophical and ideological sources. Some have likened the modern surge in ecological awareness to the growth of religious sects in the seventeenth century – the Shakers, Quakers, Diggers, Ranters, Pilgrims, Fifth Monarchists, and Levellers. Their fiercely independent spirit of egalitarian politics, their love of the Earth, their decentralist tradition, and their passionate spiritual commitment certainly number them among a long line of antecedents of an ecological world-view. Within the eighteenth and nineteenth century Romanticism expressed by Thomas Carlyle, John Ruskin, and the Romantic poets including Blake, Wordsworth, Coleridge, and Byron, we can identify a revolt against capitalism and the utilitarian, materialistic values of the time. They shared concern for the increasing effects of industrialization and urbanization on the landscape alongside a desire to hold on to traditional values and beliefs in the face of tumultuous change. Perhaps the foundation of an ecological sensibility is most clearly reflected in Ruskin who in 1859 lectured Bradford manufacturers on the potential total disfigurement of the English countryside by spreading industrialization and later carried out practical experiments in combating pollution. Ruskin called for a renewal of moral and spiritual values in society. In *The Storm Cloud of the Nineteenth Century* he noted that climatic deterioration and pollution were creating a new form of cloud, a "loathsome mass of sultry and foul fog, like smoke...a plague wind."[21] Apparently referring to the pollution from the blast furnaces at Barrow-in-Furness, what seemed to concern him most was the symbolic nature of the cloud as the material expression of moral decline instigated by industry and commerce.

Shifting Paradigms

Neil Armstrong's "small step" from Apollo 11's *Eagle* landing craft onto the Moon's surface in 1969 (the climax of a massive, politically driven, scientific and technological *coup de grâce)* served to embody an anthropocentric spirit inherent in the Baconian creed. But his "giant leap for mankind" also served as inspiration for an evolutionary stewardship approach that saw in an emancipated humanity

---

[20] Blake 1808.
[21] Cited in Cosgrove 1984, 251.

the expression of a natural evolution rendered self-conscious.[22] Some of the younger generation, raised on the exhilaration and hopefulness of space exploration and science fiction, turned their attention away from a technological future back towards Earth to confront life in all its organic richness, diversity, and creativity. In the United States campus riots and civil rights demonstrations led to the "summer of love" and onto a farm in Woodstock, where half a million turned up to "tune in and drop out" and "go with the flow" in the physical and spiritual footsteps of the beat authors and poets like Jack Kerouac and Alan Ginsberg. For many the challenge wasn't any longer technological but rather philosophical. It was really about how to get "back to nature."

A rejection of the assumption of human self-importance in the larger scheme of things has a long history that can be traced back through the wilderness/environmental/land ethics of Henri David Thoreau,[23] John Muir[24] and Aldo Leopold,[25] to the fraternal teaching of Saint Francis of Assisi (the "patron saint of ecology"), and further into the past within medieval and Renaissance cosmologies, with their images of the world that were holistic, organic, ecological, and spiritual. The medieval organic metaphor of nature also derived from a human experience in which the Earth was perceived as a living body wherein the circulation of water through the rivers and seas was comparable to the circulation of blood; the circulation of air through wind was the breath of the planet; volcanoes and geysers were seen as corresponding to the Earth's digestive system. And there were a number of Renaissance organic philosophies based on the idea that all parts of the cosmos were unified in mutual interdependence, in which everything was saturated with life, and it was impossible to distinguish between living and non-living things. Earth was a living being among humans. Even although she could also be unpredictable, wild, passionate, and dangerous, "Mother Earth" nourished and nurtured us, and so should command respect and reverence.

The organic view and the medieval cosmology stemmed from the Great Chain of Being, which had originated with the Greeks and had been transmitted to medieval writers who adapted it to their own cosmology. The Great Chain is a designed hierarchy in nature in which all matter, from rocks to angels, is in possession of a soul, and all earthly species of organic life have their appointed place on the chain, from the insects above the rock to the humans below the angels. All were joined together in a fixed hierarchy, and were interdependent. The metaphor and related ideas, which continued to influence essential assumptions framing scientific theories into the eighteenth century, placed people and nature in a mutual relationship in which each link in the chain was vital for the continued existence of the whole chain. The elimination of one link would dissolve the whole cosmic order and render the world muddled and disjointed.

The idea of a coherent cosmic order based on continuity and gradation was tied to the notion of "plenitude" or "abundance" which held that the world is filled with diverse living things such that all species that could theoretically exist do in fact exist. Fullness stemmed from a hypothetically infinite process of

---

[22] Bookchin 1990.
[23] Thoreau 1854.
[24] Muir 1901.
[25] Leopold 1949.

reproduction. The diversity of living organisms was deemed to be so great and the numbers so abundant that some feared that a single species could multiply indefinitely and eventually cover the entire Earth. This view led Malthus (in 1798) to posit that humans could theoretically fill not only Earth, but all planets in our solar system if population growth was not held in check by wars, famines, disease, and poverty, and by competition between and within species.[26]

Many today advocate a sustainable society, not because they think it would be a better place to live, but because they believe they occupy a scientifically and sociologically legitimate position. Environmental sociology offers a critique aimed at the lack of human-environmental focus in classical sociology, and has led to a new perspective that takes account of interrelationships between environment and society, environmental variables, and feedback loops from ecosystems. The New Environmental Paradigm (NEP) acknowledges that the biosphere can impose constraints on human activity.[27] It follows the post-modern ecological view of science borrowed from a number of writers in the first half of the twentieth century, including Alfred North Whitehead, Henri Bergson, and Lewis Mumford. The view draws on the work of Michael Faraday (particularly in the sense that his electromagnetic field refuted the Newtonian idea that all entities were separate and governed by fundamental mechanical laws determined by God), Albert Einstein (whose relativity theory offered an interconnected view of the Universe), evolutionary theory in biology, and in quantum theory within subatomic physics. Lovelock's Gaia hypothesis, relating aspects of Greek and Medieval cosmologies, the organic metaphor, and new physics, refers to the Earth as a self-regulating organic system, striving toward a steady-state condition favorable for the maintenance of life, while being capable of responding to changing needs for human sustenance.[28] The concept sees the Earth as a self-regulating system which is impacted upon by humans but cannot be controlled by them and in which conditions suitable for life are maintained by feedback processes involving both living things and the non-living part of the planet. According to the hypothesis, the self-regulating organic system is striving toward a steady state condition favorable for the maintenance of life, while being capable of responding to changing needs for human sustenance. In seeking homeostasis this complex system can adjust, within certain limits, to large-scale human technological interventions. But the current pattern of urbanization, resulting as it does in energy-intensive, highly-pollutant, forms of human settlement represents interventions which are spiraling out of control, causing levels of environmental degradation and social disruptions that threaten the planet's equilibrium.

We now have irrefutable scientific evidence that our activities are harming the biosphere and human life in alarming ways that may soon become irreversible. Advances in satellite technology have provided environmental data giving us crucial insights into changing geological patterns, rising sea levels, and the depletion of the ozone layer. What the evidence points toward is global environmental problems on an unprecedented scale; rapid depletion of natural resources, energy and materials, atmospheric pollution, climate change, deforestation, and dramatic loss of biodiversity. The more we investigate these

---

[26] Malthus 1798.

[27] Catton and Dunlap 1978.

[28] Lovelock 1979.

problems the more we come to realize that they are interconnected and interdependent. Scarcities of resources and environmental degradation combine with rapidly expanding populations leading to the breakdown of communities, collapsing infrastructures in cities, and to ethnic and tribal violence. Stabilizing the world population growth rate will only become possible when poverty is reduced throughout the world. The mass extinction of animal and plant species will go on as long as the developing world is burdened by huge debts. Environmentalists now describe these problems as different facets of a single crisis deriving from an out dated worldview that is no longer adequate for dealing with an overpopulated, hyper-consuming, globally connected world. Many advocate that it is the dominant attitude towards nature and the environment in Western society underpinned by a long-standing and far-reaching mechanistic paradigm that needs to change.

When Capra describes how the major problems of our times require a radical shift in our perceptions, our thinking and our values, he generalizes Kuhn's definition of a scientific paradigm to that of a social paradigm. Capra's analysis of cultural transformations defines, a set of beliefs and practices shared by a community, which forms a particular vision of reality and a basis for the way the community organizes itself. [29] Towards the end of the Second World War, Mumford foresaw not only the imperative for change but signs that we were beginning to embrace a new humanistic vision for an emerging global culture:

> An age of expansion is giving place to an age of equilibrium. The achievement of this equilibrium is the task of the next few centuries...The theme for the new period will be neither arms and the man: nor machines and the man: its theme will be the resurgence of life, the displacement of the mechanical by the organic, and the re-establishment of the person as the ultimate term of all human effort. Cultivation, humanization, co-operation, symbiosis; these are the watchwords of the new world-enveloping culture. Every department of life will record this change: it will affect the task of education and the procedures of science no less than the organization of industrial enterprises; the planning of cities; the development of regions; the interchange of world resources. [30]

Throughout the twentieth century the displacement of the mechanical by the organic has taken a variety of forms and moved at different speeds in disparate fields. It has involved revolutions, reactions and complex oscillations but primarily the basic tension was always between the parts (the mechanistic, reductionist, or atomistic) and the whole (the holistic, ecological or systemic). In dealing with our growing environmental and social problems however, the deterministic and mechanical view of the world, continues to promote a specialised instrumental approach, which relies heavily on scientific method and technological know-how. When associated with the relentless pursuit of material progress, based on a no-limits mentality, we appear blind to the fact that beyond a

---

[29] Capra 1996.
[30] Mumford 1944, 598-99.

certain threshold (carrying capacity) we will inevitably deplete the world's natural resources and overburden the biosphere with waste products incapable of being absorbed by Gaia's self-balancing system. In effect we undermine the Earth's equilibrium-seeking mechanisms. When confronted with the evidence of ecological overreach and collapse, the default response is generally (and unsurprisingly) reductionist; limited to strategies of policy reform geared to the technological solution – we continue to emphasise technological means for solving problems which are essentially ecological in origin. This almost universal and implicit assumption that all of our modern problems (rapid population growth, pollution, the threat of nuclear conflict) have a technical solution is contested. In The Tragedy of the Commons Hardin defines a "technological solution" as one that requires "a change only in the techniques of the natural sciences, demanding little or nothing in the way of changes in human values or ideas or morality."[31] If technological answers to problems associated with growth are inadequate, then it follows that more profound social, cultural, political, economic and environmental transformations will be essential to "building a better world for all," and in redefining the notion of progress. For Hardin such changes will require the recognition of the necessity of "mutual coercion" in social arrangements (e.g. pollution taxes) and a careful rethinking of the meaning of "freedom." He points to the legislation against robbery as an example of society becoming more, not less, free through mutually agreed laws. Quoting Hegel, Hardin reminds us that, "Freedom is the recognition of necessity."

Technology, the Cause, and the Cure

According to Barry Commoner pollution is an unintended by-product of the drive to increase profit by introducing technologies that increase productivity, and is intensified by the displacement of older techniques by new, ecologically faulty, but more profitable technologies.[32] In this causal relationship between pollution, economy, and modern technology, the cost of environmental degradation is borne, not by the producer, but by society as a whole. Polluters are therefore being subsidised by society. It is the kind of relationship that must be redefined in the midst of an ecological crisis. During recent decades the increasingly popular notion of sustainable development has been propelled to the forefront of our thinking, and our policy debates, because it promises to respond to an irreconcilable contradiction between growth and limit agendas, and the conflicting territorial challenges posed by society, technology, and nature. Sustainable development "is development that meets the needs of the present without compromising the ability of future generations to meet their own needs."[33] It contains within it two key concepts:

- the concept of "needs," in particular the essential needs of the world's poor, to which overriding priority should be given; and

---

[31] Hardin 1968, 1245.

[32] Commoner 1971.

[33] World Commission on Environment and Development 1987, 43.

- the idea of limitations imposed by the state of technology and social organization on the environment's ability to meet present and future needs.[34]

The urgency of the conflict steers an essential movement toward sustainability wherein the principle tension is between those who believe that technology can resolve all of our problems and those who know that it cannot. Theoretically at least it is conceivable that with sufficient investment in basic needs and infrastructure, communities in the developing world can provide the food, water, farmlands and industry needed to raise themselves above absolute poverty, with stable levels of population, and sustainable levels of energy and resource use resulting in better living standards for all. Achieving this will require us to carry out as many experiments and look to as many alternatives as possible. And perhaps it is just possible that, through careful research, innovation, and evaluation, we can develop sustainable technologies, and employ these productively at a reduced cost to the ecosystem, within our resource availability, while still maintaining our environmental and cultural integrity. But neither those who write-off science as a possible contributor to human well-being and environmental stability nor those who believe that technology will solve all of our human ills and rid us of all our environmental problems, can ever be more than half right.[35] Technology is both the cause and the cure of what ails us. Human strategy and planning is what gives it its stimulus and defines its limits. But the modern world, with its rapid growth in population, its pursuit of material wealth via increasing rates of resource consumption, and its rapid shift from rural to urban life via the process of urbanization, lacks strategies to preserve, protect, and maintain an ecological equilibrium that could lead to a better world for all. Technological solutions, no matter how clever, cannot facilitate infinite growth in a finite system. Our technological expertise has, thus far, merely shifted the problem around often at the expense of more energy and resource use, and therefore more pollution. How such expertise in the future might be allied with a human spirit and creativity that will embrace an ecological view of the world could determine how our, "wishes, habits, ideas, and goals" are not only communicated, but implemented. Better knowledge and better machines may well lead to technological advantage, but if we fail to acknowledge that many of our modern problems are ecological in origin and that technological responses must incorporate changes in human value and morality signalling a shift away from unlimited production and consumption towards a redefined relationship between society, technology and nature, then sadly not to a better world for all.

References

Bacon, Francis. 1626. *New Atlantis.* New York: Collier & Son.
Blake, William. 1808. "And Did Those Feet in Ancient Time". Cited in David E. Verdman (ed) 1988. *The Complete Poetry and Prose of William Blake.* New York City: Doubleday.

---

[34] World Commission on Environment and Development 1987.
[35] Thayer 1994.

Barbour, Ian G. 1991. *Ethics in an Age of Technology. Gifford Lectures*, Volume Two. San Francisco: HarperCollins.

Bookchin, Murray. 1990. *The Philosophy of Social Ecology.* Montreal: Black Rose Books.

Capra, Fritjof. 1986. The Concept of Paradigm and Paradigm Shift. *Re-Vision* 9 (1): 3-12

Capra, Fritjof. 1996. *The Web of Life.* London: HarperCollins.

Catton Jr., William R., and Riley E. Dunlap. 1978. Environmental sociology: A new paradigm. *American Sociologist* 13(2): 41-49.

Commoner, Barry. 1971. *The Closing Circle: Nature, Man and Technology.* New York: Knopf.

Cosgrove, Denis E. 1984. *Social Formation and Symbolic Landscape.* London: Croom Helm.

Daly, Herman E. 1977. The Steady-state Economy: What, Why, and How. In *The Sustainable Society,* edited by Dennis Pirages, 107-114. New York: Praeger.

Daly, Herman E. 1991. *Steady-State Economics,* 2nd edition. Washington, D.C: Island Press.

Descartes, René. 1637. *Discourse on the Method of Rightly Conducting the Reason, and Seeking Truth in the Sciences.* Vol. XXXIV, Part 1. New York: Collier and Son.

Dobson, Andrew. 1995. *Green Political Thought* (second edition). London: Routledge.

Douthwaite, Richard. 1992. *The Growth Illusion.* Oklahoma: Council Oak Books.

Goerner, Sally J. 1999. *After the Clockwork Universe: The Emerging Science* and *Culture of Integral Society.* Edinburgh: Floris Books.

Grierson, David. 2003. Arcology and Arcosanti: Towards a Sustainable Built Environment. *Electronic Green Journal* 1(18). Available at http://escholarship.org/uc/item/8xh5f1d1, accessed 31 March, 2014.

Hardin, Gerrett. 1968. The Tragedy of the Commons. *Science* 162 (3859), 1243-1248.

Kuhn, Thomas S. 1962. *The Structure of Scientific Revolutions.* New York: Houghton Mifflin.

Leopold. Aldo. 1949. *A Sand County Almanac.* Oxford: Oxford University Press.

Lovelock, James. 1979. *Gaia: A New Look at Life on Earth.* Oxford: Oxford University Press.

Malthus, Thomas. 1798. *An Essay on the Principle of Population.* London: J. Johnson.

Meadows, Donella H., Dennis L. Meadows D, Jørgen Randers, William W. Behrens III. 1972. *The Limits to Growth.* New York: Universe Books.

Muir, John. 1901. *Our National Parks.* Boston: Houghton Mifflin.

Mumford, Lewis. 1934. *Technics and Civilization.* New York: Harcourt Brace and Company.

Mumford, Lewis. 1944. *The Condition of Man.* New York: Harcourt Brace and Company.

Ponting, Clive. 1991. *A Green History of the World.* New York: St. Martin's Press.

Thayer Jr., R. L.1994. *World Green Heart: Technology, Nature and the Sustainable Landscape.* New York: John Wiley & Sons.

Thoreau, Henri D. 1854, *Walden.* Boston: Ticknor and Fields.

Turner, R. Kerry. 1988. *Sustainable Environmental Management: Principles and Practice.* London: Belhaven Press.

World Commission on Environment and Development (WCED). 1987. *Our Common Future.* Oxford: Oxford University Press.

Worster, Donald. 1985. *Nature's Economy: a history of ecological ideas.* Cambridge: Cambridge University Press.

Part III: Business

Editors' note: In sustainability studies business is often seen as a double-edged sword. On the debit side of the "business balance sheet" is the role of businesses in generating fossil fuel-driven technology that generates widescale greenhouse emissions as well as other forms of pollution, such as toxic sludge, persistent organic pollutants and other forms of chemical waste. To a large extent the economic infrastructures that we work and live within are those that large businesses have created. But businesses, or at least the most successful ones, are innovative organizations that offer up the possibility for change. It is difficult, for example, to imagine a transition to a low carbon economy without the active support of the world's leading energy and technology companies. Businesses may serve the public good as well as the private good of individual firms. This is a point that Robert Hinkley touches upon:

> Corporate law …casts ethical and social concerns as irrelevant, or as stumbling blocks to the corporation's fundamental mandate. That's the effect the law has inside the corporation. Outside the corporation the effect is more devastating. It is the law that leads corporations to actively disregard harm to all interests other than those of shareholders. When toxic chemicals are spilled, forests destroyed, employees left in poverty, or communities devastated through plant shutdowns, corporations view these as unimportant side effects outside their area of concern. But when the company's stock price dips, that's a disaster.

Robert Hinkley (2002)[1]

---

[1] Hinkley, Robert. 2002. "How Corporate Law Inhibits Social Responsibility", *Common Dreams: Building Progressive Community*. Available at http://www.commondreams.org/views02/0119-04.htm, accessed June 6, 2014.

Chapter 9: Economic Models for Environmental and Business Sustainability in Product Development and Manufacturing

Dariush Rafinejad and Robert C. Carlson

Introduction

Over recent decades, major global corporations and manufacturers of products have gone through multiple evolutionary phases in their outlooks toward environmental sustainability. These companies have generally evolved through stages of no-concern, to pollution control, to pollution prevention, and to resource efficiency maximization in lockstep with governmental regulations. However, the adoption of sustainable development as an imperative strategic vision is often lacking in the industrial enterprises.

Although, and as we discuss in this chapter, the issues of environmental and business sustainability have been addressed by several leading scholars, usually either environmental or business sustainability has been focused on to the detriment of the other. Most academic and popular literatures tend to state the looming sustainability problem, emphasize the need for the so-called triple bottom line objectives (profitable growth, environmental friendliness and social responsibility) and propose initiatives to mitigate environmental harm caused by human economic activities. In spite of the broad recognition of the need for sustainable products and manufacturing, progress in fulfilling triple bottom line objectives is inadequate. There is a general belief that the development of commercially successful "sustainable products" faces an insurmountable challenge in the current economic context. What is often lacking in the critique of unsustainable industrial practices is the pivotal role of the prevalent economic context which shapes industrial and financial institutions and constrains decision rules in product development and operational strategies.

The French philosopher Michel Foucault said: "A critique is not a matter of saying that things are not right as they are. It is a matter of pointing out on what kinds of assumptions, what kinds of familiar, unchallenged, unconsidered modes of thought the practices that we accept rest."[1] These unconsidered modes of thought often apply to the "sacred cows" of prevalent structures and systems - the past strategies and assumptions that are perceived to be the source of success and the essence of economic activity.

Hence, the sacred cows are not even put on the table when discussing change, even though they might be the root cause of the problem that necessitates that change. In this chapter we argue that we need to examine critically our often unquestioned assumptions of the following: growth as the overarching objective of business, the win-lose approach toward nature, survival-of-the-fittest as the desirable (and unavoidable) mechanism for development, the focus on short-term gains, globalization for exploitation of resources (of the Earth and labor) without globalization of equity, detachment from local and community issues, and consolidation/centralization for the control of resources in a zero-sum game. While many of these root causes of unsustainable development have been discussed by others (as referenced throughout this chapter), the proposed remedies are at best a modest modification of the prevalent economic system rather than the needed fundamental transformation.

Sustainable product development and manufacturing, we further argue, requires a new economic model that supports the sustainable growth of human welfare and development (rather than the sustained growth of shareholder value). This new economic model would fundamentally change the basis of competition, shift the concept of development away from raw materialism and change purchasing power as the sole indicator of success, happiness and self-actualization.

In what follows, we will first provide a brief overview of the historical regulatory and market context that has driven business decisions vis-à-vis environmental sustainability. As discussed in the next section, managerial awareness of sustainability issues has risen over the past few decades and corporations, in abidance by regulatory requirements and consumer preference, have reduced the intensity of the environmental impact of their business operations. However, the current modus operandi is unlikely to lead to environmental and business sustainability. We then review four very different and widely discussed models developed and advocated by respected and diverse authors, and argue that these models have not led to economic and environmental sustainability on a global scale due to shortcomings that have not fully addressed the root causes of unsustainability in the prevalent economic and social systems. We then propose how a new economy model of distributed capitalism can overcome these shortcomings and lead to the desired global sustainability by progressing through evolutionary phases of economic activities. Finally, the transition to sustainable development is discussed.

---

[1] Foucault 1990, 154.

Evolution of Regulations and Design for Environment, Health and Safety

Safety and health related guidelines in product design and manufacturing were imposed through codes and standards such as ASME, ASTM, NEC, NFPA[2] and others throughout the twentieth century. Many of these guidelines were developed by professional engineering organizations and adopted by local and state governments as prerequisites for issuing manufacturing or construction permits.

Modern environmental awareness grew in the second half of the twentieth century as a result of several alarming radioactive accidents, oil spills, mercury poisoning and other mishaps. The United States Environmental Protection Agency (EPA) was established in December 1970, and was charged with protecting human health and safeguarding the natural environment. Another US federal agency, the OSHA (Occupational Safety and Health Administration), was also created in December 1970. Subsequently, the United States passed new legislation such as the Clean Water Act, the Clean Air Act, the Endangered Species Act and the National Environmental Policy Act. This legislation has provided the foundations for current environmental standards. In 1972 the United Nations Conference on the Human Environment was held in Stockholm, for the first time uniting the representatives of the world's governments in negotiations on the state of the global environment. This conference led directly to the creation of the UN Environment Program (UNEP).

The common use of the terms (environmental) sustainability and sustainable development began with the publication of the Brundtland Report which characterized sustainable development as "development that meets the needs of the present without compromising the ability of future generations to meet their own needs."[3] The UN Environmental Program has organized several international conferences which have increased awareness and broadened the discourse on sustainability issues. The issues addressed include systematic scrutiny of patterns of production of toxic components, such as lead in gasoline, or poisonous waste, alternative sources of energy to replace the use of fossil fuels, new reliance on public transportation systems in order to reduce vehicle emissions, and the growing scarcity of water. An important achievement was the agreement of the UN Framework Convention on Climate Change which in turn led to the establishment of the Kyoto Protocol in December 1997. Countries that ratified this protocol committed to reducing their emissions of carbon dioxide and five other greenhouse gases by an average of 5% below their 1990 levels, or to engage in emissions trading if they wished to maintain or increase emissions of these gases. The Kyoto Protocol, which expired in 2012, was not ratified by the United States, the leading emitter of greenhouse gases (GHGs) in spite of its adoption by almost all other industrial countries of the world. However, the Kyoto requirements are now widely recognized to be inadequate in the light of recent alarming data on climate change. There is a need for much more restrictive

---

[2] ASME = American Society of Mechanical Engineers, ASTM = American Society for Testing and Materials, NEC = National Electric Code, NFPA = National Fire Protection Association.
[3] World Commission on Environment and Development 1987, 43.

measures on GHG emissions (as much as 50 to 80% reduction by 2050) if the environmental and ensuing economic calamity is to be averted.[4]

Another important international treaty regulation that significantly impacted upon product and process development in the 1990s was the Montreal Protocol on Substances That Deplete the Ozone Layer. This treaty forced industries across the globe to phase out the production of chlorofluorocarbon (CFC) chemicals (which were invented in 1929) as they are responsible for the depletion of atmospheric ozone which provides a critical health service to humans and animals by absorbing the harmful wavelength of the ultraviolet sunlight. The depletion of atmospheric ozone was a clear demonstration of how excessive human-made industrial effluents (over a very short time period – 1929 to 1987) could exceed Earth's limited capacity as a sink to absorb them. The widespread adoption and successful implementation of the "CFC ban" has been hailed as an example of exceptional international cooperation preventing a global tragedy of the commons. This case also demonstrated that global society can be prepared to make sacrifices in order to live within ecological limits.[5]

In the last decade, the European Union (EU) has led the industrialized-world in enacting laws which regulate use of hazardous substances in product manufacturing and hold the original equipment manufacturers (OEMs) responsible for recycling of the packing material and products at end-of-life, including a take-back program. The following EU directives have impacted many electronic and electrical products: RoHS - Restriction of Hazardous Substances Directive,[6] WEEE - Waste Electrical and Electronic Equipment Directive[7] and EuP - Energy-using Products.[8] The most notable of recent EU directives is REACH - Registration, Evaluation and Authorization of Chemicals (enacted in June 2007). According to REACH, the burden of proof is on the industry to ensure that chemicals put on the market do not adversely affect human health or the environment.[9] This is in contrast with Toxic Substances Control Act (TSCA) of 1976 that is the primary chemical policy in the US. TSCA places the burden of proof on the government's Environmental Protection Agency (EPA).

The EU directives have a broad implication for product design by global manufacturers who market to the EU. Because most manufacturers have a strong preference for common product designs across different market segments, the EU initiative is likely to become the de facto global standard for the transnational corporations (TNCs).

Throughout the short history of environmental protection, there has been strong opposition to government regulations by many corporations (particularly large US corporations) and by political leaders who have been concerned about the expected negative consequences of environmental regulation on economic growth. Corporations have consistently engaged in intense lobbying efforts to

---

[4] Stern Review 2006.
[5] Meadows et al 2004.
[6] European Parliament and Council 2003a.
[7] European Parliament and Council 2003b.
[8] European Parliament and Council 2005.
[9] Authorities should focus on ensuring that industry meets its obligations and takes action to reduce the harmful effects of substances of high public concern or where there is a need for community action.

dissuade law makers from enacting environmental laws rather than demonstrating imaginative leadership in sustainable product design and manufacturing.

The case of atmospheric ozone depletion and subsequent international cooperation that substantially abated the problem is a success story. It demonstrated that environmental-friendly industrial action need not necessarily harm the economy. Nevertheless when the problem was first brought to the attention of the public and environmentalists called for urgent action, the reaction from industry and government officials was a predictable denial and economic risk/benefit analysis.[10]

Over the last two decades of rapid globalization of markets and supply chains, the sensitivity to environmental and sustainability issues has risen significantly. The consequences of elevating sustainability issues to the forefront of corporate strategy have been limited to risk management, regulatory compliance, customer awareness and market opportunities.

Increased global competition has brought unprecedented pressure on corporations to increase operational efficiency and reduce costs. This has had an unintended benefit for the environment through reduction in waste. Marketers have taken advantage of this opportunity and have favorably positioned their companies as environment-friendly while reducing cost. More recently, the increased consumer awareness about global warming and other environmental degradations has created additional incentive for corporations to adopt energy efficiency measures in their operational practices.[11]

The adverse impact of traditional manufacturing and agricultural practices on the environment and human health has also created an opportunity for disruptive technologies such as organic farming and genetically-modified crops. The demand for organic food has grown at an unprecedented rate in spite of higher prices and shorter shelf life and has enticed mainstream farmers and distributors to enter the market. Riding on the sustainability coattails, Monsanto has tried to redefine the agriculture industry by the invention of genetically-modified (GM) crops. Monsanto's CEO, Robert Shapiro, positioned his company's strategy for GM crops as Growth through Global Sustainability:

> The Company has progressed from pollution prevention and clean-up to spotting opportunities for revenue growth in environmentally sustainable new products and technologies. Sustainable development is a discontinuity: the world is a closed system and we are beginning to hit the limits. The traditional model of growth in agricultural output has been to increase acreage and increase productivity through fertilizers, pesticides, and irrigation. This is not sustainable. New technology is the only answer: biotechnology and DNA-encoded information technology... Sustainability means less stuff and more knowledge and service (delivering functions to the users rather than goods.)[12]

---

[10] Meadows et al 2004, chapter 5.
[11] Denend and Plambeck 2007.
[12] Shapiro and Magretta 1997.

GM crop technology is criticized for safety concerns,[13] endangering biodiversity and indigenous farming particularly in underdeveloped regions of the world and for creating even further centralization of global industrial activities.

Other industry leaders have expressed the need for change and have tried to align their product strategies accordingly. Toyota Motor Corporation developed the highly successful Prius Hybrid car in a strategic response to an anticipated increase in consumer sensitivity to environmental and resource scarcity issues.[14] Toyota's proactive development of Prius, which exceeded regulatory emission standards, created a comparative advantage for the company by changing the rules of the game and creating competitive barriers.

Holliday et al[15] provide an excellent account of practices by several large transnational corporations and discuss an evolving paradigm in sustainable development. Stephan Schmidheiny, a Swiss industrialist and co-author of *Walking The Talk*, pioneered the formation of a business council of executives in the early-1990s which led to the formation of the World Business Council for Sustainable Development (WBCSD) in 1995. According to Schmidheiny in *Walking The Talk*: " ... the WBCSD continues to seek ways by which companies can achieve economic vitality while helping the planet toward environmental and social vigor."[16] The WBCSD website routinely documents successful cases that highlight integration of environmental and social concerns in business operations.

"No firm is yet sustainable," observes Catherine Ramus in her research.[17] According to her study, a gap between environmental policies and practices exists in many firms, even in those which are environmentally proactive. In a survey by McKinsey in February 2008, only 30% of the 1983 global companies in the survey considered climate change in their business strategy.[18] The percentage was even lower (21%) among the 470 US companies in the survey. In a 2011 survey of 4,700 executives, managers and thought leaders from around the world and from a wide range of industries, only 14% listed the threats and opportunities of sustainability among the primary business challenges facing their organization.[19]

The motivation in business for tending to environmental sustainability has been weak because the traditional strategies for competitive advantage and sustained value creation have served corporations well. Risk management, branding and improvement in operational efficiency, rather than (Brundtland's) sustainable development, have been the basis for corporate environmentalism.

Sustainable Economy: The Context for Sustainable Product and Manufacturing

In this section we review five models that have emerged from the sustainability research by economists and business management scholars and from the best

---

[13] EU directives such as directive 2001/18/EC require authorization for placing genetically modified organisms (GMO) on the market, in accordance with the precautionary principle.
[14] Carlson and Rafinejad 2007.
[15] Holliday et al 2002.
[16] Holliday et al 2002, 16.
[17] Ramus 2001.
[18] McKinsey Quarterly 2008.
[19] MIT Sloan Management Review and Boston Consulting Group 2011.

known methods practiced by industry leaders. These models represent normative frameworks of sustainable economy as the prerequisite context for sustainable product development and manufacturing. The five models are: value-based, regulatory, entrepreneurship, natural capitalism and new-economy. We will briefly review these models and discuss their contribution in creating the business context that is conducive to sustainable development. We also highlight the shortcomings of these models and argue that a significantly different model *(new economy)* is needed to fulfill the vision for sustainability. In conclusion, we explore a potential path for state change to sustainability.

Value-based Model

Walley and Whitehead's article "It Is Not Easy Being Green"[20] describes this model and captures the essence of industry's prevalent attitude toward sustainability vis-à-vis business objectives. The authors argue that environmental costs at most companies are skyrocketing, with little economic payback in sight. Win-win solutions (for shareholder value *and* the environment) should *not* be the foundation of a company's environmental strategy. As a society however, we may choose environmental goals and pay the cost. The common rallying cry of many environmental thinkers is that the environment must be integrated into everyday business decisions, yet few specify what that means. Companies would be better off focusing on the "trade-off zone," where environmental benefit is weighed judiciously against value destruction. This is the value-based approach.

In a follow up perspective article in the *Harvard Business Review*, twelve academic and industry experts review and comment on the above paper.[21] More than three decades ago, the Nobel laureate economist Milton Friedman made a similar argument suggesting that environmental protection belonged to the public sector and the primary responsibility of private corporations was to the shareholders.[22]

Reinhardt argues that business response to the environmental challenge must be based on the economic context. Environmental externalities coexist with other factors of market inefficiency that firms can profit from. Differentiation in environmental attributes, like other product attributes, must be of value to customers (who are willing to pay), be supported (with credible benefit) and be defensible. Therefore, firms must "provide environmentally preferable products and then capture the extra costs from consumers."[23]

Schmidheiny posits the free market "solution" to sustainability with appropriate government interventions to make the market efficient in accounting for the costs of environmental externalities. His proposal, which can be viewed as a hybrid of value-based and regulatory models, calls for: 1) Full-cost pricing to account for the costs of environmental externalities (getting the price right, however, is noted as the most important factor); 2) The use of economic instruments such as environmental taxes and charges and tradable permits instead of command-and-control regulations; 3) The phasing-out of perverse subsidies; 4)

---

[20] Walley and Whitehead 1994.

[21] Challenge of Going Green, Perspectives, 1994.

[22] Friedman 1970.

[23] Reinhardt 1998, 43.

Changes to standard national accounts (such as gross domestic product – GDP) to reflect environmental scarcity.[24] Schmidheiny further argues that:

> Society – through its political systems – will have to make value judgments, set long term objectives, implement measures such as charges and taxes step by step, and make midcourse corrections based on experience and changing evidence. Thus in moving toward sustainable development, it may be sufficient merely to introduce environmental charges slowly but predictably. Infrastructure planning, technology development, cultural patterns, and consumers could then anticipate the price increase and react accordingly.[25]

The value-based model is arguably the prevalent context for current business activities. Being market-driven, this model drives economic activities based on consumer preference over a short decision optimization window and on a non-cooperative and firm-centric basis. While market-driven business strategies could align with the wellbeing of consumers, they are often detached from long term environmental exigencies as is manifested in the exponential growth of the human environmental footprint over the last century and the lack of market-correcting feedback. Furthermore, it is not clear how full-cost pricing can be attained through free market economic instruments when consumers do not have equal and complete information about the short-term and long-term environmental damage caused by production and consumption and about the economic cost of recovery, cleanup or prevention.

Regulatory Model

Porter and van der Linde [26] argue that governments should enact strict environmental regulations and let competition thrive through compliance. Properly designed environmental standards can trigger innovations that enhance resource productivity (raw materials, energy and labor) lowering the total cost of a product or improving its value, thus offsetting the costs of a beneficial environmental impact. For example, pollution means waste that has resulted from inefficiencies in design or operations, embedding hidden costs throughout a product's lifecycle. Pollution, if viewed similarly to quality defects, reveals flaws in product and process design. Porter and van der Linde further argue that: "The resource-productivity model, rather than the pollution-control model, must govern decision making."[27] To accelerate progress, companies should: 1) measure their direct and indirect environmental impacts; 2) learn opportunity costs of underutilized resources and waste; 3) favor innovation-based, productivity enhancing solutions; and 4) foster relationships with regulators and environmentalists. The authors also offer a set of characteristics for "Innovation-Friendly Regulations" that do not drive up costs, including: focus on prevention

---

[24] Holliday et al 2002, 17.
[25] Schmidheiny 1992, 16-17.
[26] Porter and van der Linde 1995.
[27] Porter and van der Linde 1995, 131.

and outcomes rather than technologies; strict rather than lax phase-in deployment to allow new technologies to develop; market-incentives (pollution charges); and industry participation.

The authors try to make a business case for environmental innovation based on a cost-reduction and resource productivity argument, an opportunity that would be overlooked without regulations which would raise the priority of environmental innovation in the industry. Their regulatory-driven sustainability model views government regulations as panacea, the overarching force that brings about an integration of factors which are imperative to realization of a sustainable economy. Lack of (global) uniformity in regulations and enforcement, lack of universal standards in setting the "optimal" regulations (with self-evident benefit), strong political influences in regulatory processes, and opportunistic firm-based strategies for competitive advantage and for maximization of shareholder return, make a regulatory-based sustainability model untenable.

The underlying assumption behind the regulatory model is the view that government is the "protector" of environmental resources - "the common" property that is in danger of exhaustion because of excessive exploitation by the producers and consumers who are inherently selfish. This phenomenon, widely known as the tragedy of the commons, was put forth by Garrett Hardin. That essay was written specifically about the dangers of overpopulation and argues that "the commons" should be protected by coercive measures (laws) limiting (people's) actions that exploit the commons for self-interest. Hardin posits that the population problem has no technical solution. He disagrees with "most people who anguish over the population problem and think that farming the seas or developing new strains of wheat will solve the problem—technologically."[28] Hardin uses the example of herdsmen overgrazing a pasture open to all beyond the carrying capacity of the land because they all optimize their decision solely on the basis of self-interest. In other words, to a rational utility-maximizing herdsman there is no downside to overgrazing the shared commons until it is too late for all. "Freedom in a commons brings ruin to all" in a world that is limited, argues Hardin.[29]

Hardin further explains that while the "common sources" can be protected by privatization of ownership, the "common sinks" (e.g. the atmosphere as a sink for human-made pollution) cannot be "readily fenced." Hardin further posits: "To couple the concept of freedom to breed with the belief that everyone born has an equal right to the commons is to lock the world into a tragic course of action."[30] He argues that coercion is the only plausible strategy for protecting the commons – consciousness and sense of responsibility for the common good are not tenable. He argues that according to Darwin's theory of natural selection, "conscience is self-eliminating" because people who do not heed the conscience-call to limit breeding will produce a larger fraction of the next generation. Furthermore, we cannot expect "social responsibility" to save the commons because responsibility is a product of social arrangement and without the fear of sanctions is synonymous to conscience.

---

[28] Hardin 1968, 1243.
[29] Hardin 1968, 1244.
[30] Hardin 1968, 1246.

To adopt the regulatory model, we must ponder the basis for constructing the social arrangements, the ensuing responsibility and the enforceability of the necessary "coercive" measures (i.e. the regulations) in a democratic context. The process of identifying shared resources (the commons) and the need for a cooperative "game" on a global scale requires enlightened discourse in framing sustainability as a moral imperative and in establishing the necessary social consent. In other words, consciousness and education would be the essential ingredients for averting the tragedy of the commons in a regulatory model advocated by Porter and van der Linde.

Entrepreneurship Model

Proponents of this model advocate keeping the government out and letting the free-market forces manage natural resources and the environment through entrepreneurship [31] and innovation in technology. [32] Anderson and Leal cite numerous excellent examples where entrepreneurs envisioned profitable business opportunities in mitigating an adverse environmental impact of an industrial enterprise. For example, they narrate stories: of saving endangered birds by developing a tourism business in a logging forest; marketing fish-eggs "wastes" which were previously disposed at the peril of the local land and water; and trout fishing revival in a housing development project. [33]

The technology argument is based on the theory that environmental/resource sustainability issues can be resolved through technological innovation if the necessary investment is directed at them. And because sustainability issues give rise to entrepreneurial business opportunities, enviro-capitalists will drive the investment in the right technologies.

Meadows *et alia* see the weakness of the technology argument in the underlying simplistic assumption about the way technology works arguing that "For many economists technology is a single exponent in some variant of the Cobb-Douglas production function – it works automatically, without delay, at no cost, free of limits, and produces only desired outcomes." [34]

The paramount consideration in technological development is the objective that the technology is supposed to serve. In an economic context where unabated growth in production and consumption is the overarching goal, an enviro-capitalist is likely to invest in technologies that further promote this goal, albeit through "greener" production and consumption.

Another important dimension of human economic activities and technological solutions is the time dimension. Cumulative impact of human activities on renewable and non-renewable sources and sinks and the inherent latency in technological effectiveness in resolving problems are causes for instability in the business/ecological system behavior and for irreversible calamities. For example, many scientific and government sources around the globe have predicted dire consequences of global warming if human-generated greenhouse gas emissions are not abated drastically and very soon. In other

---

[31] Anderson and Leal 1997.

[32] Ray and Guzzo 1990.

[33] Anderson and Leal 1997.

[34] Meadows et al. 2004.

words, no technology may be able to reverse the expected melting of glaciers and rising sea level beyond a tipping point.

As will be discussed below, entrepreneurship and technological innovation are essential to an integrated sustainability framework but are effective only if they target locally-balanced production and consumption of the new economy model.

Natural Capitalism Model

This model espouses sustainability through increased resource efficiency, reduced waste, a shift to solution-oriented business and investment in nature. Lovins *et alia*[35] posit that the following four major shifts must occur in current business practices in order to achieve "natural capitalism":

1. Dramatically increase the productivity of natural resources; reduce waste and destructive flow of resources.[36]
2. Shift to biologically inspired production models. Eliminate waste by using a closed loop production system. This would require the formation of industry alliances – within and across sectors – and cooperatives in order to sell and service the byproducts. Otherwise, in order to market their operational byproducts, firms must engage in businesses that are outside their served market, or even industry.
3. Move to a solution-based business model. This results in providing "illumination, for example, instead of light bulb." Hence, manufacturers shift to a service-leasing business model and to life-cycle ownership of their products. This approach works if the local labor cost to service and repair a product is less than the manufacturing and carrying cost of the defective subsystem and component that is sourced from a "low-cost-region".
4. Reinvest in natural capital – similar to capital investment in means of production – to restore, sustain and expand the planet's ecosystems.

A few success stories from grazing (moving cattle from one place to another) and farming (avoiding single crops) are cited by Lovins *et alia* in the *Harvard Business Review*.[37] The authors argue that current metrics of business success and the taxation system encourage the waste of resources. For example, low initial cost is favored over low cost-of-ownership, spending is encouraged (use your budget or lose it), and city design ordinances encourage waste.

In order to shift current business practice toward "natural capitalism," there must be adequate incentives for industry to participate in inter-firm cooperative programs to emulate biological production models (item 2 above) and to invest in natural capital (item 4). Public corporations are driven by the financial metrics

---

[35] Lovins et al. 1999.

[36] According to the United Nations Development Programme, efficiency improvement is a staggering opportunity. The US economy could do everything it now does, with currently available technologies and at current or lower costs, using half as much energy, if it operated at the present efficiency levels of Western Europe. Source: UNDP 2003.

[37] Lovins et al. 1999.

that are set (for the most part) by investors (in global equity markets) and any of the above shifts must measure up against those metrics. Today's global supply chains of production extend over long physical distances and long networks that are inherently incongruous to the biologically inspired production models that are advocated by the proponents of this model. Furthermore, investment in natural capital by capitalists necessitates privatization and monetization of nature's services and resources which need to be sustained. This approach is likely to align access to natural resources (including existential basics of water and food) with wealth and further exacerbate global inequity. Natural capitalism and the privatization of the commons (natural capital) in the long run could drive investment decisions contrary to biodiversity and the requirements for sustainable development.

New Economy Model

As defined by the US National Academy of Engineering, sustainable development "represents the quest for an economy that exists in equilibrium with the Earth's resources and its natural ecosystems. Sustainable development brings environmental quality and economic growth into harmony, not conflict."[38]

Inspired by this definition, we argue that the prevalent economic model of global capitalism for unlimited and uneven economic growth must change to a locally-sustainable model that supports the evolutionary improvement of life quality. James Gustave Speth, a longtime leader in the environmental movement and Dean of the Yale School of Forestry and Environmental Studies, states in *Bridge at the Edge of the World* the need for a transformative change from the prevalent economic norm: "In short, my conclusion, after much searching and considerable reluctance, is that most environmental deterioration is a result of systemic failures of the capitalism that we have today and that long-term solutions must seek transformative change in the key features of this contemporary capitalism."[39]

We envision a new economic system that behaves similarly to the ecological system and is integrated with it. As such, the structure of this economy has to be distributed where there is harmony between economic activities and the environment in both spatial and temporal dimensions; a global system that comprises myriad units of linked local economies which are supported by local resources in alignment with nature's seasonal variations. In stark contrast, the current global system of production, distribution and consumption is characterized by long supply chains for physical and virtual transfer of resources from one part of the world to another creating globalized externalities.[40]

In order to conjecture on the feasibility of the above new economy model, we should revisit the definition and moral aspects of sustainability and examine the

---

[38] Richards et al. 1994, 5.

[39] Speth 2008, 9.

[40] For example, industrialization of food products is decoupling production from local consumption: soy is grown in one country, fed to pigs/chicken in another country, livestock is slaughtered and processed in a third and shipped to a fourth country for consumption. Changes in consumption in one country thus impact upon the environments of other countries across the globe.

characteristics of a sustainable model. Sustainable global economy is an economy that the Earth is capable of supporting indefinitely. This means a stable level of world population plus an environmentally-balanced use of resources so that a thriving and lasting state of high quality life for all people is created. The stable state would require recycling so that untapped resources are not exhausted and maintaining the equilibrium of the ecosystem where its operation is resilient and not perturbed to instability. Such a stable state would be feasible if human consumption is locally efficient (not wasteful) and high quality life is maintained with minimal consumption of resources. We refer to these as "locally-sustainable high-HDI[41] consumption and population levels."

According to economist Herman Daly[42] the following three rules characterize a sustainable economy: 1) For a renewable resource (e.g. soil, water, forest, fish), the sustainable rate of use is less than the rate of regeneration of its source; 2) For a nonrenewable resource (e.g. fossil fuel, high-grade mineral ore, fossil groundwater), the sustainable rate of use is less than the rate at which a renewable resource (used sustainably) can be substituted for it; and 3) For a pollutant, the sustainable rate of emission is less than the rate at which the pollutant can be recycled, absorbed or rendered harmless in nature. In the new economy model, Daly's rules on a global scale are followed through their implementation on a local and small scale.

Jouni Korhonen provides an overview of approaches to a sustainable economy and the elements of an industrial ecosystem in that economy, including: roundput (closed-cycle industrial production), diversity (presence of many actors and activities in an industrial ecosystem), cooperation ("symbiotic and cooperative relationships" between players) and locality (local product lifecycle and local actors).[43] He discusses the difficulties in implementing these approaches and concludes that they should be used not as prescriptive elements of a sustainability model but as indicators of sustainability when analyzing complex systems. Korhonen also points out that "there is risk that the ethical and emotionally loaded views about sustainability distort and hamper thorough and careful analysis" and highlights the uncertainty in environmental analysis and the difficulty in doing a "nearly complete life cycle assessment (LCA) of products."[44]

We must ask whether the ethical and moral dimension is a relevant consideration in sustainability discourse or a philosophical, personal, and emotional issue which does not belong to the realm of business and economic science; and perhaps the ethical issue also belongs to the societal collective representative that is the government. It is important to note that any economic and business model is rooted in a certain framework of purpose that forms the

---

[41] Alternate metrics to GDP have been proposed to represent a holistic condition of economic, social and environmental health. The United Nations has developed the Human Development Index (HDI) as an alternate to GDP to integrate the impact of economic and human capitals on the standard of living. HDI is calculated according to the three factors of life expectancy (health), adult literacy (education/skills) and gross-domestic-product (GDP) per capita at purchasing power parity (PPP). For the lack of a better metric, we consider the HDI to be synonymous with life quality.

[42] Daly 1990.

[43] Korhonen 2007.

[44] Korhonen 2007, 52.

foundation for analysis and establishes the criterion for optimization of action. Because of the complex and holistic nature of the sustainability concept, the discourse on ethics and dealing with uncertainty is not avoidable as it is central to the concept of development and imperative to the quest for sustainable development. We must ask what the social benefits of our economic system are. An economic system and associated business strategies that aim to maximize growth of shareholder value are fundamentally different to those that enhance quality of life equitably across spatial and temporal dimensions.

The definition of sustainability and the necessary attributes of a sustainable system can be inspired by nature and the functioning of its stable (mature) ecosystems.[45] Natural ecosystems are self-organized into a diverse and integrated community of organisms to maintain their presence in one place, make the most of what is available, and endure over the long haul. Perhaps the reason why humanity in the twenty-first century is faced with the issue of sustainability is the fact that we have deviated from the above natural sustainability model in several critical respects. We have heeded the "make the most out of what is available" doctrine, but have failed to abide by the natural law of "formation of diverse and integrated local communities" and to attend to the imperative issue of endurance over the long haul.

We have created a centralized global system, and have sought short-term benefits from what is available (more is better) at the expense of the long haul by discounting future benefits and harm. In fact, Benyus argues, humans have even gone beyond breaking natural laws. We have separated ourselves from nature, declared independence from natural laws and worse yet, we have claimed superiority over nature and striven to dominate, control and improve it.

The hallmark of the prevailing economic system is fascination with growth. Growth in shareholder value/profitability, revenue, and market share has become the most important metric of business performance today. Growth is not just a metric of success; it is framed as the imperative of survival. Consequently, market size expansion, business consolidation, increased production and promotion of consumption on the global scale have become the overarching objectives of current economic models and management strategies. The lesson of sustainability from nature is at a stark contrast with our notion of growth for survival. Nature does not have a growth objective. Its evolutionary process is aimed at sustainability as the overarching purpose in which the fittest will thrive and survive. And the fittest is not the domineering but the one that can live in harmony with diversity and does not exhaust the sinks and the sources that it needs for its own survival. In the model for human sustainability, there needs to be an additional dimension that transcends mere survival as the life purpose. Quality of life, for example in the joy of science, music, art and esthetics or in community and spirituality are important to realizing our potential and growth as humans.

Another important issue of sustainability and sustainable product manufacturing is localization versus globalization. This is also an area where we have parted from natural processes. The current economic trend of centralization of objective setting and control of supply chains (in giant transnational

---

[45] Benyus 1997.

corporations), consolidation of financial markets (in a few corporations), centralization of production (for economy of scale in agriculture and manufacturing in low cost regions), and homogenization of products and consumption (as in ubiquitous McDonald and Starbucks stores) move the economic system away from diversity, local efficiency and sustainable local ecosystems to a consolidated capitalist system.[46] The efficiency improvement as more output with less consumption of resources is commonly touted as the argument for centralization and economy of scale (large systems of production and distribution). Even some of the proponents of sustainability overlook the critical importance of localization at a scale commensurate with supporting natural ecosystems. For example, integrated renewable energy systems (solar, wind and water resources) are designed based on existing systems of distribution and consumption of electricity within political boundaries for national or regional self-sufficiency rather than natural boundaries. These large integrated system designs, enabled by advances in modern technologies in data communication and controls, have little regard for the crucial role of natural ecosystems and its key supporting functions. Such large complex integrated systems are potentially non-resilient and rely on technologies that are out of reach of the vast majority of people of the world. Furthermore, the disregard for local harmony with nature results in excessive population growth as evidenced by the creation of consumption centers such Las Vegas or Dubai that function on mostly imported resources and people.

The need for a new economy and the necessary localization of economic activities and local self-reliance has been discussed by several authors. [47] Distributed systems might be less (globally) efficient from the point of view of production capability, but they are more stable, resilient and locally efficient. Another advantage of a local (small) community of economic activities is that it has short feedback latency and hence it is self-correcting without going over a tipping point (point of no return).[48] And even if one unit of the world economic ecosystem went over a tipping point and became bankrupt the vast majority of others would thrive. No economic unit or corporation would be "too big to fail."[49] Furthermore, local economic units can be managed cooperatively for sustainable

---

[46] Today's consolidated capitalist system has the following characteristics: a) few large global corporations in each industrial sector (or sometimes across industrial sectors) often dominate the entire supply chain through vertical/horizontal integration or market power through economy of scale; b) centralized strategy and decision making; and c) strong support from national governments.

[47] Ring 1997; Wackernagel and Rees 1997; Milani 2000; McKibben 2008; Speth 2008; Korten 2009.

[48] The quality, validity and accuracy of information to appropriators of common resources are critical to timely decisions making about the institutional and operational rules that affect benefits and costs to the appropriators and sustainability of the common resource. To Ostrom, "One should expect individuals to be willing to adopt new rules that will restrict their appropriation activities when there are clear indicators of resource degradation, generally perceived to be accurate predictors of future harm, or when leaders are able to convince others that a 'crisis' is impending" (Ostrom 2008, 208).

[49] The US (and world) economic crisis of 2008-2009 amply demonstrated the lack of resilience of very large financial services corporations.

sharing of natural resources (the commons) without privatization or external government intervention.

In her seminal work, political scientist Elinor Ostrom who won the 2009 Nobel Prize in economic sciences, critiques government coercion (regulations) and market solutions (privatization) which are often cited as the (only) solutions to the tragedy of the commons dilemma. She demonstrates that relatively small natural common pool resources (CPRs) can be sustainably utilized through cooperative arrangements among local actors. They self-organize effective institutions and decision rules for appropriation and provision of the CPR and for monitoring compliance to the rules. Through extensive empirical evidence, Ostrom demonstrates that governance of natural resources through collective action can be successful when the CPR scale is relatively small[50] (enabling information symmetry among the individuals) and when external public (government) instruments are supportive.

The distributed economy model envisions local production and delivery of life essentials (such as food, water and energy from local CPRs) to support local consumption so as to be technologically efficient. It also includes local decision-making within the bounds of global cooperative rules and global information/technology sharing for efficient production at the local level.[51] This system stabilizes local population, production and consumption at a level that Earth's resources can sustain locally. For other products and services of modern life such as electronic consumer goods and global transportation that do not abide by a locally-sustainable model, cooperative institutional and operational rules must be devised on a global scale to bound economic activities by sustainable sharing of Earth's common goods.

Distributed capitalism supports biodiversity and diversity in human culture and esthetics. It strengthens communities and enhances quality of life through caring for others and for nature. What is needed is development of theoretical models which define the parameters and system characteristics of a distributed economy and demonstrate its feasibility.

Transition to Sustainability

The new economy model represents an evolutionary stage in development of the framework for human economic activities. These economic activities, vis-à-vis environmental sustainability, have evolved through the three phases of pioneering economic development, environmental protection and eco-efficiency. The impetus for evolving from one phase to the next has been either to correct the adverse consequences of activities in the current state or to take advantage of the prospective opportunities of the next higher state.

> *Phase 1.* Pioneering development: during this phase the assumptions of boundless natural resources and negligible impact of human activities on the ecosystem led to an economic system that facilitated maximization

---

[50] The largest CPR among the successful cases in Ostrom's book involves 15,000 appropriators. See Ostrom 2008.
[51] Korten calls for revised intellectual property rules. See Korten 1999.

of production and consumption. Moreover, because governmental regulations for protection of the environment were minimal, this phase led to significant and tangible adverse impact on the environment, health and safety of populace by the second half of the twentieth century.

Phase 2. Environmental protection: in this phase the objective of maximizing economic growth was further pursued at an exponential rate. The environmental impact of industrial activities, however, was moderated by end-of-pipe pollution reduction through governmental regulations. By the 1990s an excessive human ecological footprint and the potential scarcity of nature's sources and sinks were recognized as potential threats to economic growth.

Phase 3. Resource-to-value efficiency (eco-efficiency): the current state of the world's economic development is, we argue, at Phase 3. In this phase, business strategies and government actions are aimed predominantly at slowing down deterioration of nature's sources and sinks through conservation and improvements in operational efficiency in ubiquitous global production and consumption. In this phase, nature's capital (i.e. the value of natural resources and services) is not integrated into the economic system and the necessary coordination for protection of global common goods is lacking. Therefore, global population growth and increase in GDP per capita will continue to accelerate, thus increasing the human ecological footprint beyond Earth's sustainable capacity. In this phase, self-interest of business is the overarching driver of economic activities and hence global competition for access and control of limited resources is likely to intensify and lead to conflict and increased inequity.

The new economy represents *Phase 4* of economic development. We call this phase Distributed Capitalism and envision it as a globally linked system of local economic units that harmonize human activities with nature's resources and achieve sustainable, global and equitable human development. At this stage human development means enhancement of global HDI and happiness, resources are exploited without disturbing the equilibrium of the ecosystem in its ability to provide the necessary source and sink services, and human dominance over nature is substituted with a strategy of mutual interdependence with nature.

Our economy will not evolve to the sustainable state of *Phase 4* through natural adaptation. A deliberate global strategy must be devised. The case of the CFC ban that reversed the ozone depletion trend in 1990s is an example of such a focused strategy; a successful international cooperation that was orchestrated by the United Nations. Another example of widespread cooperation among different players is the quality transformation of products, services and manufacturing in the United States in 1970s and 1980s. Unlike the CFC ban, quality transformation was more than a cooperative problem solving effort and led to ubiquitous cultural and behavioral change in business practice. The US semiconductor manufacturers, who were struggling in the early 1980s to compete with high quality products from Japan, caught up in quality practices and recovered a major portion of lost market share by the end of the decade, thanks to Sematech, a government-industry funded organization that led the industry in a broad quality improvement collaboration.

Although quality transformation occurred on a national scale, it nevertheless demonstrates that cooperation among competitive businesses, their suppliers and the government can be effective in shifting business culture and operational practices in a profound way. A similar cooperative effort is under way in the EU among chemical manufactures to facilitate implementation of the REACH directives.[52] While recent corporate initiatives are positive, they contribute limited improvement to the modus operandi and merely slow down the prevalent unsustainable economic activities. Sustainability and adoption of the distributed capitalism model require a major business transformation, a metamorphosis of economic goals, framework, assumptions and activities. This transformation does not mean regression or slowing of historical improvement in the average human development index (HDI) in economically developed regions. Sustainability transformation and the necessary technological, economic, and social innovations are the imperative next steps in the human development process. On the one hand, transition to a sustainable world changes the status quo in socio-cultural, economic and institutional (power) structures and modes of behavior and hence will be resisted by those who perceive a loss. On the other hand, visionary business, social and intellectual leaders will see change as the opportunity and the imperative of business sustainability.

If environmental and economic sustainability is to be attained, long term, systemic and deliberate transition strategies must be implemented across virtually all business enterprises and governments worldwide and numerous innovative business processes must be adopted. The probability of success in change management depends on four factors: 1) the degree of dissatisfaction with the status quo; 2) the persuasiveness of the vision and quality of the model for change; 3) the quality of the implementation process; and 4) the perceived cost and risk of change. The first two factors are largely unmet. While dissatisfaction with sustainability risks has risen globally, more education is needed to create dissatisfaction with the current economic system to the level that is necessary to trigger a profound change to a sustainable state. Furthermore, the vision for a sustainable world, such as our proposed concept of distributed capitalism, is only in its infancy and requires much development. Comprehensive and defensible theoretical models must be developed for economic and socio-cultural structures and for associated supporting institutions and public policy in the new state. And all concerned social actors, including universities, businesses, governments, and citizens, must engage in education, research and action aimed at developing the vision for a sustainable state and groundbreaking models for operationalizing the vision.

---

[52] Carlson et al. 2008.

References

Anderson, Terry. L. and Donald R. Leal. 1997. *Enviro-Capitalists, Doing Good While Doing Well*. Lanham, Maryland: Rowman and Littlefield Publishers, Inc.

Benyus, Janine, M. 1997. *Biomimicry – Innovation Inspired by Nature*. New York: William Morrow and Company.

Carlson, Robert, C. and Dariush Rafinejad. 2007. *Development of Prius Hybrid Vehicle – A Case Study*. Stanford University, Management Science and Engineering Department.

Carlson, Robert, C., Feryal Erhun and Dariush Rafinejad. 2008. WACKER Invests in Sustainable Production of Solar Silicon for the Photovoltaic Industry – A Case Study. Stanford CA: Stanford University.

Challenge of Going Green, Perspectives. *Harvard Business Review*, July-August 1994. Reprint 94410.

Daly, Herman. 1990. Toward Some Operational Principles of Sustainable Development. *Ecological Economics* 2: 1-6.

Denend, Larry and Erica Plambeck. 2007. *Wal-Mart's Sustainability Strategy*. Stanford University, Graduate School of Business, Case Study: OIT-71.

European Parliament and Council. 2003a. Directive 2002/05/EC on the Restriction of the Use of Certain Hazardous Substances (ROHS) in Electrical and Electronic Equipment. Brussels

European Parliament and Council. 2003b. Directive 2002/96/EC on Waste Electrical and Electronic Equipment (WEEE). Brussels.

European Parliament and Council. 2005. Directive 2005/32/EC on the Eco-Design of Energy-using Products (EuP). Brussels.

Foucault, Michel. 1990. *Politics, Philosophy, Culture: Interviews and Other Writings, 1977-1984*. London: Routledge.

Friedman, Milton. 1970. The Social Responsibility of Business Is to Increase Its Profits. *New York Times Magazine*. September 13, 1970: 1-6.

Hardin, Garrett. 1968. The Tragedy of the Commons. *Science* 162(3859): 1243-8.

Holliday, Charles O. Jr., Stephan Schmidheiny and Philip Watts. 2002. *Walking the Talk: The Business Case for Sustainable Development*. San Francisco: Berrett-Koehler Publishers.

Korhonen, Jouni. 2007. Environmental Planning vs. Systems Analysis: Four Prescriptive Principles vs. Four Descriptive Indicators. *Journal of Environmental Management* 82(1): 51-9.

Korten, David C. 2009. *Agenda for a New Economy – From Phantom Wealth to Real Wealth*. San Francisco: Berrett-Koehler Publishers.

Lovins, Amory, B., Hunter, L. Lovins, and Paul Hawken. 1999. A Roadmap for Natural Capitalism, *Harvard Business Review* Reprint 99309. Available at http://salient.nohomepress.org/wp-content/uploads/2008/03/hbr-rminatcap.pdf, accessed March 8, 2014.

McKibben, Bill. 2008. *Deep Economy: The Wealth of Communities and the Durable Future*. New York: Holt Paperbacks.

McKinsey Quarterly. 2008. *Climate Change Considerations in Business, a Survey of Global Companies*. February Online Newsletter. Available at http://www.mckinsey.com/insights/mckinsey_quarterly accessed March 28, 2014.

Meadows, Donella, Jorgen Randers, and Dennis Meadows. 2004. *Limits to Growth: The 30-Year Update*. White River Junction VT: Chelsea Green Publishing.

Milani, Brian. *Designing the Green Economy*. 2000. Lanham, MD: Rowman & Littlefield Publishers.

MIT Sloan Management Review and Boston Consulting Group. 2011. SMR405, (53) No.1. Cambridge MA: Sloan Management Review Association, MIT Sloan School of Management.

Ostrom, Elinor. 2008. *Governing the Commons – The Evolution of Institutions for Collective Action*. Cambridge: Cambridge University Press.

Porter, Michael and Claas van der Linde. 1995. Green and Competitive, Ending the Stalemate, *Harvard Business Review* reprint 95507. Available at http://hbr.org/1995/09/green-and-competitive-ending-the-stalemate/ar/1 , accessed March 8, 2014.

Ramus, Catherine A. 2001. Organizational Support for Employees, Encouraging Creative Ideas for Environmental Sustainability. *California Management Review* 43(3): 85-105.

Ray, Dixy Lee and Lou Guzzo. 1990. *Trashing The Planet*. Washington DC: Regnery Gateway.

Reinhardt, Forest L. 1998. Environmental Product Differentiation: Implications for Corporate Strategy. *California Management Review* 40(4): 43-73.

Richards, Deanna J., Braden R. Allenby, and Robert A. Frosch. 1994. *The Greening of the Ecosystems*. Washington DC: National Academy of Engineering, National Academy Press.

Ring, I., 1997. Evolutionary strategies in environmental policy. *Ecological Economics* 23(3): 237-250.

Schmidheiny, Stephan, 1992. *Changing Course – Perspective on Development and the Environment*. Cambridge MA: MIT Press.

Shapiro, B. Robert and Joan Magretta. 1997. Growth Through Sustainability: An Interview with Monsanto's CEO Robert B. Shapiro. *Harvard Business Review* 97110. Boston, MA: Harvard Business School Publishing.

Speth, James Gustave. 2008. *The Bridge at the Edge of the World: Capitalism, the Environment, and Crossing from Crisis to Sustainability*. New Haven: Yale University Press.

Stern Review. 2006. *The Economics of Climate Change: A UK Government Report*. Cambridge: Cambridge University Press.

United Nations Development Program (UNDP). 2003. *Human Development Report 2003*. Available at http://hdr.undp.org/en/content/human-development-report-2003, accessed March 28, 2014.

Wackernagel, Mathis and William E Rees. 1997. Perpetual and Structural Barriers to Investing in Natural Capital: Economics from an Ecological Footprint Perspective. *Ecological Economics* 20: 2-24.

Walley, Noah and Bard Whitehead. 1994. It Is Not Easy Being Green. *Harvard Business Review* reprint 94310 (5). Boston, MA: Harvard Business School Publishing.

World Commission on Environment and Development. 1987. *Our Common Future*. Oxford: Oxford University Press.

Chapter 10: Sustainable Corporate Strategy: Who's Sustaining What?

Michael L. McIntyre and Steven Murphy

Introduction

In order to address the idea of sustainable corporate strategy it is necessary to be specific about both the idea of sustainability and the idea of corporate strategy. Further, a useful discussion of sustainability and corporate strategy needs context. In our view, the appropriate context is the role of the corporation in society. In other words, we think it is useful to ground the discussion of sustainable corporate strategy in a discussion of what society is looking for when it gives the corporation its social license to operate. This approach brings focus to the discussion of what, exactly, one is attempting to sustain and why when one uses the term sustainable corporate strategy.

Corporate strategy is about organizing the corporation (the "firm") to achieve its goals. There is a significant focus on the establishment and exploitation of comparative advantages in key elements of what the firm does. The firm uses its comparative advantages to develop value propositions that are as good as, or better than, those of its competitors.[1] Firms can identify sources of comparative advantage in many ways, but the usual approach involves an environmental scan,[2] assessment of the firm's industry and the nature of competition within the industry,[3] and an evaluation of the firm's capabilities. The firm can evaluate its comparative advantages in terms of its ability to exploit them, the comparative advantages' relative ability to contribute to the firm's goals, the extent to which the advantages are differentiated from the advantages possessed by competitors, and the extent to which competitors can replicate the firm's advantages and thus neutralize their benefits.[4] The firm can also identify important gaps in its capabilities and seek to address them. Firms do not necessarily have to develop

---

[1] Porter 1980.
[2] For a discussion of the PEST (Political, Economic, Sociological, and Technological) approach, see Grundy 2006.
[3] Porter 1980.
[4] Wernerfelt 1984; Barney 1991; Grant 1991.

any or all of their comparative advantages internally, nor must they address gaps using solely internal resources. While in many cases firms emerge and grow over the long term using home grown comparative advantages, they can also acquire them from elsewhere if they are available, reasonably priced, and the firm has the resources to pay the prices at which they are available.

Two additional matters need to be addressed to provide a foundation for discussion. The first is to specify the goals of the firm. We do so as follows: achieving financial viability while maintaining its social license to operate. The second is acknowledgement of the fact that the firm uses resources to carry out its strategy. The firm's resources include the assets under its control, the input factors to production that it owns, and its labor force. The implication of the firm's use of and control over resources is that it makes the firm a governance structure in society. This begs the questions: "Why would society wish to delegate supervision of some of its resources and people to such governance structures? What are the implications of this delegation for issues of sustainability in society?" We will address these matters in the next section. First, we consider the idea of sustainability.

Sustainability and the Corporation

The idea of sustainability in the context of corporate strategy is interesting because it has only limited relevance if the firm is primarily concerned with economic performance. Business units and their chosen strategies exist in a context, and in a randomly evolving world, the context is ever changing. There is therefore always a requirement to ensure that the initially specified strategy evolves so that it stays well adapted to its environment most of the time. In a strict sense, therefore, the specific details of a strategy are sustained only for as long as it takes for enough change to occur in the business environment to render the strategy unsuitable. Having said this, it is usual for a business unit to implement a strategy with a view to making enough money over a specific time frame to justify the investment in the business. Since justification depends on the realization of planned cash flows over the planning horizon, and the typical maintained hypothesis is that the cash flows are an artefact of the business unit's strategy, one can imagine that it is desirable if the strategy can be sustained over the planning horizon. It saves management the trouble of repeatedly reinventing the business unit. The point is the strategy does not likely need to be sustained for an indefinitely long period of time to achieve the business unit's economic goals.

A firm with more than one business unit invokes a corporate level strategy that amounts to managing the firm's portfolio of business units. The multi-business-unit firm is able to cycle through business units indefinitely, retiring some and adding others as time unfolds, so it is easy to imagine such a firm having an indefinitely long life. Although this may be so, it is not necessary for it to be so, and even less necessary for it to have an infinite life, for it to achieve economic legitimacy. If, after the passage of time, the firm finds itself without the comparative advantages it needs to carry out its corporate level strategy, it can sell its resources to another firm and terminate its existence. It can do this and simultaneously be a financial success over the course of its existence.

Consequently, there is not necessarily an economic requirement for corporate strategy in multi-business-unit firms to be indefinitely sustainable.

Figure 10.1 presents a decision tree for firms assessing the viability of strategies. Clearly there are economic frictions as capital and labor are redeployed, but the issue isn't one of sustainability versus nonsustainability. The issue is one of deciding which activity will get the resources, and which firm will act in a governance role over them: the existing firm or a different one. These decisions depend on the relative costs and benefits of the alternatives. This being said, there is considerable focus on business unit and firm level sustainability. This is because management's job is typically considered to include scanning the environment and the competition to find new opportunities, but sustainability that arises from this, from an economic point of view, is a "nice to have" and not a "need to have." In addition, managers may strive for long term sustainability because they don't like the alternative: explaining to prospective employers that the best idea they had in their old job was to sell the firm's assets to someone who had better ideas than they did.

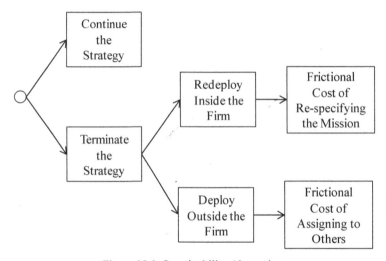

Figure 10.1: Sustainability Alternatives

So far we have focused on ideas of sustainability that are linked to economic performance: that is, the nature of sustainability needed to achieve economic goals. Since it seems clear there isn't necessarily an economic need for corporate strategies that are sustainable for particularly long periods of time, what is it that sustainable corporate strategy must be getting at? We now explore this in the context of the social role of the corporation.

The Social Role of the Corporation

As we stated at the beginning, both the idea of sustainability and the idea of corporate strategy ought to be considered in the context of the socially constructed idea of, and role, for the corporation itself. We also indicated that the fact that the firm exerts influence over its resources and employees makes it a governance structure in society, and we posed these questions: "Why would society wish to delegate supervision of some of its resources and people to such a governance structure? What are the implications of this delegation for issues of sustainability in society?" We consider these matters next.

Williamson[5] examines the firm as a governance structure under the rubric "The Economics of Governance." He further indicates that he does so in the spirit of eunomics, which Fuller defines as "the science, theory, or study of good order and workable arrangements."[6] While we would agree that there are consistencies between ideas in economics and ideas in eunomics, in this chapter the distinction is important. If one is concerned with sustainability for the firm, one is concerned with sustainability of a socially constructed phenomenon, one that was ostensibly constructed for the benefit of society. The firm must be economically and socially viable, so one is necessarily concerned with governance as contributing to good order of both a social and an economic nature, and as contributing to arrangements that are both socially and economically workable. We see the term eunomics as the more general term because it spans social and economic dimensions, and is therefore the more relevant term for the purposes of this chapter.

Williamson views governance more narrowly than we do in other ways as well. He sees it as part of the process of working out the management of transactions, with spot transactions in markets constituting the benchmark case. He sees transactions that are other than spot transactions emerging for various reasons, of which transaction costs are the most important. The object of his analysis is the exchange transaction as is evident from the following quote:

> ...the economics of governance is a lens of contract construction, broadly in the spirit of James Buchanan's (2001 p. 29) observation that 'mutuality of advantage from voluntary exchange... is the most fundamental of all understandings in economics'.[7]

We view the idea of a transaction more generally. For us, a transaction is the elemental unit of human interaction, sometimes including economic interaction but also including all other elemental interactions between one individual and another, between an individual and an organization, and between organizations, one type of which is the firm. We similarly view governance more generally as being the various social processes, including economic processes that govern our more broadly defined notion of transaction as will be evident from our more detailed discussion below.

Because our approach is eunomic rather than economic, and because we define "transaction" more generally, which is consistent with a eunomic approach, we will ultimately be able to consider the firm, and sustainable corporate strategy for the firm, in a relatively complete context. Our approach also provides proper context within which to consider governance of the firm and

---

[5] Williamson 2005.
[6] Fuller 1954, 477.
[7] Williamson 2005, 1.

governance within the firm. This is important because corporate strategy, sustainable or not, is the product of these governance forces.

Although our purpose is different to that of Williamson, we begin with the same benchmark case as in the Williamson construct: an economic general equilibrium in which production, consumption and investment are mediated by a price system. Williamson seeks to explain why there are different methods for transacting, markets versus hierarchical methods, whereas we are interested in establishing the origins and social purpose of the firm to build ideas of sustainable corporate strategy. We start with the benchmark case to put up a comparator case which, in our view, is a useful counterpoint to observed social organization, that is, social organization that requires socially constructed governance mechanisms beyond the creation of a price mediated economy, one of which is the firm. By doing this, we establish that the role of the firm is to act as a mechanism that addresses some market failures in society's pursuit of its goals. We argue that a clear position on the role of the firm provides a solid background against which to consider corporate strategies in general, and ideas of sustainable corporate strategies in particular.

The idea of meliorism is that society or the world can be improved through human effort. Our view is melioristic in the senses that we consider a broadly defined governance structure that evolves through human action and interaction, and we consider sustainable corporate strategy in the context of this evolution. In our governance construct, the idea of sustainable corporate strategy is an important component of, and plays a significant role in, meliorism and eunomics.

We argue that a holistic discussion of sustainable corporate strategy must recognize that in the corporate form, the firm, is a governance structure, and that a shared view of the meaning of sustainable corporate strategy requires a shared view of governance. To move toward a shared view of governance, we pose this question: "What is governance about?" We propose that governance structures are, or should be, created by the community, for the community. Thus, for our purposes, we propose that one should look for an answer that is grounded in the concept of community. Corporate strategy originates from and affects the community.

Like McMillan and Chavis,[8] we view a community as a group of individuals coexisting with a sense of shared values and belonging, organized to enhance quality of life for its members. This can include many things – building prosperity together, offering interesting activities that members of society can undertake with low barriers to participation, creating an environment for the exchange of ideas, preserving the physical environment the community inhabits, and working productively to enable the community to afford and enjoy a wide range of goods and services.

An important characteristic of successful communities is that they provide individual community members with the ability to respond to self-actualization.[9] Thus, communities have a dual goal:

---

[8] McMillan and Chavis 1986.
[9] Maslow 1943.

a) Nurturing success of individuals within the community according to each individual's standards; and,

b) Instilling in members of the community a sense of responsiveness to the community's shared values.

c)

Consistent with McMillan and Chavis, who find that a sense of community depends on feelings of membership, influence, integration and fulfillment of needs, and a shared emotional connection among community members, we assert that achieving this dual goal builds strong, vibrant and self-reliant communities. In a highly abstracted sense then, governance is the set of social, political and economic phenomena that enable each individual within a community to succeed according to his or her individual standards while ensuring individual respect for the community's shared values. For brevity, we use the term social order to denote this condition.

To be clear, we do not suggest that social order is easy to achieve, to measure or perhaps even to describe. Nor do we think it is easy to create the institutions and processes that help society move toward social order. Finally, we do not necessarily think that it is easy to build in and deploy an optimal amount of flexibility and responsiveness in social institutions and processes to enable them to remain consonant with society as society evolves. Whereas Hobbes describes life in the state of nature as "solitary, poor, nasty, brutish, and short"[10]; we would characterize social order as difficult, stubborn, amorphous, elusive and transient. It is never solitary, and it is often considerably more than poor, although not always.

The notions of flexibility and responsiveness introduced in the preceding paragraph allude to a fundamental challenge for corporate strategy: the firm is an important institution in a randomly evolving society, within which the firm seeks financial success, relevance and consonance with society's goals. How does the firm balance the costs of maintaining flexibility and responsiveness against the benefits of relevance and consonance with societal goals? How does the firm mould its corporate strategy to do so?

To set the stage for our discussion, let us consider our counterpoint scenario. It is the case in which governance, the set of social, political and economic phenomena that support social order, is limited to a frictionless pure exchange economy in which prices float freely, markets clear, and each economic agent acts optimally relative to his or her own objective function. In such a setting, and subject to the existence of certain supporting conditions, market exchange, prices, and individual optimization lead naturally to a social optimum in the sense contemplated by Walras, and Arrow and Debreu.[11] Because the price system governs what occurs in the economy, and the economy is at a social optimum, there is no need for any other governance structure. Coase expresses this sentiment in the context of the formation of firms in the following way: "...having regard to the fact that if production is regulated by price movements, production could be carried on without any organisation at all..."[12]

---

[10] c. f. Hobbes 1651.

[11] Walras 1874; Arrow and Debreu 1954.

[12] Coase 1937, 388.

Thus, economic agents (or in our context, members of the community) express their preferences through their exchange decisions (buying and selling; consumption and investment), and allocations of resources in the community (who consumes what, individual wealth, and allocations to productive capacity) are all reflections of the sentiments of community members expressed through the exchange system. In a sense, economic agents vote with their money.

When we observe modern Western democracies we do not see social organization implemented exclusively by individuals engaged in selfish optimization in a price mediated economy. Rather, we see social organization that is a hybrid of:

a)   Individuals engaged in selfish optimization; and,
b)   Institutions that play important roles in social organization.

It is not a surprise to modern economists that communities are not exclusively governed by pure exchange economies. In order for a pure exchange economy to lead to a social optimum it is necessary for every element of the community and all interaction among community members to be priced, and for markets to function perfectly at all times. We know these conditions do not hold in observed communities. The entire range of human interaction, whether for exchange or otherwise, is not necessarily priced. In addition, there are a number of well-known market failures. Some noteworthy ones are:

a)   Externalities – costs or benefits for one economic agent that arise out of the selfish optimizing decisions of another economic agent;
b)   Unequal bargaining power between transacting economic agents;
c)   Asymmetric information between transacting economic agents;
d)   Natural monopoly or monopsony conditions;
e)   Appropriability problems;
f)   Bounded rationality;
g)   Various free rider problems;
h)   Various other economic frictions, not the least of which fall under the broad heading of transaction costs.

A pure exchange economy is therefore unlikely to achieve the governance goal we have set for ourselves in this chapter: to enable each individual within a community to succeed according to his or her individual standards while ensuring that each community member respects the community's shared values. Instead, communities need alternate governance structures to achieve what pure exchange economies cannot. Thus, we see the formation of various governmental and nongovernmental organizations in society that provide governance and need to be governed themselves. Coase explains this in the context of the formation of firms for the purpose of production:

> Outside the firm, price movements direct production, which is coordinated through a series of exchange transactions on the market. Within a firm, these market transactions are eliminated and in place of

the complicated market structure with exchange transactions is substituted the entrepreneur-coordinator, who directs production.[13]

Using the more general term "organization", to the extent that price movements outside the organization direct production inside the organization, an exchange economy will contribute to governance of the organization's activity. To the extent that an entrepreneur-coordinator within the organization directs production without market exchange transactions and without influence by market prices outside the organization, the entrepreneur-coordinator (or equivalently, the organization) is the governance mechanism that is operating in substitution for market governance. This begs the question: What forces exist to ensure that Coase's entrepreneur-coordinator deploys the factors of production and adopts behaviors internal to the organization that are consistent with a naturally occurring, socially optimal general equilibrium, assuming the latter could be specified? This can be put another way if one imagines that an organization can completely specify in its mission statement what the entrepreneur-coordinator intends to do. The questions then are:

a) Is the organization's mission statement consistent with the community's wishes; and,

b) Is the entrepreneur-coordinator following the organization's mission statement?

In a community that evolves randomly over time, as presumably all do, these questions must be addressed repeatedly, and the mission statement and the entrepreneur-coordinator's behavior adjusted as necessary to retain alignment with our hypothetical naturally occurring, socially optimal general equilibrium.

Within the community, individuals and organizations pursue their respective goals, so community success depends on the consonance of the values and behaviors of individuals, organizations, and the community as a whole. We must therefore expand our earlier definition of governance to include the set of forces that encourages a positive response to the two key questions above. The more complete definition is that governance is the set of forces that aligns individual and organizational behavior in a way that fosters feelings of membership, influence, integration and fulfillment of needs, and a shared emotional connection among community members, while supporting individuals and organizations in the pursuit of their respective goals. In this context, sustainability is whatever achieves these outcomes.

Our view renders governance the object of eunomics because it is in the spirit of seeking good order and workable arrangements in the community. It also renders governance melioristic because it seeks ongoing improvement in governance to meet the needs of a changing community.

While we have suggested above that organizations are a response to market failure, and have provided a list of commonly discussed causes or symptoms of market failure, we are, of course, not the first to attempt to explain the existence of organizations in observed economies. Some have proposed that they occur as

---

[13] Coase 1937, 388.

economic agents seek monopoly status to the detriment of social welfare, or that they are an optimal response to the technology of production. Williamson, in particular, argues that organizations form because they minimize transaction costs that would occur in a pure exchange economy.[14] Others argue that organizations offer benefits of scope and scale, reduced coordination costs, increased bargaining power, and realization of the benefits of scale in information gathering and processing. While there might be useful outcomes from a discussion that pins down the motivating forces behind the existence of organizations, the more important matter for the purposes of this chapter is that they exist, and they develop and implement strategies that have an effect on the communities within which they operate. As a consequence, there are social alignment problems that can either support or stand in the way of social order.

It is worth noting that in this conception of governance, good governance, or any kind of governance for that matter, is not the goal in and of itself. Rather, the goal is social order. The need for governance mechanisms to take steps toward it, and the concomitant task of attempting to make them good ones, are a reflection of the fact that society is not able to achieve social order by means that are relatively more dependent on its members acting individually. When we discuss sustainable corporate strategy then, we are discussing the ability of the firm to contribute to sustained social order of some kind.

Our conceptions of society, governance and the role of the firm in society exist in the context of a social alignment mechanism that includes a feedback loop intended to provide ongoing improvement to the alignment process through time. To facilitate discussion, we present a graphic depiction illustrating the social alignment mechanism in Figure 10.2.

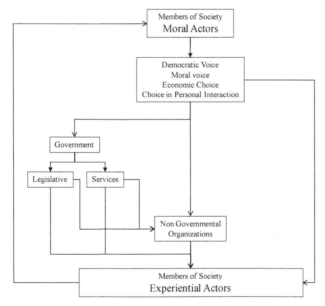

Figure 10.2: The Social Alignment Mechanism

---

[14] Williamson 2005.

In Figure 10.2 members of society appear wearing two hats. In the first instance, they have membership in the group of moral actors, wearing the hat of influencers in society. In the second instance, they appear as experiential actors, wearing the hat of individuals who experience the socially constructed reality within which they live their lives. Their experiential selves feed back to themselves in their moral actor role, under which they assume responsibility for shaping society. We use the characterization "moral actor" because it seems appropriately broad. The term economic agent and other candidate terms seemed too narrow.

In our conception, members of society in a moral actor role appear at the top of the process. Through the deployment of whatever capabilities they have, they exert influence on the conduct of the society they populate. We include organizations in both a government and nongovernmental role. The firm is a member of the set of nongovernmental organizations. Organizations both govern, and need to be governed. They are instruments of the governance process. They govern to the extent that they influence the outcomes members of society experience. That is, they affect members of society as experiential actors. They need to be governed in the two senses addressed in questions a) and b) above.

The lines in the chart denote channels of influence. There are probably numerous ways in which moral actors influence society. We have proposed four, which in our view capture a sufficiently wide range of influences for the purposes of this chapter. We briefly elaborate on each below:

a) *Democratic voice*: There are typically numerous opportunities for members of society to participate in democratic processes. By democratic processes we mean the processes related to running for elected office, exercising the right to vote in the various circumstances where this is available, and exercising the right to influence the political process once an individual or individuals are elected. This includes voting as a shareholder in cases where one owns shares in a corporation that carry voting rights.

b) *Moral voice*: Individuals in civil society have the opportunity to express their views in a way that exercises intellectual leadership. They can use moral leadership to sway public opinion. Individuals are also able to use their moral voice to organize civil action, and even in some cases, civil disobedience.

c) *Economic choice*: Individuals can make financing and expenditure decisions that have an economic effect on individuals and organizations in society.

d) *Choice in personal interaction*: Individuals can adopt individual and group behaviors toward others that may influence the behavior of others. For example, an individual or a group of individuals can decide to withdraw interaction from members of society who, in the view of the individual or group, are acting objectionably.

The lines of influence for moral actors extend to governmental organizations, nongovernmental organizations, and to members of society in general. The fact that this third line of influence, to members of society in general, exists suggests the possibility that the system can function completely without government, and without nongovernmental organizations. To those who would ask, "Why would you have government?" we would respond, "Only if you need to."

The government category is divided between a legislative and enforcement branch, although the idea of enforcement is suppressed in Figure 10.2, and a service delivery branch. This reflects practice in society: typically governments in civil society do much more than legislate and enforce. They often provide many services. For example, many governments own and operate large systems of parkland.

The government category includes all organizations that report to elected officials. Thus, it includes national and local governments and their agencies and crown corporations. The nongovernmental organization category includes all organizations in the private sector, and so, subsumes both for profit and not for profit enterprise. Thus, what is typically referred to as corporate governance and the governance of NGOs is notionally attached to the nongovernmental organization category.

The lines of influence of government extend to both nongovernmental organizations and to members of society as experiential actors. This is because nongovernmental actions are subject to legislation and enforcement, and are often users of government provided services. Similarly, members of society as experiential actors are subject to legislation and enforcement, and are often users of government provided services.

The fact that members of society appear in our model as both moral actors influencing society and experiential actors that bear the outcomes of society means that sustainable corporate strategy amounts to what members of society as experiential actors are willing to tolerate. The goal of the firm ought to be whatever this is. Certainly, firm goals include an element of economic performance: presumably society would prefer to see its resources deployed effectively to their highest and best use. Having said this, society may be intolerant of approaches that abuse resources simply because they are unpriced or because their present value in economic terms is near zero. Similarly, society may be intolerant of approaches that make known society itself inherently finite, even though the approach might be economically beneficial over planning horizons that make economic sense.

We can now return to the questions posed at the beginning of this chapter: "Why would society wish to delegate supervision of some of its resources and people to such a governance structure? What are the implications of this delegation for issues of sustainability in society?" We have argued that society has selected the corporate form because it is a device that enables society to meet its goals. We suggested that when the corporate form responds to economic exigencies, there is little sense in which it requires its approach (ie its strategies) to be sustainable for very long. Thus, it appears that while the corporate form is an effective governance structure for output and effective resource utilization in the short term, it may well be a counterproductive governance structure for sustainability in the longer term. In the next section we discuss society's expectations concerning sustainability, and the relationship of those expectations to the corporate form.

Societal Notions of Sustainability

The term sustainable usually implies some notion of continuation for an indefinitely long period of time. In the context of development, the World Commission on Environment and Development (the Brundtland Commission) define sustainability as follows:

> We define sustainable development in simple terms as paths of progress which meet the needs and aspirations of the present generation without compromising the ability of future generations to meet their needs.[15]

A sustainable corporate strategy that follows the Brundtland concept would be a corporate strategy that can be carried out over the long term without precluding subsequent generations from continuing to carry it out. Society may be intolerant of corporate strategies that fail on the Brundtland definition if society perceives such failing strategies as contributing to making society itself inherently finite.

Adherence to the Brundtland idea of sustainability rules out many corporate strategies over the very long term even though those strategies might be economically viable over shorter planning horizons. For example, it rules out strategies with finite planning horizons and it rules out strategies that depend on unbounded resource extraction. It also rules out strategies that depend on selling into markets that depend on unbounded resource extraction. Certainly, financial analysis of a corporate strategy respects resource limitations, but it does so only if the resource limitation manifests within the planning horizon of the analysis. The fact that planning horizons are almost always finite means that in practical effect, financial analysis responds only to a subset of resource limitations: those that fall within planning horizons. Finally, the Brundtland idea of sustainability rules out strategies that result in an unbounded negative environmental load on the planet. As with resource limitations, financial analyses respect only the subset of environmental limitations that a) fall within planning horizons and b) are priced. This is not to say that corporations ignore resource limitations and environmental consequences. It is simply that if corporations do so, they are doing so for other than financial reasons.

Let's pause here to think about the dynamics of sustainability and economic reaction. To facilitate discussion, we assume that there is finite planetary capacity to absorb human activity and that it is a critical event if we reach that capacity. At the moment, we are accelerating in population and economic growth, so if there is finite planetary capacity, we are accelerating toward it. Think of this as consuming the space between now and the critical event not proportionately through time, but in exponentially larger chunks. The corporate form, and market based economies will continue to conduct economic activity that is economically viable in the short term even, although it may not be economically, environmentally or socially viable beyond economic planning horizons. An absence of economic viability will lead to change in behavior in a market economy, but only when this absence is explicitly manifested, that is, a

---

[15] World Commission on Environment and Development 1987a, 4. Similar wording appears in the main report of the commission: World Commission on Environment and Development 1987b.

realization point is reached at which the market economy concludes it must react. So what we need to know is whether or not society can change course quickly enough at, and after, the realization point in order to avoid the critical event. Do we really know whether or not economics will warn us soon enough? Do we really know if economics can deliver change quickly enough after the realization point to avoid the critical event?

A challenge for economics is that even if planning horizons are very long, there are very significant limitations to the attention it pays to long dated financial consequences. Table 10.1 shows that the present value of a payoff of $1.00 made thirty-one years hence is about one cent if the financial discount rate is 16 per cent. One can debate whether a payoff of one cent on the dollar is small enough for the firm to ignore, or if it must be smaller, and one can debate whether 16 per cent is a discount rate that would actually be applied, but the fact remains that at some point in the not too distant future, financial considerations, positive or negative, cease to be a material consideration on a present value basis.

Table 10.1: Present Values Over the Long Term

Financial Discount Rate	Years to Payoff	Present Value of a Payoff of $1.00
16%	31	$ 0.01
12%	41	$ 0.01
8%	60	$ 0.01

In the same way that financial discount rates compress future payoffs to almost nothing, percentage growth rates explode future values quite rapidly. Table 10.2 shows that doubling periods can be rather short even at modest growth rates.

Table 10.2: Doubling Periods under Exponential Growth

Annual Percentage Growth	Years to Double
1%	70
2%	35
3%	23
4%	18
5%	14
6%	12

The issue for business and society is that most of the time, progress is measured using percentage growth rates: measures of exponentially rapid change in society. We are not in a position to settle the questions of exactly how long the doubling period is, what exactly is doubled if there is a doubling period, and how many doubling periods society and the planet can sustain. What we can say is that a paradigm that is inextricably linked to doubling is a questionable paradigm, and is so a fortiori if one believes the planet offers finite resources and has a finite ability to tolerate human population and human activity.

At present, world population is about 6.9[16] billion and growing somewhere in the range of 1.18 per cemt to 1.34 per cent[17] per year. Estimated world GDP for 2009 is estimated at about US$58 trillion (2008, US$61trillion).[18] The current growth estimate is about 3.4 per cent per year.[19] World GDP is estimated by the IMF to reach US$82 trillion[20] (in constant 2008 US dollars) by 2015 at which point world population is estimated to rise to 7.3 billion.[21] It is difficult to assess the planet's tolerance for growth. What is, perhaps, easier to assess is this. If a long term goal is worldwide social welfare improvement and distributional equity, let us ask what happens if the GDP per capita for all individuals below the level of the twentieth best nation on Earth (currently the United States of America) is raised to that level. World GDP would have to be five times what it is now to achieve this at current population levels. Over a 100 year period, with population growth of 1.18 per cent and growth in the benchmark base level GDP per capita (the current rate in the United States) of 3 per cent, world GDP in real terms would have to be 316 times what it is now to reach this distributional goal.

It seems inconceivable with current knowledge that Earth can sustain this level of activity, so in the absence of monumental change along many dimensions (technological, social), something has to give if we are to achieve ideas of distributional justice cross sectionally and longitudinally. At present, we have limited cross sectional distributional justice. There are huge gaps between rich and poor when considered both within and between nation states. There is limited longitudinal distributional justice as it is doubtful that hundreds of years from now the planet will be in as good shape across important dimensions as it is now. So the current best guess is that the lives of future citizens may not be as pleasant as ours as a result of our actions.

Another way to look at the matter of economic development in the context of limits to economic growth is to consider redistribution of current GDP. The United Nations (UN) attempts to capture the idea of human development in most countries on Earth using its Human Development Index. The index is based on measures of a long and healthy life (life expectancy at birth, life expectancy index), knowledge (mean years of schooling, expected years of schooling) and standard of living (gross national income per capita, gross national income index). The UN publishes its Human Development Index for 169 of the approximate 208 countries on Earth and divides the measured countries into the quartiles identified in Table 10.3.

---

[16] United Nations 2008a.

[17] United Nations 2008b. 1.18% is the indicated world population growth rate as at 2010. 1.34% is the population-weighted-average of individual country population growth rates as at 2010.

[18] International Monetary Fund 2010.

[19] This is the GDP-weighted-average growth rate based on individual country growth rates presented in Figure 20, International Monetary Fund 2010.

[20] International Monetary Fund 2010.

[21] United Nations 2008a.

Table 10.3: World GDP by Human Development Quartile

Level	Number of Countries	Population (billion)	2008 GDP		Equalizing Shift		
			Trillion US$	Per Capita (thousand US$)	GDP (trillion US$)	% of World GDP	Multiple of Sector GDP
Very High	42	1.1	43.0	40.7	-33.2	-55	.75
High	43	1.1	9.1	8.7	.7	-	-
Medium	42	3.6	8.0	2.2	25.5	40	3
Low	42	1.1	.8	.8	9.4	15	10
World		6.9	60.9	9.0			

Source: Human Development Index data obtained from United Nations Human Development Program, Human Development Reports, Statistics. Downloaded from http://hdr.undp.org/en/statistics/hdi/ on December 1, 2010.

While the table highlights the distributional inequality mentioned above, it also illuminates a distributional opportunity. The average per capita GNP for the world is about US$9,000. Raising all countries to this level requires a shift of US$33.2 trillion in GDP from the very high category to the medium and low categories, with only a small accretion to the high category. This is a little over half the world's GDP and three quarters of the very high category's GDP. Such a shift translates to a tripling of GDP in the medium category and a tenfold increase in the low category. Countries with GDP per capita in the US$9,000 range include Argentina (US$8,358), Malaysia (US$8,197), Brazil (US$8,311), Romania (US$9,518), Uruguay (US$9,610) and Mexico (US$9,964),[22] all of which are in the high category in Table 10.3.

We realize that there is more to human development than simply raising GDP in poor jurisdictions, that it is difficult to achieve, that a large shift from rich to poor nations is very difficult to achieve, and that large increases to GDP in poor countries are likely to have over-arching social and political consequences. Nevertheless, we argue that consideration of the shift shown in Table 10.3 is a useful reference point.

For comparison to our first hypothetical (setting minimum per capita GDP no lower than the current level of the twentieth best nation) let us consider it relative to the current average level, again under conditions of 3 per cent real growth per year and 1.18 per cent population growth per year over a 100 year period. This second hypothetical implies a real economy at the end of 100 years that is sixty-two times the level of the current economy. In its 100th year alone, the year over year change in the size of the economy is an amount that is 2.5 times the total size of the current world economy.

These results are motivated by the doubling rates attached to both the population and economic growth assumptions. Thus, even if we set a much more modest distributional goal, it seems unlikely that the planet can tolerate growth rates over the long term in the absence of very radical changes in social

---

[22] Individual country GDP per capita information was obtained from United Nations Statistics Division, Demographic and Social Statistics, Social Indicators. Downloaded from http://unstats.un.org/unsd/demographic/products/socind/inc-eco.htm. on November 26, 2010. Individual country GDP per capita information was also obtained from the International Monetary Fund 2010.

development, political development, governance, consumption of material goods, food production and distribution, recycling, water management, waste management, energy generation and usage, climate management, and perhaps technology in general. Even if one could sustain the argument that raising the welfare of 4.7 billion people at the expense of a welfare decline for 1.1 billion people is a worthwhile social calculus, it remains that the corporate form is not much interested in distributional justice and so is a dubious choice of social instrument if this is society's goal. We identify the corporate form as a social instrument because society has elected to use it as an instrument in the economic sphere of the larger society.

Returning to the notion that society has turned to the corporate form as a governance structure to deal with problems that could not otherwise be solved, we can now see the need for emphasis on sustainable corporate strategy, that is, corporate strategy that meets the Brundtland condition.

Sustainable Corporate Strategy

If society subscribes to the Brundtland conception of sustainability, then the firm as a social instrument must also. In other words, the corporate form's social license to operate ought to be conditional on corporate strategy meeting the Brundtland condition. This is our conception of sustainable corporate strategy.

We see this idea as different from ideas of corporate social responsibility.[23] This is because there are many ways in which a firm can act in a socially responsible manner, while failing to meet the Brundtland idea of sustainability. For example, a firm can extract a nonrenewable resource while reducing the harm it does to the surrounding environment. This is certainly socially responsible, but if the resource is finite, then the strategy does not meet the Brundtland-Khalid condition, especially if the extraction is growing exponentially.

We argue that rewriting the firm's social license to operate to require Brundtland sustainability amounts to a change from the existing economy to a 3R economy: Recycle, Renew, Re-Birth. Clearly, one route to separating exponential economic growth from exponential growth in extraction of finite resources is to engage in significant recycling. Similarly, we must separate exponential economic growth from exponential dependence on nonrenewable resources, and link it to dependence on renewable resources. The idea of Re-birth is in the sense that society must shed an old paradigm in which the wellbeing of members of society is closely linked to material consumption (ie to objects) and give birth to a new paradigm in which ideas, introspection, and inquiry play a larger role: an idea economy in which creativity is paramount. For it is creativity and ideas that will give rise to the Re-birth and sustainability of corporate strategy.

Conclusion

Society has adopted the corporate form as a powerful engine of economic growth, and a platform for the profitable deployment of capital. If economic success is the measure, then it has served society well. If long term sustainability of society is

---

[23] For a detailed history and discussion see Engel 1979.

the goal, perhaps it has not. As we have discussed, corporate strategy that serves an economic master has only a tenuous connection to sustainability of any kind. As a result, the corporate form is naturally set up have a tenuous connection to more important conceptions of sustainability: specifically the notion articulated by the World Commission on Environment and Development.[24] Thus, if society has evolved to the point that Brundtland sustainability is of paramount importance, then there is a fundamental conflict between the corporate form as a socially constructed reality and society itself? The corporation's social licence to operate is in jeopardy.

We therefore argue that sustainable corporate strategy is strategy that extends the corporate form's social license to operate for the foreseeable future. This is necessarily corporate strategy that meets the Brundtland condition. In order to achieve this, society will have to move from its existing economic conception to a 3R economy: Recycle, Renew, Re-birth. If the corporate form can invoke strategy that divorces economic growth from exponential pressure on finite resources, and the planet's potentially finite ability to absorb human population and human activity, then it will survive. It seems that part of the coming Re-birth is a separation of social welfare from material consumption and migration toward an idea economy in which creativity is paramount. In addition to thinking in terms of the firm requiring comparative advantages in order to survive in its industry, we must also think in terms of the corporate form offering comparative advantages to society in order for the corporate form to survive. Ultimately, sustainable corporate strategy is about sustaining the relevance of the corporate form as a governance structure in society.

References

Arrow, Kenneth J. and Gérard Debreu. 1954. Existence of an Equilibrium for a Competitive Economy. *Econometrica* 22(3): 265–290.

Barney, Jay. 1991. Firm Resources and Sustained Competitive Advantage. *Journal of Management* 17(1): 99-120.

Buchanan, James. 2001. Game Theory, Mathematics, and Economics. *Journal of EconomicMethodology* 8(1): 27-32.

Coase, Ronald H. 1937. The Nature of the Firm. *Economica* 4(16): 386-405.

Engel , David L. 1979. An Approach to Corporate Social Responsibility. *Stanford Law Review* 32(1): 1-98.

Fuller, Lon. 1954. American Legal Philosophy at Mid Century. *Journal of Legal Education* 6(4): 457-85.

Grant, Robert M. 1991. The Resource-based Theory of Competitive Advantage: Implications for Strategy Formulation. *California Management Review* 33(3): 119-135.

Grundy, Tony. 2006. Rethinking and reinventing Michael Porter's Five Forces Model. *Strategic Change* 15(5): 213-229.

Hobbes, Thomas. 1651. *Leviathan.* Edwin Curley (ed.) 1994. Cambridge MA: Hackett Publishing.

---

[24] World Commission on Environment and Development 1987b.

International Monetary Fund. 2010. International Monetary Fund, World Economic and Financial Surveys, World Economic Outlook Database. Available at http://www.imf.org/external/pubs/ft/weo/2010/02/weodata/index.aspx, accessed November 26, 2010.

McMillan, David W. and David M. Chavis. 1986. Sense of Community: A Definition and Theory. *American Journal of Community Psychology* 14(1): 6-23.

Maslow, Abraham H. 1943. A Theory of Human Motivation. *Psychological Review* 50(4): 370-396.

Porter, Michael E., 1980. *Competitive Strategy*. New York: Free Press.

United Nations. 2008a. United Nations, Population Division, Department of Economic and Social Affairs, World population prospects: The 2008 Revision. File 1: Total population (both sexes combined) by major area, region and country, annually for 1950-2050. Available at http://esa.un.org/unpd/wpp2008/peps_stock-indicators.htm, accessed November 26, 2010.

United Nations. 2008b. Population Division, Department of Economic and Social Affairs, World population prospects: The 2008 Revision. File 20: Average annual rate of population change by major area, region and country, 1950-2050 (percentage). Available at http://esa.un.org/unpd/wpp2008/peps_period-indicators.htm, accessed November 26, 2010.

Walras, Léon. 1874. *Elements of Pure Economics: Or the theory of social wealth.* 1954 translation of 1926 edition, Homewood, Ill.: Richard Irwin.

Wernerfelt, Birger. 1984. A Resource-Based View of the Firm. *Strategic Management Journal* 5(2): 171-180.

Williamson, Oliver E. 2005. The Economics of Governance. *The American Economic Review* 95(2): 1-18.

World Commission on Environment and Development, 1987a. Report to UNEP's 14th Governing Council Session, June 8, 1987, Nairobi, Kenya.

World Commission on Environment and Development. 1987b. *Our Common Future*. Oxford: Oxford University Press.

Chapter 11: From Corporate Social Responsibility to the Democratic Regulation of Transnational Corporations

David Humphreys

Introduction

Those who work in the public education sector and are funded from the public purse should contribute to public debate on matters of public importance.[1] As Edward Said argued in his 1993 Reith Lectures, the public intellectual is "someone whose place it is publicly to raise embarrassing questions, to confront orthodoxy and dogma (rather than to reproduce them), and to be someone who cannot easily be co-opted by governments or corporations."[2] To Said, the public, broadly defined, is the intellectual's natural constituency. In this respect scholars can arguably find no more pressing contemporary public welfare issue on which they should speak than global environmental degradation.

One of the drivers of environmental problems is the transnational corporation. This publication analyses the increasing ease with which corporations are now able to evade public oversight. It seeks to move beyond analysis to address the normative question of what policies, political systems and governance structures are necessary if corporations are to be brought under democratic public control. Grappling with normative issues raises the question of whether public scholars should engage in advocacy; and if so on whose behalf should we claim to speak? In line with Said I would suggest that academics should be free to engage in advocacy providing that the advocacy in question is expressly intended to promote human rights or otherwise enhance the public interest. The public interest is, of course, an essentially and perpetually contested concept. The spatial scale at which we analyze "the public" may vary

---

[1] This chapter was published earlier in Jonathan H Westover (ed.) 2013. *Socially Responsible and Sustainable Business around the Globe: The New Age of Corporate Social Responsibility.* Champaign Il: Common Ground Publishing.
[2] Said 1996, 1.

significantly between communities, while in sustainability discourse the notion of "the public" should include the needs of generations that have yet to be born.

This publication first outlines why we should study the role of transnational corporations in environmental governance. It then introduces and critiques the idea of corporate social responsibility (CSR), differentiating between CSR and corporate accountability. The final substantive section presents an original model for the democratic regulation of corporations.

The Transnational Corporation and Environmental Governance

A transnational corporation is a company that operates in two or more countries, either in terms of activities or in terms of having subsidiaries or affiliates in different countries. The last twenty years has seen the emergence of a burgeoning literature within the academic and activist communities on corporations, the power they wield and how they may evade public oversight, especially in the majority world of Latin America, Africa and Asia.[3]

The modern business corporation was originally conceived to serve public needs. There are three attributes of the corporation that affect its role in environmental governance. First, corporations have a legal personality. The legal right of a corporation to exist and raise money through issuing shares is provided by public authorities in a charter. In 1886 a US court case, *Santa Clara County v. Southern Pacific Railroad*, ruled that a corporation should be considered a legal person with the same constitutional rights as a US citizen, thus giving to corporations a status that is equal to people.[4] Second, corporations have a fiduciary responsibility to act in the interests of their shareholders. The doctrine of shareholder primacy was established in the Michigan court case, *Dodge v. Ford Motor Company* 1919. Third, the shareholders of corporations have a limited liability, a principle first introduced in England in 1851. If a corporation is sued shareholders are liable only for their initial investment, and not for any damages greater than this. For limited liability corporations it is the personality of the corporation that is liable, not its directors or managers. Corporations can divide themselves into legally separate entities, such as holding companies and subsidiaries, so that if successfully sued any damages will be confined to a small "firm" within a larger organizational structure. This makes it more difficult for plaintiffs to obtain legal redress for any environmental damage corporations may cause.

Sustainability, however defined, cannot be attained unless corporations actively promote it. If the world's largest economies are measured by either the annual GDP of a country, or the annual turnover of a corporation, then over half of the world's largest economies are corporations. Transnational corporations are major users of natural resources and owners of industrial and agricultural land. They generate a significant percentage of global carbon dioxide emissions. The 200 largest corporations control approximately 30% of global Gross Domestic

---

[3] For example, Bakan 2004; Bollier 2003; Christian Aid 2004; Clapp and Utting 2008; Cromwell 2001; Derber 2002; Drutman and Cray 2004; Gates 2001; Korten 1995, 1999, 2006; Litvin 2003; Monk 2008.

[4] Korten 1995, 1; Drutman and Cray 2004, 63.

Product (GDP), while the largest 500 control approximately 80% of foreign direct investment and 70% of world trade.[5] Environmental standards may be driven down as countries compete to offer more attractive investment terms to transnational corporations, the so-called "race to the bottom." Yet many corporations operate with no effective control from their shareholders or from governments; indeed increasingly it is corporations that influence governments. Some corporations effectively conduct their own diplomacy, with political and legal departments that specialize in negotiating with governments. [6] Big businesses have become standard setters rather than standard takers. Corporate alliances such as the World Business Council for Sustainable Development, the European Round Table of Industrialists and the US Council for International Business have prepared the first drafts of international legal agreements tabled in intergovernmental negotiations. [7] The increased influence of corporations in international negotiations has led to what may be termed the "privatization" of the United Nations, in which international law increasingly reflects the interests of business.[8]

Deregulation has freed corporations from public oversight. Most now function with little, if any, commitment to the spaces and communities within which they operate. They increasingly operate not only beyond local community control but beyond the control of national governments, wielding enormous power yet protected from the worst consequences of their action by limited liability law. They have rights, including some important rights in international law, but without concomitant responsibilities. If corporations are growing in size and political influence then how can local communities regain democratic control over them?

From Corporate Social Responsibility to Corporate Accountability

In the contemporary neoliberal era, mandatory national and international environmental legislation is eschewed in favor of voluntary and market-based initiatives. [9] Corporate social responsibility (CSR) fits comfortably within neoliberal logic. CSR refers to the wide range of voluntary initiatives by corporations, sometimes in conjunction with other stakeholders, "to promote ethical corporate behavior and minimize the negative impacts of business activity on society and the environment." [10] CSR is based on the assumption that corporations themselves best know how to improve their social and environmental performance. There are various forms of CSR including voluntary statements of principles from individual businesses, schemes agreed jointly by several corporations (such as the UN Global Compact), and market-based certification and labelling schemes. While it originated from business, CSR has

---

[5] Elliott 2004: 117.
[6] Ross 2007: 216.
[7] Humphreys 2001; Derber 2002.
[8] Lee, Humphreys and Pugh 1997.
[9] Harvey 2003.
[10] Clapp and Utting 2008, 1

been accepted by many developed governments and by the European Union as the dominant approach to business (self-)regulation.[11]

Companies may adopt CSR schemes for a variety of reasons, including improving resource and energy efficiency, public relations, demands from stakeholders and to ward off tougher mandatory regulation. Some corporations adopt CSR schemes for more altruistic and ethical reasons. CSR schemes vary enormously but typically include commitments against sweatshop or child labour and the promotion of gender equality, fuel efficiency, waste management, nature conservation and carbon dioxide emission reduction. Proponents argue that CSR helps to fill a global regulatory vacuum by exerting a normative pull on business that raises standards. Corporations that take a lead in CSR can set new standards that laggard businesses will later adopt in order to avoid reputational damage.[12] To John Ruggie, schemes such as the Global Compact signify "the emergence of a new advocate for a more effective global public sector: business itself."[13]

But CSR has many critics. CSR may be seen as a contradiction in terms; under the law the corporation has no social responsibility, only the fiduciary duty to maximize its shareholders' interests. Corporations often evade even the voluntary commitments they make.[14] Commitments are usually vague, often taking the form of generalized principles. Time bound and quantifiable targets and verifiable performance benchmarks are the exception rather than the rule. Because of the voluntary nature of CSR schemes there is no mandatory verification, no obligation for independent auditing and no penalty for corporations that fail to adhere to their commitments.

One argument against joint CSR schemes such as the Global Compact focuses on the relationship between the strength of standards and the number of participating corporations. The standards of such schemes have to be weak to induce participation. Schemes with strong standards will have low uptake; they will achieve significant impacts for the limited number of corporations that adopt them, but low impact overall. Schemes with weak standards will achieve high uptake, but because they promote only minor changes in corporate practices, again there is only a limited overall impact. Uptake is uneven, both by sector and across space, with some types of business and some countries more active on CSR than others. There remain regulatory gaps through which unscrupulous businesses can slip.

However, CSR schemes have had some impact in ratcheting up standards for some firms. But whether CSR schemes may be considered effective depends on how "effectiveness" is defined. If effectiveness is defined as behavioral change that would not otherwise occur then some CSR schemes should be considered effective. But if effectiveness is defined as solving the problem at hand, such as maintaining or enhancing environmental quality, then CSR must be seen as fundamentally ineffective, as environmental problems have worsened since CSR was first adopted in the early-1990s. CSR leaves the onus on responding to environmental problems to corporations; it is an agency response to systemic problems that can only be solved with a coordinated system level response. CSR

---

[11] European Commission 2002.
[12] Assadourian 2006.
[13] Ruggie 2003, 115.
[14] Christian Aid 2004.

did not emerge primarily as a response to global environmental and social problems, and has thus failed to grapple meaningfully with them. CSR schemes focus primarily on changes that will benefit the corporation, such as providing predictability in an uncertain policy environment and improving resource efficiency. But there is no reason to conclude that simply improving corporate efficiency will arrest broader systemic problems, such as nature conservation and decarburization of the global economy. Foundational questions such as does sustainability require a reduction in international trade, a macro-level restructuring of global financial markets or a worldwide reduction in resource-intensive modes of production modes are not admitted into CSR discourse.

CSR constitutes a weakening of the public sphere at the national and international levels. Governments have, in effect, passed responsibility for standard setting from the public to private sectors. At the international level the Global Compact enhances the status of the UN secretary-general, but bypasses, and thereby diminishes, intergovernmental bodies within the UN system. It presumes that the era of state regulation has ended, supplanted by a new era of global governance in which the state is just one actor amongst many. As David Coleman has argued, by appealing directly to corporations to adopt global standards and thus, in effect, inferring that states and international organizations are the laggards at standard setting, the Global Compact has subverted the idea of global public regulation.[15]

So there are problems with both the means of CSR (often untransparent and profoundly undemocratic) and the ends (simply not fit for the purpose of arresting environmental degradation and achieving sustainability). CSR is a surface phenomenon: it does not penetrate deep into corporate governance structures. Under limited liability law corporations can impose environmental externalities on society by, for example, clearfelling forests, polluting waterways and dumping toxic waste, while internalizing the financial benefits, leading to huge imbalances between private profits and social welfare. In a neoliberal policy environment it is often "rational" to degrade the environment for private gain. The work of Arild Vatn on rationality is relevant here. Vatn argues that rationality is best understood as a plural concept that encompasses the "I" rationality of self-interested utility maximizing states and corporations, and "we" rationality that focuses on cooperation and collective gains.[16] Institutions are guided by different types of rationality, and whatever notion of rationality is internalized in an institution will shape the behavior of those who work within it. As Elinor Ostrom[17] shows, institutions can be designed that prevent commons tragedies, at least at the local level. But the modern corporation is the result of the progressive weakening of collective "we" rationality and the strengthening of individualistic "I" rationality leading to a weakened public sphere and the utilitarian/neoliberal theory that the common good is best realized by the interaction of buyers and sellers in "free markets."[18]

To Vatn, therefore, the behavior of corporations is in large part determined by the particular socio-historical context in which it operates. The voluntarism of

---

[15] Coleman 2003, 353.
[16] Vatn 2008.
[17] Ostrom 1990.
[18] Vatn 2008, 74.

CSR generates a free rider problem: businesses that do not sign up to CSR, or sign up but do not implement, avoid environmental costs that more principled businesses incur. A corporation motivated by a strong notion of "we" rationality will inevitably suffer a comparative disadvantage relative to those governed by a ruthless and instrumentalist "I" rationality. From this argument it follows that if the notion of rationality that governs corporations has evolved as it has due to changes within the socio-political environment that have led to neoliberalism, what is needed in order to promote corporate behavior that serves the common good is a fundamental shift in this broader environment.

This is what corporate accountability advocates. Proponents of corporate accountability seek to stipulate the duties and obligations of corporations, emphasize the answerability of corporations to stakeholders and publicly accountable authorities and reassert the role of public policy in governance.[19] The next section develops an original model for corporate accountability. Drawing from and building upon earlier work[20] the model seeks to reconcile two pressing governance challenges: to uphold tough international standards that provide a framework within which corporations can compete fairly, while passing the final decisions on corporate activities back to local communities.

Towards the Democratic Regulation of Transnational Corporations

Local responses can promote and enhance environmental, cultural, social and economic sustainability, but even the best local responses will ultimately fail if they take place within a neoliberal economic system that places the rights of investors above those of communities. To be effective, local responses need to be embedded within an international framework dedicated to the enhancement of what John McMurtry calls life values.[21] The democratization of globalization and the democratization of local spaces should thus be seen as symbiotic and mutually reinforcing processes.

Central to the model proposed in this section is the democratic regulation of the corporation. The social and environmental consequences of corporate practices should no longer be subordinated to the quest for profit. The fiduciary duty of corporations to maximize shareholders' interests – a duty that from a legal standpoint currently trumps all moral and ethical considerations – should be replaced by an explicit legal responsibility for the corporation to act *pro bono publico* (for the public good). To infuse accountability into corporate practices, the notion should be revived that the right of the corporation to operate is conditional upon its satisfying public needs as stipulated in public charters. The final arbiters on the standards to which corporations should adhere should be local public bodies.

Under the Westphalian system of international law that has emerged over the last four centuries, states both make international law and are subject to it. However, the state is no longer the main international actor. To ensure minimum global standards a new corpus of international law is needed that regulates

---

[19] Clapp and Utting 2008, 17.
[20] Humphreys 2006.
[21] McMurtry 1999, 2003.

corporations rather than states. States, as the legitimate representatives of national publics in international politics, will negotiate this corpus of law, which will stipulate the standards to which corporations should adhere if they are to be permitted to trade and invest internationally. This new body of law should be agreed free of interference from corporations, which, as regulatees rather than regulators, would be denied access to the legislative process.

The first step would be the negotiation of a Convention on Transnational Corporations outlining the responsibilities and duties of corporations. This would set a raised plateau of tough obligations and standards for corporations. Whereas an individual state adopts a treaty by signing and ratifying it through its domestic legislature, a corporation would adopt the convention by endorsing it through its board of directors and shareholders' meetings. No corporation could trade or invest internationally without first adopting the convention. By doing so corporations would recognize that their right to engage in transnational economic activity is conditional upon the observance of obligations to the global public. These obligations would, at a minimum, include commitments to uphold environmental quality, respect the precautionary principle, undertake environmental and social impact assessments, allow independent environmental and social auditing and respect community rights and traditions. Corporations ratifying the convention would receive an international charter permitting them to trade internationally. In terms both of intent and content, the convention would be the opposite of the contemporary corpus of international trade law, which outlines the rights of investors rather than their duties.

To invest in a country, a corporation would then need to obtain a country-level charter from a public authority in the host country. This would stipulate the terms and standards that the corporation should observe. The granting or withholding of country-level charters would be a matter of public debate, with the final decision being made by the national government. The conditions stipulated in a country charter could be stronger, but not weaker, than those outlined in the convention. A corporation would need a charter from every country in which it operated. Inevitably the contents of charters would vary as different corporations would be admitted to different countries under different circumstances to meet different publicly-defined needs.

A corporation with a country-level charter would then be free to find a locality in which to invest. Publicly accountable groups at the sub-state level would have the right to determine which corporations invested in their space. The conditions would be stipulated in a local charter. The provisions contained in a local charter could strengthen, but not weaken, those laid out in the country-level charter. Actors at the sub-state level charged with upholding the public interest will come in many guises, reflecting the rich diversity of local cultures, economies and commons regimes that are to be found in most countries. They will include, for example, democratically elected local councils and more traditional local community governance structures. Irrespective of the form of governance at the local level, a local group or authority would be under no obligation to issue a charter if it did not consider this to be in the interests of its public.

The revitalization of the charter as an instrument for public accountability will stimulate citizen engagement and participatory democracy. Citizens' views would now count. Corporations would have to adjust to the needs of local communities rather than, as now, communities being forced to adjust to intrusions from big business playing by international trade rules. In a post-neoliberal world, a new mode of doing business would emerge based upon the needs of the local economy. This will benefit, for example, those communities that wish to conserve local fishing and forest resources and those that do not want large superstores or "global brands" disrupting the cultural distinctiveness of the local economy. If local communities exercise greater control over their economies then a greater share of the income from the local economy will flow to the local level where it can be used to fight poverty and social exclusion.

But the local level cannot entirely dominate. Global targets should be set on public good provision. Targets would be necessary on, for example, forest cover and greenhouse gas emissions. Multilevel coordination between the different layers of governance would then be needed to ensure that these targets were met. Agreement would need to be reached at the intergovernmental level on the responsibilities of individual countries with respect to global targets. Similarly, within countries, coordination will be necessary between the state and sub-state authorities on the responsibilities that local spaces should make to public good provision.

Corporations adopting the Convention on Transnational Corporations would be obliged to ensure its implementation by all subsidiary companies and to undertake to trade only with businesses that respect and implement the convention. This would reinforce standards amongst market players. Standards adopted by one business would thus be passed along supply chains which would act as transmission belts along which rules are passed and enforced. Insisting that a corporation know its supply chains is not unreasonable, but it would call for mandatory international certification schemes. Some reform of international financial markets would also be necessary, including a minimum period for holding shares. This would promote shareholder activism in corporate governance, encouraging shareholders to monitor the social and environmental performance of the business. There is no public interest in share dealers buying shares and selling them at the click of a mouse a few minutes later to earn a "margin." A minimum period of, say, five years would ensure that shareholders took an active interest in corporate management and invested only in socially responsible businesses.

If the courts were to find that the convention had been violated, then a corporation would be liable to penalties, such as fines, suspension of the international charter or, in severe cases, the revocation of the international charter and the dissolution of the corporation's assets. Similarly, if a corporation were to violate a national or local charter it would answer to national and local authorities, and may suffer financial penalties or the withdrawal of its charter. Where a corporation causes environmental damage due to unsustainable, unscrupulous or negligent corporate practices, then it should be required to make good (as far as possible) on the damage caused and to pay a financial penalty. Should the corporation have insufficient funds to pay these costs, the courts should have the power to dissolve the corporation and seize its assets. Limited liability legislation should be repealed and the courts should have the power to seize the private assets of company directors.

Some will argue that this is draconian. In response four points should be made. First, in most countries the law can be used to pursue assets gained from illegal and immoral activities, such as drug smuggling and prostitution. There is no reason to deal differently with those who enrich themselves from the degradation or asset stripping of nature. Second, the principle of limited liability has enabled corporations to shifts risks and costs from the corporation to society as a whole. The repeal of limited liability legislation will mean that those whose policies generate environmental and social externalities will be those who pay for them. Third, the risk of financial penalties accruing to corporations and directors would provide a strong financial incentive for corporate policies that promoted long-term sustainable practices. Fourth, financial penalties would co-opt the world's financial markets into promoting environmental, cultural and social sustainability. Financial investors want profitable businesses. If knowledge emerged that a corporation had engaged in unsustainable practices that violated a charter, the result would be a fall in the value of the corporation's stock as the markets reacted in anticipation of heavy fines or the dissolution of the corporation. Financial investors would thus have an incentive to monitor corporate environmental practices. This would help to repair the problem of what Robert Nadeauhas termed the absent feedback loop, whereby the depletion of natural capital simply does not register with corporations and international financial markets.[22]

A further objection may be that international rules outlining the obligations of corporations are "impractical." Yet business leaders and their political allies have consistently pressed for international rules on neoliberal objectives, such as trade, investment and intellectual property rights. On what basis can it be claimed that international rules for sustainability and human rights are, in some way, impractical? It may also be claimed that a multilevel system of charters would be "bureaucratic" and "protectionist," both of which are pejorative terms for neoliberals. But if bureaucratization is the price of accountability, then let these charges be made, as the social and environmental costs of allowing corporations to penetrate new spaces without effective public oversight have already proved too large. And it is hardly protectionist to point out that local and, sometimes, national businesses are at a significant disadvantage when competing with powerful transnational corporations. A nested system of public charters would help to equalize the huge power asymmetries between local communities and corporations. Indeed, on what basis would anyone seek to deny public authorities the right to decide which corporations can have access to their economies and which should not?

It may also be claimed that if environmental and social standards are set "too high" then corporations will stop investing in local communities. But given that a key driving force of environmental degradation is investment that is not constrained by strong safeguards, this argument should be dismissed. Irrespective of the regulatory environment, corporations will continue to search for investment opportunities. They will soon learn to adjust to the needs of local communities if the alternative is not to invest at all. With effective public safeguards in place, only environmentally and socially irresponsible investment will be blocked.

---

[22] Nadeau 2003.

Finally, it may be argued that it is utopian, even naive, to argue that corporate-driven environmental degradation can be reversed using the same institutions – the state and intergovernmental organizations – that so far have failed. In response, it should be asked that if publicly accountable bodies, such as the state and intergovernmental organizations, are not going to represent the interests of citizens and promote public goods, then who, precisely, will do this? The problem is not so much with publicly accountable bodies, but with their penetration by corporate interests. The solution is to reform and strengthen public bodies rather than to agree with the neoliberal approach, premised on the notion that a weakened state is "inevitable" under globalization, of searching for new forms of governance which hand more power to business, such as CSR. Far from being abandoned, public government and intergovernmental organizations should be reclaimed, revitalized and democratized.

Conclusions

Public authorities have both the right and the duty to govern business in the public interest, yet economic decision making is in large part driven by corporations whose sole duty is to serve private shareholders. This represents an unreconciled tension in global governance, one that often passes unrecognized. The model of corporate accountability proposed here would reconcile this tension by establishing as paramount the collective public interest to which all other considerations, including corporate policy and shareholder value, should be firmly subordinated. It would reinvigorate the principle that corporations should be free to operate if, and only if, they provide benefits for the public good. It would render CSR obsolete, restore full political authority to a dynamic, inclusive and democratically accountable public domain and reconnect corporations to the communities within which they were previously embedded.

There will, admittedly, be huge costs to corporations if they are to adopt more sustainable and responsible practices. But the environmental and social costs that have arisen from predatory and exploitative corporate practices – including massive displacement of communities, species loss and climate change – have also been vast, although they have not been accurately chronicled. Simply put, nature and communities are currently paying the costs of corporate enrichment. In any case, the governance reforms advocated here will be minimal and small scale compared to the massive interventions and curbs on corporate activity that will be necessary in, say, another half-century if global environmental degradation continues unchecked.

Under the system of nested and differentiated governance proposed here, countries and communities would not adopt identical rules, although they would adopt minimum rules. The system would promote universal values and standards while respecting and promoting the diversity and right to self-determination of countries and communities. It would seek to govern corporations in the interests of environmental sustainability, at both the local and global levels. It would enable local spaces to sustain their cultural and social diversity while also providing a non-discriminatory framework within which corporations can compete in global markets. As such, the proposed model could provide an embryonic framework within which the interconnected fundamentals of environmental, social, cultural and economic sustainability can flourish and thrive. It is in this spirit, and in line with Said's entreaty to scholars to seek to speak on behalf of the public at large, that this proposal is offered for criticism and debate.

References

Assadourian, Erik. 2006. Transforming Corporations. In *State of the World 2006*, edited by Worldwatch Institute, 171-189. New York: WW Norton.

Bakan, Joel. 2004. *The Corporation: The Pathological Pursuit of Profit and Power.* London: Constable.

Bollier, David. 2003. *Silent Theft: The Private Plunder of Our Common Wealth.* London: Routledge.

Christian Aid. 2004. *Behind the mask: The real face of corporate social responsibility.* London: Christian Aid.

Clapp, Jennifer, and Peter Utting. 2008. Corporate Responsibility, Accountability, and Law: An Introduction. In *Corporate Accountability and Sustainable Development: Ecological Economics and Human Well-being*, edited by Peter Utting and Jennifer Clapp, 1-33. Oxford: Oxford University Press.

Coleman, David. 2003. The United Nations and Transnational Corporations: From an Inter-nation to a "Beyond-state" Model of Engagement. *Global Society: Journal of Interdisciplinary International Relations* 17(4): 339-357.

Cromwell, David. 2001. *Private Planet: Corporate Plunder and the Fight Back.* Oxfordshire: Jon Carpenter Publishing.

Derber, Charles. 2002. *People Before Profit: The New Globalisation in an Age of Terror, Big Money, and Economic Crisis.* London: Souvenir Press.

Drutman, Lee and Charlie Cray. 2004. *The People's Business: Controlling Corporations and Restoring Democracy.* San Francisco: Berrett-Koehler.

Elliott, Lorraine. 2004. *The Global Politics of the Environment* (second edition). London: Palgrave Macmillan.

European Commission. 2002. COM(2002) 347 final. Communication from the Commission concerning Corporate Social Responsibility: A business contribution to Sustainable Development. July 2.

Gates, Jeff. 2001. *Democracy at Risk: Rescuing Main Street from Wall Street.* Cambridge MA: Perseus.

Harvey, David. 2003. *The New Imperialism.* Oxford: Oxford University Press.

Humphreys, David. 2001. Environmental Accountability and Transnational Corporations. In *Governing for the Environment: Global Problems, Ethics and Democracy,* edited by Brendan Gleeson and Nicholas Low, 88-101. Basingstoke: Palgrave.

Humphreys, David. 2006. *Logjam: Deforestation and the Crisis of Global Governance.* London: Earthscan.

Korten, David C. 1995. *When Corporations Rule the World.* London: Earthscan.

Korten, David C. 1999. *The Post Corporate World: Life After Capitalism.* San Francisco: Berrett-Koehler.

Korten, David C. 2006. *The Great Turning: From Empire to Earth Community*. San Francisco: Berrett-Koehler.

Lee, Kelley, David Humphreys, and Michael Pugh. 1997. "Privatisation" in the United Nations System: Patterns of Influence in Three Intergovernmental Organisations. *Global Society: Journal of Interdisciplinary International Relations* 11(3): 339-357.

Litvin, Daniel. 2003. *Empires of Profit: Commerce, Conquest and Corporate Responsibility*. London: Texere.

McMurtry, John. 2003. *Value Wars: The Global Market Versus the Life Economy*. London: Pluto Press.

McMurtry, John. 2013. *The Cancer Stage of Capitalism: From Crisis to Cure* (second edition). London: Pluto Press.

Monks, Robert A.G. 2008. *Corpocracy*. Hoboken NJ: John Wiley and Sons.

Nadeau, Robert L. 2003. *The Wealth of Nature: How Mainstream Economics Has Failed the Environment*. New York: Columbia University Press.

Ostrom, Elinor. 1990. *Governing the Commons: The Evolution of Institutions for Collective Action*. Cambridge: Cambridge University Press.

Ross, Carne. 2007. *Independent Diplomat: Dispatches from an Unaccountable Elite*. London: Hurst and Company.

Ruggie, John Gerard. 2003. Taking Embedded Liberalism Global: The Corporate Connection. In *Taming Globalization: Frontiers of Governance*, edited by David Held and Mathias Koenig-Archibugi, 93-129. Cambridge: Polity.

Said, Edward. 1996. *Representations of the Intellectual: The 1993 Reith Lectures*. New York: Vintage Books.

Vatn, Arild. 2008. Sustainability: The Need for Institutional Change. In *Corporate Accountability and Sustainable Development: Ecological Economics and Human Well-being*, edited by Peter Utting and Jennifer Clapp, 61-91. Oxford: Oxford University Press.

Part IV: Art

Editors' note: Creative works are integral to human interactions with the natural world. These works help us to understand our origins and where we are today, and they may play an important role in our future interactions with the environment. Jerzy Kosinski was a Polish-American novelist with award-winning books and a tumultuous legacy. He is reported to have said the following:

> The principles of true art is not to portray, but to evoke.

Jerzy Kosinski[1]

Image 2: *Displaced Enchantment*, 2012, by Spencer S. Stober, pencil drawing, 9x12 inches. Carl Jung challenged us to imagine the birth of our species—when we as a group first became aware—fading in and out of consciousness over time with faith and reason as facilitators. Has faith displaced our enchantment with Nature? Are we treading on reason at the expense of Natural enlightenment? Permission to reprint has been granted by Spencer S. Stober (co-editor).

---

[1] BrainyQuote.com, Xplore Inc, 2014. Jerzy Kosinski quote available at http://www.brainyquote.com/quotes/quotes/j/jerzykosin308851.html, accessed July 6, 2014.

Chapter 12: Sustainability, Aesthetics, and Future Generations: Towards a Dimensional Model of the Arts' Impact on Sustainability

Alisa Moldavanova

Sustainability has been an important topic in many disciplines for over two decades, and its urgency keeps rising. [1] At the same time, the conceptual understanding of sustainability remains quite vague, which poses a challenge for sustainability research. The term sustainability has been widely used in the context of ecological, economic, and social studies. [2] In ecological economics, it is often used interchangeably with the term sustainable development, defined as "development that meets the needs of the present without compromising the ability of future generations to meet their own needs." [3] Some scholars use context-specific definitions of sustainability, implying particular meanings of the word based on the subject of study. [4] However, the contextual approach to defining sustainability is criticized on the grounds that it is not the meaning of sustainability that changes with respect to context; rather, our understanding of the context itself is being shaped by sustainability.

Proponents of the multidimensional view of sustainability generally identify three main dimensions of this concept – environmental, economic, and social (or sociopolitical) – and some also add the technological aspect of sustainability. [5] For example, Edwards and Onyx attempt to reconcile three imperatives: the ecological imperative to live within global biophysical carrying capacity; the social imperative to ensure the development of democratic systems of governance to effectively promote public values; and the economic imperative to ensure that

---

[1] Asheim et al. 1999; Budd et al. 2008; Dasgupta 2008; DeCanio and Niemann 2006; Hasna 2007; Helm and Ferenc 1999; Howarth 2003; Krautkraemer, 1998; Matarasso 2001; McCarthy et al. 2007; Nurse 2006; Padilla 2002; Shearman 1990; Tubadji 2010.

[2] Daguspta 2008; Helm and Fenrenc 1999; Howath 2003; Padilla 2002.

[3] World Commission on Environment and Development 1987, 43.

[4] Brown et al. 1987.

[5] Shearman 1990.

basic needs are met worldwide. [6] There is the recognition that the multidimensional treatment of sustainability often involves resolving conflicts between multiple dimensions and reconciling competing values embedded in each of the important aspects of sustainability. [7] Consequently, sustainability may be viewed not only as an end result of natural or human-led actions but also as a process, or as means to creating and maintaining system equilibrium.

Sustainability also has an ethical dimension, although the concept of an ethic of sustainability is less developed in the sustainability literature. A normative view of sustainability implies treating sustainability as a form of intergenerational equity or fairness. [8] The question of intergenerational equity constitutes a growing normative concern, and our obligation to future generations requires looking beyond the short-term impacts of current public policies.

Some recent scholarship suggests that the traditional dimensions of sustainability do not fully reflect the complexity of contemporary societies, and a fourth dimension, cultural sustainability, should be added to ensure a more holistic understanding of this concept. [9] For instance, according to Nurse, culture should be the central pillar of sustainable development, which "is only achievable if there is harmony and alignment between the objectives of cultural diversity and that of social equity, environmental responsibility and economic viability." [10] Likewise, David Trosby unifies cultural and economic systems under his framework of "culturally sustainable development." [11] He argues that bringing economy and culture together provides a workable model for policy analysis.

Overall, there is a shared agreement that both multidimensionality and the systemic approach to development are a central strength of the idea of sustainability, and at least in theory there is an acknowledgement that the questions of social justice, peace, democracy, self-reliance, ecology, climate change, and quality of life are closely connected. [12] Among the four dimensions of sustainability (environmental, economic, social, and cultural), it is the last aspect that appears to be least examined by scholars. However, understanding how culture contributes to the long-term sustainability of communities and societies is key to a holistic understanding of sustainability itself and, therefore, worthy of scholarly attention.

Art and sustainability is a much-neglected area in sustainability research. In an attempt to fill this gap, this chapter argues that the ethic of long-term sustainability can be informed by aesthetics and arts in their embodied, institutionalized form. Configuring the long-term impact of aesthetics and its role for shaping the values of future generations, as well as studying the resilience potential of art organizations, is valuable for an understanding of the ethic of long-term sustainability. This argument is further developed by taking a deeper look into the multidisciplinary sustainability research, explaining how culture fits

---

[6] Edwards and Onyx 2007.

[7] Hasna 2007; Tubadji 2010.

[8] Barry 1997; Catron 1996; Frederickson 2010; Moldavanova 2013; Throsby 1995.

[9] Haley 2008; Nurse 2006; Packalén 2010; Tubadji 2010; Throsby 1999.

[10] Nurse 2006, 33.

[11] Throsby 1999.

[12] Brocchi 2010.

into a holistic, multidimensional concept of sustainability, and suggesting what makes aesthetics valuable for the ethic of long-term sustainability.

Adding Culture to the Dimensions of Sustainability

The idea of looking at culture and aesthetics as the core for the ethic of sustainability can be traced to the works of the earlier philosophers of art, who justified the powerful role of art in society, and argued that values communicated through aesthetic experiences exceed the boundaries of art institutions. [13] According to Collingwood, the aesthetic consciousness is "the absolutely primary and fundamental form of all consciousness and all other forms emerge out of it." [14] Moreover, art is a collaborative enterprise that involves the creator, performers, and spectators. Adorno further looks at the works of art as social monads, which are reflective of the socio-historical process. He emphasizes the repetitive character of "culture industry" (a term developed with Max Horkheimer), and criticizes popular culture for producing standardized cultural goods, which lead to the standardization of society.[15] Unlike mass culture, which is temporal, works of high art are capable of concentrating past and future in the present, through the conflict presented in them, thus appearing as timeless. They are capable of cultivating socially significant values – creativity, freedom, and human happiness.

Wittgenstein stressed the importance of the aesthetic dimension of life and recognized the symbiotic relationship between ethics and aesthetics by precisely stating that "ethics and aesthetics are one and the same."[16] This interdependence of ethics and aesthetics "is rooted in the fact that the ethical, as a way of understanding life in its absolute value, expresses itself in aesthetic form, while aesthetic form expresses the ethical as an individual, yet universal, aspect of the artistic act."[17] Based on Wittgenstein's philosophy, aesthetics is the expression of an ethical perspective on the world that is capable of putting current life events in a long-term perspective, and ethical values and ideas can be uniquely expressed and communicated through works of art.

In contemporary scholarship, the actual significance of cultural values, arts, and creativity keeps rising in importance alongside socio-economic, scientific, and technological concerns, [18] and it is the values and ideals embedded in the cultural dimension of sustainability that contribute to sustainable thinking in the longer-term. When adding culture to the dimensions of sustainability, one should be aware that culture itself is a very complex term, and there is a multitude of definitions associated with it.[19]

When considered as the fourth pillar of sustainable development, culture is often viewed as a complex mix of tangible and intangible resources: artifacts, cultural products, milieu, values, symbols, identity, patterns of behavior, ways of

---

[13] Adorno 2004; Adorno 2001; Collingwood 1964; Wittgenstein 1984.
[14] Collingwood 1964, 115.
[15] Adorno 2001.
[16] Wittgenstein 1957, 86.
[17] Stengel 2004.
[18] Haley 2008.
[19] Williams 1983.

life, and civilization traits with social, political, and economic dimensions.[20] For instance, Nurse argues that it is critical to move beyond talking about the preservation of "the arts," "heritage," and "cultural identities," when discussing sustainability.[21] Packalén too calls for the adoption of a broader understanding of culture that is composed of the traditional elements that make up cultural policy (such as theater, film, music, architecture, literature, etc.) and cultural norms, values, assumptions, traditions, and practices.[22] He further argues that it is very important to discuss the potential of art and culture in shaping a desirable future because "responsibility, ethics, and aesthetics go together."[23]

It is possible to develop the arguments of these scholars by suggesting that values within a culture of sustainability could be successfully promoted through aesthetics and art, since multidimensionality and other ideas are naturally embedded in the process of creating, interpreting, and experiencing art. Therefore, this chapter looks at art and aesthetics as the core of the cultural dimension of sustainability. It claims that the ethic of long-term sustainability could be informed through aesthetics and art institutions for two main reasons: first, because the cultural dimension of sustainability has the greatest potential among all of the other dimensions to reflect the long-term aspect of sustainability; and second, because aesthetics is essential for ethics, and normative sustainability as intergenerational equity can be greatly informed by institutions specifically created to promote and preserve important values for the future generations. The analysis in this chapter results in a dimensional model of the arts' impact on individuals and societies. This model is an important contribution towards filling the gap in art and sustainability studies by connecting various dimensions of the arts' impact to sustainability.

Aesthetics and Art as the Core of the Cultural Dimension of Sustainability

There are two main streams of scholarly discourse related to the relationship between sustainability and art: first, the issue of the self-sustainability of art institutions; and second, the role of aesthetics and institutions of art for the sustainability of communities and societies as a whole. It should be noted that neither of these issues is completely independent from one another. It is necessary for arts institutions to be resilient and sustainable in the first place, in order to be able to serve as a catalyst of a broader, multidimensional idea of sustainability. Overall, the issue of the sustainability of art institutions implies both sustaining particular art institutions through the pressures of current economic recession in the short-term and addressing the question of long-term institutional survival and the legacy of art institutions across generations.

Despite the timeless significance of art and aesthetics, institutions of art are nevertheless among the first to face the pressures of economic downturns, and there have been several attempts to study sustainability responses within the arts

---

[20] Tubadji 2010.
[21] Nurse 2006.
[22] Packalén 2010.
[23] Packalén 2010, 118.

sector.[24] For example, the study of the sustainability strategies developed in eleven metropolitan areas in the United States concludes that the most common immediate sustainability strategies adopted by the managers of local arts institutions include: changing fundraising strategies by diversifying revenue sources, changing operating strategies in order to reach wider audiences, enhancing collaboration with other arts institutions, promoting wider community goals by engaging in cross-disciplinary projects, and lobbying the state and local government for greater financial support.[25] The study also identifies two tendencies observed within the arts sector: reconsidering the old non-profit fundraising model and shifting to the search for powerful individual investors, and greater collaboration with other art organizations, in some cases leading all the way to institutional consolidation.

As public support for the arts declines, the structure of private donations and the profile of private donors are changing as well, which raises some concerns about the future. Thus, an expert on museum ethics, Erik Ledbetter, predicts that the decline of the middle class and the reemergence of a true American plutocracy will have serious consequences for the business models currently favored by art managers.[26] Economic recessions that impact traditional income streams, in combination with tight federal, state and municipal budgets, increases the reliance of art institutions on powerful private donors, who in return expect greater influence over the strategic aspects of their work. A situation where the same individuals control both the strategic priorities and the economic resources of an institution may lead to a potential conflict of interest in the long-run.

Professional arts organizations have been trying to assess the long-term tendencies within the changing environment for the arts, and to develop some response strategies. One example is a recent project by the American Alliance of Museums – the Center for the Future of Museums.[27] This interactive virtual platform, hosted on the web site of the AAM, serves as a discussion forum for experts, art managers, and individual artists concerned about the long-term survival of museums. A recent survey conducted by the Center demonstrates that race and ethnicity become inescapable categories for examining demographic change of the visitors' profile. However, with the incorporation of technology into the practices of art institutions, a generational experience already seems more determinant than race or ethnicity. Audiences of art institutions continue to age, while younger people are more likely to prefer more participatory forms of cultural engagement.[28]

The current situation in the arts sector in the United States is comparable with the situation that existed before the New Deal reforms stepped in and took responsibility for promoting cultural life.[29] The difference is that, besides the traditional direct governmental intervention, today's art institutions have many more tools and possible channels of response available to them, including a

---

[24] Genoways 2006; Hooper-Greenhill 1999; McCarthy et al. 2001; 2004; Paris 2006; Foster 2010; Seaman, 2005.
[25] McCarthy et al. 2007.
[26] Ledbetter 2011.
[27] Center for the Future of Museums 2014.
[28] Farrell and Medvedeva 2010; NEA 2004; 2006; 2008.
[29] Cherbo et al. 2008; Wyszomirski 1995.

greater number of market solutions, as well as greater networking and cross-disciplinary opportunities. On the positive side, economic recession forces artists to rely less on the corporatized and commercialized forms of art infrastructure that have dominated since the late twentieth century, and fosters their greater reliance on community resources, which often implies serving a broader public.

With the passing of the Industrial Age and the arrival of the Knowledge Economy, there was a recognition that the future of the arts depends upon both a deep understanding of their past and present and a comprehensive awareness of business practices.[30] However, while shifting to new business models generally enhances the resilience capacity of cultural institutions, it also has potential negative long-term outcomes. Arts organizations are forced to rely more on economic arguments and borrow the vocabulary and methodological tools from other disciplines in order to make their case to potential contributors.[31] This may diminish the intrinsic value of arts organizations and the alternative, non-economic values that they promote.

In an effort to address concerns regarding the future, arts institutions are looking for ways to better know and serve their publics, thus building a base of current and future supporters. Cultural institutions employ a multitude of methods to improve outreach and ensure responsiveness to the public, including greater engagement in more popular, experimental forms of art, greater dialogue with the public through social media, and arranging multidisciplinary events that spark visitors' curiosity.[32] Increasingly, institutions also emphasize their educational role and strive to serve as a learning environment for the public.[33] The public value of arts institutions becomes increasingly important, and by serving the public interest in numerous ways, arts institutions build up their resilience capital, which appears important for their long-term survival.

At the same time, some responses to recession have generated debate within the professional community. As an example from the museum community, according to a survey of thirty U.S. leading museums, there is an ideological split within the professional community, with some institutions focusing on revenue generation while others stress the museum's role as a free community resource.[34] It appears that particular management choices often depend on institutional size and location. Museums in major cities that attract tourists are more likely to charge entry fees, while their counterparts in areas with fewer tourists, which mostly rely on local visitors, tend to be free. Consequently, those institutions that charge only for special exhibitions on average make less than 1 percent of the total operating budget through ticket sales, while museums that issue a general entry fee earn an average of 9.5 percent of their budgets in this way.

Decreased funding from the National Endowment for the Arts and the declining budgets of state art agencies have led some scholars to consider the importance of arts to individuals. In the long-run individual appreciation for the

---

[30] Falk and Sheppard 2006.
[31] McCarthy et al. 2007.
[32] Falk and Sheppard 2006.
[33] McCarthy et al. 2007; McCarthy et al. 2004; Falk and Sheppard 2006; Hooper-Greenhill 1999; Paris 2006; Suchy 2006; Seaman 2005.
[34] Stoilas and Burns 2011.

value of art will be key to the long-term sustainability of the cultural sector.[35] A study conducted in Kentucky demonstrated that the value individuals place on art significantly affects support for proposed increases in art events and exhibits.[36] However, regardless of the personal value they place on art, individuals are generally willing to provide support in order to avoid decreases in already-existing performances and exhibits. This means that there is a certain threshold, or a baseline, below which individual support for the arts is unlikely to fall. The main question is: what is the major factor that ensures the existence of such a baseline for the art? This chapter argues that the answer lies within the arts sector itself, and can be found in the uniqueness of the long-term role of art. The values promoted through art institutions to future generations build some sort of "unseen" intrinsic endowment for the arts, thus contributing to the resilience capital of arts institutions and ensuring their long-term sustainability.

The Long-term Significance of Aesthetics and Art

There is an ongoing theoretical discussion regarding the importance of arts and culture for more sustainable thinking. However, in practical terms, this role for art is underappreciated, and may be even not well understood or conceptualized by artists and their professional communities. [37] This chapter conceives of aesthetics and the arts sector (the embodied, institutionalized form of aesthetics) as the core of the cultural dimension of sustainability, and argues that, although the cultural aspect of sustainability is perhaps least examined by scholars, understanding how culture contributes to the long-term sustainability of communities and societies is key to a holistic understanding of sustainability itself.

The issue of cultural sustainability is becoming especially urgent now, since many societies are facing a crisis of identity and lack a clear vision of the future in the post-industrial era. Such a point of view is shared by several scholars of cultural sustainability.[38] For instance, Brocchi argues that there is a natural link between genuine democracy, cultural diversity, and sustainability, since a culture of sustainability is able to address many important limits of sustainable development by treating such dogmas as "economic growth" and "free competition" with skepticism.[39] Packalén also believes that current lifestyles and ideas regarding what a good life is can be reevaluated in the light of aesthetics, serving as a counterweight to technocratic tendencies in society.[40] Ideas regarding the role of art for sustainability give central importance to the creative human being, who is capable of bringing art into the discourse on sustainable development.

It is possible to distinguish three major roles for aesthetics and the arts sector as the core of the cultural dimension of sustainability: instrumental (art for economic and social development), semi-instrumental (art for sustainable thinking and sustainable action), and intrinsic (art as a public good and value in

---

[35] Clark and Kahn 1988.
[36] Thompson et al. 2002.
[37] Bachmann 2008.
[38] Brocchi 2010; Haley 2008; Kagan and Kirchberg 2008; Nurse 2006; Packalén 2010.
[39] Brocchi 2010.
[40] Packalén 2010.

itself). Table 12.1 summarizes these roles and explains their relationship to sustainability. These three dimensions of arts' impact have varying influence on sustainability: the instrumental role of the arts is focused on using the arts for resolving immediate sustainability concerns. On the other hand, arts organizations that fulfil semi-instrumental and intrinsic roles are capable of short and longer-term, or intergenerational, impact. However, these three functions do not always exist in separation. According to Nurse, the cultural industry serves as a catalyst for regional and national identity formation, a key driver of the new digital, intellectual property-focused economy, and an economic sector with substantial growth potential.[41]

Table 12.1: Dimensional Model of Arts Impact on Sustainability Instrumental Semi-Instrumental Intrinsic

Instrumental   Semi-Instrumental   Intrinsic

$\longleftrightarrow$

	Instrumental	Semi-Instrumental	Intrinsic
Definition	Art as an instrument for social and economic sustainable development	Art as a source of social values	Art as a public good and value in itself
Dimensions of impact	Social/individual Direct Explicit	Social Indirect Implicit	Individual/social Indirect and direct Implicit
Examples	Culture-based development Art for urban revitalization Sustainable architectural design Art as a vehicle for attracting creative class Extrinsic benefits of arts participation	Sustainable thinking and sustainable action Intergenerational impact of the arts: transmitting values to future generations, fostering thinking about the future Arts' contribution to social discourses Art as ecology Prominence of public education in the arts	Art appreciation Unity of aesthetic experience Aesthetics and ethics Intrinsic benefits of arts participation Intergenerational impact of the arts: developing and promoting timeless values

The dimensional model recognizes that the impact of the arts is not dichotomous, but can be described as a combination of three facets: 1) social or individual; 2) direct or indirect, and 3) implicit or explicit. Each dimension of arts' impact could then be as the combination of these facets. Thus, the *instrumental role* implies that the arts could have direct, tangible, and very explicit effects on both

---

[41] Nurse 2006.

individuals and communities. For example, at the local level, culture-based development initiatives often result in improved job opportunities for individuals as well as a better quality of life and overall social climate within communities.

The impacts of both the *semi-instrumental* and *intrinsic roles* are more implicit, particularly since both of these roles impact values, perceptions, emotions, and modes of human thinking, rather than produce immediate tangible results. The intrinsic role of the arts impacts individuals, by evoking individual level emotions via aesthetic experiences, as well as communities, by contributing to the aesthetic values of a given society. While the *intrinsic role* of the arts results in both direct and indirect benefits, such as aesthetic pleasures from experiencing the arts and ethics sentiments inspired by art, the *semi-instrumental role* connects arts with other social discourses, and thus indirectly impacts social values via those discourses (i.e. arts contributing to environmental, economic, social and political discourses). Due to the implicit and indirect nature of the social impact that it poses, the *semi-instrumental role* of the arts in formalized arts organizations is embedded via the function of public education that bridges arts and societies.

Instrumental Role of the Arts

The *instrumental role* of the arts and creative sector in European cultural capitals has long been recognized, and there are studies attempting to quantify the contribution of arts to the gross domestic product and the impact of cultural industry on many areas of domestic economy (including but not limited to entertainment, tourism, and recreation).[42] Public attention to the arts and creative sector in the United States has been rising as well, as evidenced by a number of public forums, scholarly works, and professional conferences dedicated to the role of arts as part of the creative economy.[43]

Indeed, the arts sector in the U.S. is one of the most dynamic segments of the modern post-industrial economy. According to some estimates, there are 564,560 arts-centric businesses in the United States, employing 2.7 million people and representing 4.2 percent of all businesses and 2.0 percent of all employees.[44] In terms of the public participation in the arts, a study conducted by the National Endowment for the Arts found that about 39 to 41 percent of the adult population in the United States attend a "benchmark" art exhibition at least once a year, which can include opera, symphonic orchestra, theater, ballet and other dance forms, classical and jazz music performances.[45]

The role of nonprofit arts organizations in the American context is particularly important for understanding the *instrumental role* of the arts. In present-day America, the nonprofit arts industry is an important sector of the economy and a defining aspect of contemporary American life, which generates US$ 36.8 billion in economic activity annually, supports 1.3 million jobs, and returns US$ 3.4 billion in federal income taxes and US$ 790 million in local

[42] Caust 2003; Cunningham 2002; Holden et al. 2007.
[43] Tepper 2002; Lloyd 2010; Florida 2002; Cooke and Lazzeretti 2008.
[44] Cherbo et al. 2008.
[45] NEA 2006.

government revenues. Additionally, a number of nonprofit arts institutions have experienced dramatic growth during the last decades of the twentieth century. Half of the 8,200 American museums have come into existence since the 1970s, and the number of opera companies with budgets over US$100,000 has grown from 29 in 1964 to 209 by 1989.[46]

In the second half of the twentieth century, the increase in the number of cultural industries accelerated, the boundaries between culture and economics, and between art and commerce continued to shift, and cultural industries began to emerge as an issue in local policymaking.[47] The role of cultural industries in reinvigorating post-industrial economies has become much more influential than traditional concerns regarding the high arts and their future. This idea encompasses numerous elements: place-marketing, stimulating a more entrepreneurial approach to the arts and culture, encouraging innovation and creativity, finding a new use for old buildings and derelict sites, and stimulating cultural diversity and democracy.[48]

There is evidence that the engagement of arts industries in the redevelopment and reinvigoration of cities improves these cities' images, helps to attract tourists, fosters a better cultural climate, serves as a symbol of good taste and excellence, enhances quality of life, attracts people from other creative professions, and socially stabilizes downtowns.[49] With the recognition of these factors, cultural policy has moved from the margins to the very center of the mainstream urban regeneration agenda.[50] According to Strom, the co-mingling of arts and local economic revitalization builds on earlier American traditions of civic promotion, but it also represents a reframing of arts policies and their role in the larger community.[51]

Despite hopes regarding the role of arts in urban revitalization, there is also some skepticism about the results of such programs. For instance, the strategy of urban revitalization in American contexts, based on privileging the arts in a society that lacks a national consensus about the importance of the arts, would entail considerable risk. This is not the case in places like Germany, where the national government has made the arts part of its primary strategy for reinventing Berlin.[52] Additionally, although culture-based urban projects increasingly stress equity concerns by developing outreach programs to make sure that different population groups have comparable access to arts, there is evidence that successful neighborhood redevelopment may lead to a significant increase in property values and gentrification.[53] Therefore, along with the benefits of using arts as a strategy for urban redevelopment, there are also some concerns that need to be addressed.

Overall, the economic contributions of the arts sector to employment, GDP, and exports have been recognized as the root of the instrumental function of

---

[46] Cherbo and Wyszomirski 2000, 5.
[47] Hesmondhalgh and Pratt 2005.
[48] Hesmondhalgh and Pratt 2005.
[49] Florida 2002.
[50] Wilks-Heeg and North 2004.
[51] Strom 2003.
[52] Cherbo et al. 2008.
[53] Markusen and Gadwa 2010.

culture.[54] This perspective is useful; however, there are some theoretical problems with looking at art from a purely instrumental perspective. The main limitation of this idea stems from its grounding in the fragmented vision of culture. Thus, in the mono-dimensional view, culture is defined according to one specific channel of impact on sustainable development.[55] Studies based on a mono-dimensional definition of culture suggest that the cultural impact is a robust factor that can boost socio-economic development;[56] however, they do not acknowledge the significance of culture in and of itself.

The multidimensional conceptualization of culture takes into account the multitude of culture's channels of impact and prioritizes different channels on the basis of the strength of their impact and importance on a macro level.[57] It eventually became a platform for the studies of culture-based development, which interprets culture as living culture and cultural heritage, both of which utilize culture as a resource for generating social well-being and economic welfare – the two components of sustainable development.[58] The idea of culture-based development recognizes the critical role of cultural transformation, and presupposes that culture functions as an institution with a dual role – it is capable of replacing natural resources as the material of economic growth, and it also defines our value systems.[59]

The developmental model of sustainability has been incorporated in many international development programs and domestic policy agendas. However, it has several weak points as well, including the widening gap between sustainable goals and real development, the unwillingness of political actors to make change in the name of sustainability, too much reliance on technological solutions to sustainability issues, and difficulty in coming to terms with the limits of economic growth, among others.[60] Therefore, this framework has faced challenges on both ideological and implementation levels.

Semi-Instrumental Role of the Arts

The *semi-instrumental* view states that artistic thinking and appreciation of beauty could shape our way of thinking about the long-term future, and could be used instrumentally to support such a vision; however, it also recognizes that art and sustainability is about understanding the art in and of itself.[61] According to this logic, all art is considered contemporary, and in some way will always covey a vision for the future. Thus, it becomes the mission of artists and their organizations to allow the ordinary public to become part of professionally facilitated art actions. In the long-run this provides an avenue for the public to rethink their future.

---

[54] Nurse 2006.
[55] Tubadji 2010, 184.
[56] Tubadji 2010, 185.
[57] Tubadji 2010, 186.
[58] Tubadji 2010, 195.
[59] Matarasso 2001.
[60] Brocchi 2010.
[61] Bachmann 2008.

The role of art for social change outside of its own domain has been recognized in the literature, and became especially prominent during the late 20th century. It presupposes that social processes may occur through artistic processes that could potentially generate social change. Thus, an artist as an entrepreneur, who functions within social conventions, both benefits from some advantageous characteristics of art as a social process and has to put up with the conventional barriers of the art world, as well as the institutional and material barriers of the outside environment.[62] By overcoming these barriers art entrepreneurs foster the process of social change; therefore, art is inseparable from the society, and is part of the same ecology.

Art as ecology is an example of the *semi-instrumental role* of the arts; it emphasizes the importance of a synthesis of art and science, nature and culture. According to this line of thought, sustainability has the potential to find, through art, ethical value.[63] By contributing to reflexive thinking modes, artists and designers may become key change agents in sustainability.[64] This approach to art allows expanding the problem of sustainability from a mere question of sustaining art institutions to integrating art, as an equal partner, into the inter-disciplinary understanding of sustainability itself. In this regard, art is valuable for sustainability as a strategy and a process of moving toward the future, aimed at the creation of "an ecologically and socially just world within the means of nature without compromising future generations."[65]

After the events of 9/11, views regarding the social role of art and the significance of cultural policy gained a new perspective.[66] In particular, many artists in the United States and in other Western democracies could neither stay indifferent to, nor separate themselves from, the shocks American society has endured. It has been a time of unprecedented involvement by the arts in the public and social domain, and the works of artists not only reflect the social pain and suffering caused by the tragic events of 9/11, but they also became symbols of social solidarity, and provided psychological relief and healing. Thus, the 9/11 attacks resulted in immediate, visibly evident increases in expression of national identification and unity throughout the United States, and artists and their creations played important role in fostering the common national identity.[67] What happened in the domain of art in the U.S. had shaped cultural processes around the world and increased the inclusion of art into numerous socio-political agendas including security issues and the phenomenon of terrorism.[68]

Evidence of the arts' *semi-instrumental role* is often found in the missions of arts organizations, which emphasize the use of art as a way of fostering transformative thinking. This means including arts in broader social discourses, including the discourse on sustainability. As an example from museums, an increasing number of museum exhibitions are designed to educate people about

---

[62] Kagan and Kirchberg 2008.

[63] Haley 2008.

[64] Dieleman 2008.

[65] Kagan and Kirchberg 2008, 15.

[66] Bleiker 2006; Li and Brewer 2004; Robert 2003.

[67] Li and Brewer 2004.

[68] Bleiker 2006.

environmental sustainability.[69] These exhibitions cover such topics as global climate change, environmental awareness, sustainable clothing and food, and the preservation of natural resources. They indicate the potential for a unified aesthetics of natural and artistic beauty to stimulate critical thinking about intergenerational sustainability. Increasingly, museums also engage in exhibitions that highlight the relationship between culture and the social and political dimensions of sustainability.

The tendency to connect art with other social discourses is prevalent in the performing arts as well. There are many examples illustrating the involvement of performing arts organizations in social projects, public policies, and long-term community development initiatives.[70] For instance, when the issue of poverty was on the global political agenda in the 1990s, music was used as a vital element in campaigns against poverty organized by civic alliances between trade unions, nongovernmental organizations, and youth movements.[71] Indeed, music has been particularly recognized as a universal language of cross-cultural communication.[72] During the mid-1990s the arts joined the environmental discourse.[73] These examples demonstrate that the performing arts could be very powerful agents of social change.

Based on the studies of arts participation conducted by the National Endowment for the Arts,[74] performing arts organizations realize that stronger engagement with the arts results in higher levels of social capital and stronger and more sustainable communities overall. Hence, the performing art organizations of the future are likely to view themselves as civic organizations that perform socially important functions in addition to their primary cultural purpose. Performing civic functions, in addition to aesthetic and cultural roles, appears important for enhancing the institutional resilience of individual performing art organizations and achieving the long-term sustainability of the sector as a whole.[75]

Arts' ability to promote important social values and contribute to the multidimensional social discourse is the product of the distinct instrumental value of arts-based learning, often implemented via an institutionalized public education function.[76] The value of arts-based learning stems from the fact that through arts beauty and human aspirations can be learned through direct experience. Experience-based learning is particularly important for instilling values, and although specific values are not always mentioned explicitly in mission statements of formalized arts organizations, they are nevertheless embedded within them and are reflected in the ways that art managers understand the purpose of their institutions.

The role of art as symbolic capital is important because it capitalizes on the strength and long-term significance of art institutions. Thus, a balanced economic

---

[69] Moldavanova 2013.

[70] Causey 2006; Higgins 2012; McCarthy et al. 2004; Ramnarine 2011.

[71] Causey 2006; Higgins 2012; McCarthy et al. 2004; Ramnarine 2011.

[72] Higgins 2012.

[73] Ramnarine 2011.

[74] NEA 2006.

[75] Moldavanova 2013.

[76] Genoways 2006; Falk and Dierking 2000; Packer and Ballantyne 2002.

development and cultural policy that seeks to develop artistic or cultural capital needs to incorporate initiatives aimed at supporting the intangible, symbolic contributions of artists to local communities.[77] The *semi-instrumental* role of arts is about the capacity of arts institutions to produce both tangible and intangible benefits for society that stem from the value of art itself as well as the values promoted through the arts. Thus, the social significance of the arts is directly linked to the sustainability of art institutions.

Intrinsic Role of the Arts

The *intrinsic role* of art is based on the assumption that art can be a source of personal and collective values and principles. According to this line of thinking, sustainability within the arts relates to the processes by which art activities are carried out: search and research, learning and working, developing reflexivity of different types, appealing to a diversity of human qualities, and exceeding the limited types of rationality embedded in scientific discourses, common rules, and routines.[78] The *intrinsic role* of art recognizes that responsiveness to the public is not the only purpose of art institutions. According to Hein, "fascination with things whose value is intrinsic, with anything that is an 'end in itself,' although seems archaic in today's world where nearly all activity is engaged to some purpose, is nevertheless very important because it promotes alternative modes of coherence."[79] The process and practice of making art is valuable in itself because we can inform ourselves through it.[80]

The scholarship regarding the intrinsic significance of the arts is quite scarce, which is explained by the preoccupation of scholars with studying the numerous *instrumental* benefits of arts participation (such as an opportunity to socialize, escape from everyday routines, the improved quality of life, arts as a source of therapy, etc.),[81] and a focus on the wide range of economic, social, cultural and other *semi-instrumental* contributions of the arts to communities. Another explanation is that arts advocates themselves are reluctant to emphasize the *intrinsic* aspects of the arts experience because such arguments do not resonate well with donors.[82] However, *intrinsic significance* is an important element of the intergenerational sustainability of the arts, and is worth of separate attention.

The intrinsic significance of the arts is about the unique ways in which arts contribute to a society through their aesthetic value, the ability to make communities more aesthetically appealing, and promotion of a wider range of values. For example, the distinct impact of the performing arts is based on their aesthetics, in particular the unique ability of the live performing arts to evoke strong emotions, which are enhanced by the interaction between performers and their audience. Aside from the emotional and aesthetic appeal, which is usually short-lived, performing art experiences are capable of influencing certain long-

---

[77] Currid 2009.

[78] Kagan and Kirchberg 2008.

[79] Hein 2006, 1.

[80] Dewey 1934.

[81] Belfiore 2002; Guetzkow 2002; McCarthy et. al 2004.

[82] McCarthy et. al 2004.

lasting values and moral attitudes.[83] For example, to be a successful performer – whether a musician or a dancer – one needs to invest a lot of time, energy, persistence, and personal dedication to this occupation on an everyday basis. Hence, when young people look at performing artists as role models, they are likely to pick up this attitude, and adopt the commitment to excellence as their own approach to pursuing important life goals.

These individual-level values are foundational for the intergenerational sustainability of the arts because people are likely to practice values and attitudes acquired through the art experience in their lives and transmit them to the future generations. Such values and moral attitudes also enhance the prospects of bonding connections between individuals and the arts, which necessitates the lasting relationships.[84] At the same time, the attitudes and values developed by experiencing art are not the same as the individual level benefits of art participation that are widely described in the literature.[85] These values and attitudes are not about the instrumental value of art, they are more about the unexpected impact of art on an individual, which is intrinsic in nature.

Intrinsic significance is a key to the intergenerational sustainability of arts institutions due the ability of these institutions to remain relevant for past, present, and future generations. Aside from shaping individual attitudes and values, the aesthetic experience is important in and of itself as a critical reflection of reality and as a source of alterative thinking about everyday life. Arts organizations that produce and promote this kind of experience are likely to build lasting relationships with their audiences and sustain the arts institutions in the long-run.

Finally, the ability of art to evoke thoughts of the future is important for the purpose of intergenerational justice, which is concerned with the legacy of the current patterns of living for future generations.[86] According to Bachmann, since life, love, death, and everything in-between are the natural themes of art, the intergenerational theme is also reflected in the arts and creative thinking.[87] Thus, the *intrinsic role* of art reflects a way of thinking that is favorable to long-term sustainability. By being able to develop and promote timeless values, art created today is able to exceed the boundaries of current generations. Therefore, the value of art should never be reduced to merely perceiving art as an event, entertainment, or a mere instrument for something else.

Conclusion

Art and sustainability is a much-neglected area within sustainability studies. To fill this gap, this chapter examined relevant bodies of literature and offered a dimensional model of the arts' impact on individuals and societies, and connected various dimensions of the arts' impact to sustainability. This chapter conceptualized art and aesthetics as the core of the cultural dimension of sustainability, and suggested that an ethic of sustainability can be greatly

---

[83] Moldavanova 2013.

[84] Moldavanova 2013.

[85] Hager and Winkler 2012; McCarthy et. al 2004; Seaman 2005; Swanson et al. 2008.

[86] Catron 1996; Throsby 1995; Frederickson 2010; Moldavanova 2013.

[87] Bachmann 2008.

informed through aesthetics and arts organizations for two main reasons: first, because the cultural dimension of sustainability has the greatest potential among all of the other dimensions to reflect the long-term aspect of sustainability; and second, because aesthetics is essential for ethics, and intergenerational sustainability can be greatly informed by institutions specifically created to promote and preserve important values for the future generations.

As arts organizations struggle to address the consequences of economic recession and find new models of conducting their temporal business, their very existence and preservation contributes to the long-term sustainability of communities and societies as a whole. The primary question is how well the multitude of existing and emerging resilience strategies can coordinate with the strategic missions of art organizations and the values promoted through their programs. When a manager of an art museum or a symphonic orchestra makes a decision in response to the pressures of recession, does she consider the impact of such a decision on the mission of her institution? In case of a conflict between values embedded in the institutional mission and the decision to be made, does the mission get altered, or does it alter the outcome of the decision? These important questions exist at the intersection of ethics and management, and have a very prominent intergenerational impact.

It is possible to suggest two avenues for further research. First, the values and ideals embedded in the strategic priorities of art institutions and promoted through their programs contribute to building resilience capital, and serve as the foundation of long-term institutional survival. Therefore, ethics of long-term sustainability can be informed by studying institutional legacy and values that are being transferred to future generations through the aesthetics and institutions of art. Second, by fulfilling their institutional missions through both short- and long-term strategies and acting "sustainably," managers of art organizations ensure institutional endurance, thus vouching safe the interests of future generations. Consequently, understanding the resilience potential of arts organizations becomes crucial for configuring the long-term impact of aesthetics and its role for future generations.

References

Adorno, Theodor. 2001. *The Culture Industry: Selected Essays on Mass Culture.* London: Brunner-Routledge.

Adorno, Theodor. 2004. *Aesthetic Theory.* London: Bloomsbury Academic.

Asheim, Geir B., Wolfgan Buchholz, and Bertil Tungodden. 1999. Justifying Sustainability, Working Papers, Norwegian School of Economics and Business Administration. Available at http://econpapers.repec.org/RePEc:fth:norgee:5/99, accessed March 8, 2014.

Bachmann, Günther. 2008. Gatekeeper: A Foreword. In *Sustainability: A New Frontier for the Arts and Cultures*, edited by Sacha Kagan, Volker Kirchberg, 8-13. Frankfut am Main: Verlag fur Akademische Schriften.

Barry, Brian. 2002. Sustainability and Intergenerational Justice. *Theoria* 44(89): 43-64.

Belfiore, Eleonora. 2002. Art as a Means of Alleviating Social Exclusion: Does It Really Work? A Critique of Instrumental Cultural Policies and Social Impact Studies in the UK. *International Journal of Cultural Policy* 8(1): 91-106.

Bleiker, Roland. 2006. Art after 9/11. *Alternatives: Global, Local, Political* 31(1): 77.

Brocchi, Davide. 2010. The Cultural Dimension of Sustainability. In *Religion and Dangerous Environmental Change: Transdisciplinary Perspectives on the Ethics of Climate and Sustainability*, edited by Sigurd Bergmann, Dieter Gerten: 145-176. New Brunswick: Transaction Publishers.

Brown, Becky J., Mark E. Hanson, Diana M. Liverman, and Robert W. Merideth. 1987. Global Sustainability: Toward Definition. *Environmental Management* 11(6): 713-19.

Budd, William, Nicholas Lovrich Jr., John Pierce, and Barbara Chamberlain. 2008. Cultural Sources of Variations in U.S. Urban Sustainability Attributes. *Cities* 25(5): 257-67.

Catron, Bayard L. 1996. Sustainability and Intergenerational Equity: An Expanded Stewardship Role for Public Administration. *Administrative Theory & Praxis* 18(1): 2-12.

Causey, Matthew. 2006. *Theatre and Performance in Digital Culture: From Simulation to Embeddedness*. New York: Routledge.

Caust, Jo. 2003. Putting the "Art" Back into Arts Policy Making: How Arts Policy Has Been "Captured" by the Economists and the Marketers. *The International Journal of Cultural Policy* 9(1): 51-63.

Center for the Future of Museums. 2014. American Alliance of Museums. Interactive site available at http://www.aam-us.org/resources/center-for-the-future-of-museums, accessed March 8, 2014.

Cherbo, Joni and Margaret, Wyszomirski, eds. 2000. *The Public Life of the Arts in America.* New Brunswick: Rutgers University Press.

Clark, David and James Kahn. 1988. The Social Benefits of Urban Cultural Amenities. *Journal of Regional Science* 28(3): 363-77.

Collingwood, Robert. 1964. *Essays in the Philosophy of Art.* Bloomington: Indiana University Press.

Cooke, Philip and Luciana Lazzeretti. 2008. *Creative Cities, Cultural Clusters and Local Economic Development.* Cheltenham: Edward Elgar Publishing.

Cunningham, Stuart D. 2002. From Cultural to Creative Industries: Theory, Industry, and Policy Implications. *Media International Australia Incorporating Culture and Policy: Quarterly Journal of Media Research and Resources* 102: 54-65.

Currid, Elizabeth. 2009. Bohemia as Subculture; "Bohemia" as Industry. *Journal of Planning Literature* 23(4): 368.

Dasgupta, Partha. 2008. Discounting Climate Change. *Journal of Risk and Uncertainty* 33: 141-169.

DeCanio, Stephen J., and Paul Niemann. 2006. Equity Effects of Alternative Assignments of Global Environmental Rights. *Ecological Economics* 56(4): 546-59.

Dewey, John. 1934. *Art as Experience.* New York: Putnam.

Dieleman, Hans. 2008. Sustainability, Art and Reflexivity: Why Artists and Designers May Become Key Change Agents in Sustainability. In *Sustainability: A New Frontier for the Arts and Cultures*, edited by Sacha Kagan, Volker Kirchberg, 108-146. Frankfut am Main: Verlag fur Akademische Schriften.

Edwards, Mel, and Jenny Onyx. 2007. Social Capital and Sustainability in a Community under Threat. *Local Environment: The International Journal of Justice and Sustainability* 12(1): 17-30.

Entman, Robert. 2003. Cascading Activation: Contesting the White House's Frame after 9/11. *Political Communication* 20(4): 415-32.

Falk, John and Lynn Dierking. 2000. *Learning from Museums: Visitor Experiences and the Making of Meaning*. Lanham: Altamira Press.

Falk, John, and Beverly Sheppard. 2006. *Thriving in the Knowledge Age: New Business Models for Museums and Other Cultural Institutions*. Lanham: Altamira Press.

Farrell, Betty and Maria Medvedeva. 2010. *Demographic Transformation and the Future of Museums.* Washington, D.C: Center for the Future of Museums.

Florida, Richard. 2002. *The Rise of the Creative Class: And How It's Transforming Work, Leisure, Community and Everyday Life*. New York: Basic Books.

Foster, Kenneth J. 2010. Thriving in an Uncertain World: Arts Presenting Change and the New Realities. Report 24. Published electronically, January 2010. Available at http://www.apap365.org/knowledge/knowledge_products/Documents/Thriving_In_An_Uncertain_World.pdf, accessed March 8, 2014.

Frederickson, H. George. 2010. *Social Equity and Public Administration: Origins, Developments, and Applications*. Armonk: M.E. Sharpe, Inc.

Genoways, Hugh, ed. 2006. *Museum Philosophy for the Twenty-First Century*. Lanham: Altamira Press.

Guetzkow, Joshua. 2002. How the Arts Impact Communities: An Introduction to the Literature on Arts Impact Studies. *Centre for Arts and Cultural Policy Studies Working Paper Series* 20. Princeton: Princeton University Press.

Hager, Mark, and Mary Winkler. 2012. Motivational and Demographic Factors for Performing Arts Attendance across Place and Form. *Nonprofit and Voluntary Sector Quarterly* 41(3): 474-96.

Haley, David. 2008. The Limits of Sustainability: The Art of Ecology. In *Sustainability: A New Frontier for the Arts and Cultures*, edited by Sacha Kagan and Volker Kirchberg, 194-208. Frankfut am Main: Verlag fur Akademische Schriften.

Hasna, Abdallah. 2007. Dimensions of Sustainability. *Journal of Engineering for Sustainable Development: Energy, Environment, and Health* 2(1): 47-57.

Hein, Hilde. 2006. Assuming Responsibility: Lessons from Aesthetics. *Museum Philosophy for the 21st Century*: 1-11.

Helm, Carsten, Thomas Bruckner and Ferenc Tóth. 1999. Value Judgments and the Choice of Climate Protection Strategies. *International Journal of Social Economics* 26(7/8/9): 974-1021.

Hesmondhalgh, David and Andy Pratt. 2005. Cultural Industries and Cultural Policy. *International Journal of Cultural Policy* 11(1): 1-13.

Higgins, Kathleen. 2012. *The Music between Us: Is Music a Universal Language?* Chicago: University of Chicago Press.

Holden, John. 2007. Publicly-Funded Culture and the Creative Industries. London: Demos.

Hooper-Greenhill, Eilean. 1999. *The Educational Role of the Museum*. London: Routledge.

Howarth, Richard B. 2003. Discounting and Sustainability: Towards Reconciliation. *International Journal of Sustainable Development* 6(1): 87-97.

Kagan, Sacha and Volker Kirchberg. 2008. *Sustainability: A New Frontier for the Arts and Cultures*, edited by Sacha Kagan and Volker Kirchberg. Frankfut am Main: Verlag fur Akademische Schriften.

Krautkraemer, Jeffrey. 1998. Nonrenewable Resource Scarcity. *Journal of Economic Literature* 36(4): 2065-107.

Ledbetter, Erik. 2011. *Museum Ethics in a Gilded Age*. Center for the Future of Museums. Blog post, August 23, 2011. Available at http://futureofmuseums.blogspot.com/2011/08/museum-ethics-in-gilded-age.html, accessed March 8, 2014.

Li, Qiong and Marilynn Brewer. 2004. What Does It Mean to Be an American? Patriotism, Nationalism, and American Identity after 9/11. *Political Psychology* 25(5): 727-39.

Lloyd, Richard. 2010. *Neo-Bohemia: Art and Commerce in the Postindustrial City*. New York: Routledge.

Markusen, Ann and Anne Gadwa. 2010. Arts and Culture in Urban or Regional Planning: A Review and Research Agenda. *Journal of Planning Education and Research* 29(3): 379.

Matarasso, François. 2001. Recognising Culture: A Series of Briefing Papers on Culture and Development. London: Comedia.

McCarthy, Kevin, Elizabeth Ondaatje, Laura Zakaras and Arthur Brooks 2004. *Gifts of the Muse: Reframing the Debate About the Benefits of the Arts*. Santa Monica: RAND Corporation.

McCarthy, Kevin, Arthur Brooks, Julia Lowell and Laura Zakaras. 2001. *The Performing Arts in a New Era*. Santa Monica: RAND Corporation.

McCarthy, Kevin, Elizabeth Ondaatje, and Jennifer Novak. 2007. *Arts and Culture in the Metropolis: Strategies for Sustainability*. Santa Monica: RAND Corporation.

Moldavanova, Alisa. 2013. *Sustainable Public Administration: The Search for Intergenerational Fairness*. PhD Dissertation, University of Kansas, Proquest, UMI Dissertations Publishing number 3591643.

NEA. 2008. Reading on the Rise: A New Chapter in American Literacy. Available at http://www.nea.gov/research/readingonrise.pdf, accessed March 8, 2014.

NEA. 2006. The Arts and Civic Engagement: Involved in Arts, Involved in Life. Available at http://arts.gov/publications/arts-and-civic-engagement-involved-arts-involved-life-0, accessed March 10, 2014.

NEA. 2004. Reading at Risk: A Survey of Literary Reading in America, Research Division Report. Available at http://arts.gov/publications/reading-risk-survey-literary-reading-america-0, accessed March 10, 2014.

Nurse, Keith 2006. Culture as the Fourth Pillar of Sustainable Development. *Small States* 11: 28–40.

Packalén, Sture. 2010. Culture and Sustainability. *Corporate Social Responsibility and Environmental Management* 17(2): 118-21.

Packer, Jan and Roy Ballantyne. 2002. Motivational Factors and the Visitor Experience: A Comparison of Three Sites. *Curator: The Museum Journal* 45(3): 183-98.

Padilla, Emilio. 2002. Intergenerational Equity and Sustainability. *Ecological Economics* 41(1): 69-83.

Paris, Scott G. 2006. How Can Museums Attract Visitors in the Twenty-First Century. In *Museum Philosophy for the Twenty-First Century*, edited by Hugh Genoways, 255-266. Lanham: Altamira Press.

Ramnarine, Tina. 2011. The Orchestration of Civil Society: Community and Conscience in Symphony Orchestras. *Ethnomusicology Forum* 20 (3): 327-351.

Seaman, Bruce. 2005. Attendance and Public Participation in the Performing Arts: A Review of the Empirical Literature. Atlanta: Andrew Young School of Policy Studies Research Paper Series.

Shearman, Richard. 1990. The Meaning and Ethics of Sustainability. *Environmental Management* 14(1): 1-8.

Stengel, Kathrin. 2004. Ethics as Style: Wittgenstein's Aesthetic Ethics and Ethical Aesthetics. *Poetics Today* 25(4): 609-25.

Stoilas, Helen and Charlotte Burns. 2011. To charge or what to charge? *The Art Newspaper* (227), September 26. Available at www.theartnewspaper.com/articles/To+charge+or+what+to+charge%3F/24451, accessed July 4, 2014.

Strom, Elizabeth. 2003. Cultural Policy as Development Policy: Evidence from the United States. *International Journal of Cultural Policy* 9(3): 247-63.

Suchy, Sherene. 2006. *Museum Management: Emotional Value and Community Engagement.* Taipei: INTERCOM: International Committee on Management.

Swanson, Scott, J. Charlene Davis, and Yushan Zhao. 2008. Art for Art's Sake? An Examination of Motives for Arts Performance Attendance. *Nonprofit and Voluntary Sector Quarterly* 37(2): 300-23.

Tepper, Stephen. 2002. Creative Assets and the Changing Economy. *The Journal of Arts Management, Law, and Society* 32(2): 159-68.

Thompson, Eric, Mark Berger, Glenn Blomquist, and Steven Allen. 2002. Valuing the Arts: A Contingent Valuation Approach. *Journal of Cultural Economics* 26(2): 87-113.

Throsby, David. 1999. Cultural Capital. *Journal of Cultural Economics* 23(1-2): 3-12.

Throsby, David. 1995. Culture, Economics and Sustainability. *Journal of Cultural Economics* 19(3): 199-206.

Tubadji, Annie 2010. See the Forest, Not Only the Trees: Culture Based Development (CBD). Conceptualising Culture for Sustainable Development Purposes. In *Culture as a Tool for Development:Challenges for Analysis and Action*, edited by Florent Le Duc, 180-205. Paris: ARCADE.

Wilks-Heeg, Stuart and Peter North. 2004. Cultural Policy and Urban Regeneration: A Special Edition of Local Economy. *Local Economy* 19(4): 305-11.

Williams, Raymond. 1983. *Culture and Society, 1780-1950*. New York: Columbia University Press.

Wittgenstein, Ludwig. 1984. *Culture and Value*. Chicago: University of Chicago Press.

Wittgenstein, Ludwig. 1957. *Tractatus Logico-Philosophicus*. Mineola: Dover Publications.

World Commission on Environment and Development. 1987. *Our Common Future*. Oxford: Oxford University Press.

Wyszomirski, Margaret. 1995. From Accord to Discord: Arts Policy During and after the Culture Wars. In *America's Commitment to Culture: Government and the Arts,* edited by Kevin Mulcahy and Margaret Wyszomirski, 1-46. Boulder: Westview Press.

Chapter 13: A Call to Experience the Earth Collectively

Elizabeth More Graff and Wolfram Hoefer

Art as Potential Catalyst for Increased Environmental Stewardship via Shared Experience

As introduction and framework, we will explore the role of art as an indicator of environmental conditions and as precursor to public consciousness. Following a brief history of landscape representation in American painting and land art in association with land development, values and movements, we will present a case study of the current, international and travelling environmental art exhibit, Earth From Above by Yann Arthus-Bertrand, and the process intended to bring the exhibit to America. We will present the planning for the United States premiere as a catalyst for increased sustainable design and stewardship via collective cultural experience and its subsequent postponement as a reflection on the need for positive change where business, environment, and culture meet. Further, we will discuss strategies that include much needed support and funding for innovative community engagement models that connect academic and professional practice, such as Emerge Studio.

Individual human relationships with the Earth are subject to many factors influencing a people, era, and region. For example, perception, scale, spirituality, and commerce are just a few factors that play a role in environmental attitudes. Fundamentally, however, a visual record encapsulating such cultural factors can be traced via art. The artist is a seer, translator, poet, and visionary of both the perceived real, the potentially grave, and the ideal. Often at the fore of social commentary, the artist traditionally has the role of reflecting what is of a certain time, along with a bold statement about such conditions. Thus, it is fitting that we can track environmental consciousness and the dynamics of society through artistic representation. Analyzing art as a portal to environmental consciousness brings to the fore the significance of beauty and inspiration in daily life as global citizens at a time of increasing and logarithmic pressure on the carrying capacity of the Earth. This discussion provides the context for a critical appreciation of the environmental art exhibit, Earth From Above in the U.S.

However, art with a message and not just for personal enjoyment is mostly seen in relation to conceptual art and land art as instruments for political action; aspects of such a politically-minded positioning of art can be traced as early as the work of landscape painters of the Hudson River School of the mid-19th century. The second context of environmental consciousness is within the environmental movement and its reflection in the arts.

This interdependence between art, public realm and environmental awareness was exceptionally evident in the project "The Gates" in New York City in 2005. The creators of this public art event in Central Park, Christo and Jean-Claude, have evolved from the land art of the 1960s in that they were seen and experienced by many people directly. Placing hundreds of structures draped with textiles in a well-known urban park made visitors see the park in a different light; additionally, book publications and other forms of media contributed significantly to exposure. The work of Yann Arthus-Bertrand is also within the context of the public realm, inspiring a changing perception of the world. The photographs are representations of the beauty of the Earth and are accessible via book publications, but most significantly, as an experience of a large-scale, outdoor travelling exhibition. This aesthetic approach to changing environmental attitudes places this originally French exhibition in the context of American environmental history. Coincidentally, the United States' initial attempts to preserve nature were driven by aesthetic motives. For example, the beauty of the American landscape inspired the emergence of the National Park System. "Earth From Above" is a current case study that looks at the planet's fragility to deliver a poignant and hopeful educational message within these perceptions and historical context.

American Land Development: Art as Indicator of Environmental Conditions & Consciousness

> "No (hu)man can lose what (they) never had," Izaak Walton said. Nor can (we) find what (we have) never lost. That is the paradox of the wilderness. It was only when we had already lost it that we could begin to see the value in it.
>
> - Paul Gruchow, *The Necessity of Empty Places*[1]

Historically, there has been a doomsday approach to environmental movements in America and often witness trends of apathy toward sustainability. Art can be a path to present and evaluate positive change in terms of global stewardship from an optimistic avenue. But, firstly, it is necessary to understand some of the obstacles, opportunities and ideas inherent in land development that have historically been barriers to environmental responsibility, like perceptions of wilderness, manifest destiny, and the blatant obliteration of nature that ensued since European settlement. For example, the term "wilderness," derived from *wylder ness* meaning nest of a wild beast, symbolized settlers' fear for the unknown territory. Nature came to represent everything evil, savage, dangerous, and unruly. In accordance with their understanding of Christian doctrine,

---

[1] Gruchow 1988, 91.

"wilderness" identified the land beyond human control as spaces of heathen.[2] Many of today's historic cities were built on two major proponents: methods of transportation of goods and the pioneering practice of taking Native American sites as beneficial for starting settlements.[3] With the notion of economic progress combined with a fear and need to control wilderness, came the desecration of Native American culture and all that came before. Reflecting a continuing obstacle to environmental stewardship is the perpetuating image of "wilderness" as reflected in a current definition, "a wild and uncultivated region, as of forest or desert, uninhabited or inhabited only by wild animals; a tract of wasteland."[4]

A brief history of American land development from early colonial river settlement westward across the Great Plains sets the context for understanding current resistance to environmental responsibility. The scattered spatial organization depicted within artists' depictions of frontier towns resulted from rapid, unchecked development that we may now equate to the commercial strip and challenges of orienting oneself in a lack of identifiable cohesion of such physical layout. Settlement diaries speak to such disorientation.[5]

The Public Land Survey System has had much to do with how we perceive land in terms of property ownership and divided the landscape without regard to natural resources and features such as landform.[6] In other words, the method of division along a grid is dominant over ecosystem, elevation or character. At the same time, allowing an organic development of property ownership to unfold, as witnessed in the initial port towns like Boston, brings its own brand of confusion and inefficiencies that inspired the regimented and unrelenting system to be developed originally.

Interestingly, the landscape painters of the 1800s gathered their data for their compositional content by joining surveying expeditions through the great wilderness. With sketchbook in hand, they documented landscapes that they later painted in their studios, often in urban environments.[7] They were seeing a landscape and bringing it back to a public who were largely not privileged to witness these places first hand, as train travel for passengers beyond goods was yet to become prevalent. Coincidentally, today, one of the premises of the Earth From Above exhibit is to bring landscape images to the public who, statistically speaking, seldom experience the planet beyond their hometown.

Looking at environmental art as an indicator of environmental conditions and consciousness, we will begin looking at the Hudson River School of the 19th century. Thomas Cole's painting, *View on the Catskill-Early Autumn* of the early 1880s, romanticizes pastoral nature that depicts the association of people's relationship with God.[8] At this time, towns were forming on the basis of covenants between the community and God. Every member signed a treaty of agreed upon principles and rules for development.[9] Landscape paintings of this

---

[2] Stilgoe 1982, 10.
[3] Butzer 1990.
[4] Dictionary.com 2014.
[5] Hamer 1990.
[6] Kunstler 1993, 30.
[7] Cole 1836-1837.
[8] Cole 1836-1837.
[9] Smith 1966.

time period reflect a utopian idealism in the portrayal of nature as created by God in all of its glory. Some current planned communities with required associations and regulations, are not dissimilar in concept, no longer with a binding agreement with God but with an agreed upon way of life and maintenance standards.

Simultaneously, we have the cumulative town type based on economic progress and manifest destiny leading the development of the American landscape westward. This predominant town type existed and disappeared by the thousands, but most of the lasting metropolitan cities derived from this model.[10] Concurrently, by the mid 19th century, the Industrial Revolution came to the United States with the advent of the steam engine and witnessing a shift from river transport to train transport, opening the country up for growth at a quickening pace. The rapid development of towns depending on the use of natural resources brought pollution and human health issues along with it.

Significantly, a shift in the tone of landscape painting can be seen in the work of Sanford Robinson Gifford, a contemporary to and inspired by Cole. In "Hunter Mountain, Twilight" of 1866, he portrays romanticized nature as backdrop to clear-cutting in the foreground, sending a public message to pay attention to what is happening to the landscape and its resources with an implied call to action (see Figure 13.1).[11]

Figure 13.1: Hunter Mountain, Twilight, 1866, by Sanford Robinson Gifford (1823-1880), Oil on canvas, 30 5/8 x 54 1/8in. (77.8 x 137.5cm), Daniel J. Terra Collection, 1999.57.[12] Permission to reprint has been granted by the Terra Foundation for American Art, Chicago / Art Resource, NY to Wolfram Hoefer; image printed with permission and fees supported by Rutgers University and Wolfram Hoefer.

Simultaneously, we start to see planning efforts develop including Frederick Law Olmsted's park systems and parkways for horse and buggy transport, arising from the understanding that getting from point to point can be an aesthetic one.

---

[10] Lingeman 1980.
[11] Gifford 1866.
[12] Gifford 1866.

However, with independent personal transport combined with property ownership, came the conception of the American Dream and reinforced the notion of a fierce and rugged individualism that persists today and underlies resistance to sustainable behavior.

At the turn of the 20th century, in an untitled watercolor landscape by landscape etcher and painter Henry Farrer, beyond the great light and clouds of the sky, we can see in his representation an underscore of the idea that beyond America's cities, there is an awaiting vastness (see Figure 13.2). This early image of the open road carries with it the stereotypical fragment in the psyche of America that there is plenty of land left, plenty for everyone, plenty to develop. At this time, we begin to witness utopian planning models, like Ebenezer Howard's Garden City, spawning a movement that began in England and briefly debuted in America, to address unchecked development.[13] Though very limited in its actual application, Garden City remains one of the few responsible models of planned development.

Figure 13.2: Henry Farrer, American, 1843-1903, Untitled (landscape), ca. late 19th century Watercolor, 22.8 x 28.8 cm (image), The Fine Arts Museums of San Francisco, gift of Davis and Langdale Company, 1981.2.25.[14] Permission to reprint has been granted by The Fine Arts Museums of San Francisco to Wolfram Hoefer; image printed with permission and fees supported by Rutgers University and Wolfram Hoefer.

Looking to our artist peers, we see Georgia O'Keefe revealing detailed depictions of found natural objects. Giving a detailed glimpse of natural objects close up, her work used scale to remind us to stop and see the beauty in front of us every day. In addition to giving us a glimpse at American life and the introduction of garden

---

[13] Howard 1946; first published in 1902.
[14] Farrer 1843-1903.

varieties, she is more forthright in revealing a shift of view inward and of the self and looking in detail at a micro-scale at nature. Perhaps a reflection of awakening, her work stands in contrast to her predecessors depicting broad, atmospheric landscapes.

The advent of the automobile and the combustion engine amplified land development and spawned massive highway infrastructure that moved ahead of train transport. In 1904, construction of Route 1 from Maine to Florida began; and, in 1940, the Pennsylvania Turnpike brought goods via trucking, tourists, and modern settlers westward. Trucking remains the major mode of transportation of goods across the US and a major contributor to the country's carbon footprint. Later, the Dwight D. Eisenhower Interstate Highway System of 1956 reflects and contributes to post-World War II catering to the American Dream and beginnings of sprawl via segregated use, unchecked development, and focus on the automobile.[15]

As impetus and result, the preservation of nature and set boundaries are created for the first time since European settlement. Thus, the experience of the country's awe-inspiring natural landscapes can be a personal and spatial one for the public beyond artists' renditions in museums or reproduced in books. With the Organic Act of 1916, the United States established the National Park Service whose mission statement is "...to promote and regulate the use of the...national parks...which purpose is to conserve the scenery and the natural and historic objects and the wild life therein and to provide for the enjoyment of the same in such manner and by such means as will leave them unimpaired for the enjoyment of future generations." [16] The necessity of protection and maintenance was underscored by the increase of natural places as tourist attractions bringing customers via the newly built railroad system.

The introduction of the car further allowed settlement patterns that would avoid the negative impacts of urbanization. A unique residential development in Radburn, NJ designed by Stein and Wright in 1929 brought Garden City principles to the community of Fairlawn to welcome the "motor age" while providing amenities of open space, community service, and economic viability to address quality of life and sustainability in light of changing complexities of modern life.[17] Radburn is a covenanted community, not bound to God, but bound to community ideals. Still thriving as a community, there is a long waiting list to live there and the validity of the concept is even more compelling as antithesis to conventional development. However, Radburn is considered a highly successful early experimental suburban exception; predominantly, the postwar housing boom created a rather unfortunate phenomenon of suburbia, a grotesque cartoon of the American dream, driven by an inexhaustible housing market longing for immediate profit. The outcome of that development is clearly represented in Arthus-Bertrand's image of Highlands Ranch on the outskirts of Denver, Colorado with which he underscores, "These networks of low-density suburbs make their residents totally dependent on their cars, one of the chief sources of greenhouse gases. This dependence is one reason that Americans generate the highest emissions of greenhouse gases on the planet. Although North Americans

---

[15] US Department of Transportation 1956.
[16] National Park Service 1916.
[17] Gatti 1969-1989.

are only 5 percent of the world's population, in 1998, they produced almost a quarter of human-generated carbon dioxide"[18](see Figure 13.3). Since, trends of development and ways of living in North America, despite economic slowdown, have only increased. Additionally, obstacles to widely embraced alternative energy and other means of extracting fuel, such as fracking, not only contribute to increased greenhouse emissions, but also demonstrate further injury to the Earth.

Figure 13.3: Highlands Ranch, Outskirts of Denver, Colorado, United States (39°44' N, 104°59' W), 2003 © Yann Arthus-Bertrand / Altitude.[19] Permission to reprint has been granted by Altitude to Elizabeth More Graff; image printed with permission and fees supported by WSU and editors.

Art in the public realm reflects this conflict between the desire to live a better life away from the negative effects of urbanization and the impact that this increased use of undeveloped land has on the environment. A main approach was to increase environmental awareness by exposing people to the beauty of nature through art. By the 1960s, we can reflect upon Georgia O'Keefe's sketches that portray a portal from within nature, looking out. At the same time, and providing a view from within the National Parks out to the public, photorealistic work of Ansel Adams reveals Yosemite National Park, as he was inspired to represent since his first visit as an adolescent.[20] His art serves as inspiration for environmental dialogue and highlights the healing capacity of both nature and art.

With the dawning of the land art movement in the late 1960s came the experience of art brought from inside museum walls directly into the public landscape and, technically, more accessible. For example, people may not have had the means to travel to Michael Heizer's "Nine Nevada Depressions" in the

---

[18] Arthus-Bertrand 2014.

[19] Arthus-Bertrand 2003.

[20] National Park Service 2014.

Black Rock Desert created in 1968, but nonetheless, any socio-economic barriers that may be implied by museum walls are eliminated.[21] Nancy Holt redefined museum in the mid-1970s with "Sun Tunnels" in the Great Basin Desert of Utah, making a broader connection to the universe and human context by aligning massive concrete pipes drilled with holes that create light patterns with equinoxes and solstices.[22] Of course, this is happening simultaneously with the fascination with space travel.

Akin to the methods of Walter de Maria, Richard Long's sculptures since the 1960s find inspiration in the tradition of walking as experiencing, immersing, and interacting with the landscape. He poetically conveys his intentions as, "Art about mobility, lightness and freedom; Simple creative acts of walking and marking; about place, locality, time, distance and measurement; Works using raw materials and my human scale in the reality of landscapes."[23] In method and meaning, such work is in direct contrast to automobile culture, and once again, interjects a different pace as way of experiencing the world.

Christo and Jean-Claude have accomplished much in assisting the public re-see what can often become commonplace in nature and the built environment. Their work reminds us that what is awesome is still awesome, but can get lost, invisible or lose its grandness among the demands and routines of everyday life. For example, in 1969, "Wrapped Coast" in Little Bay, Australia covered one million square feet of coastline with erosion control mesh, exemplifying their guiding philosophy and artistic vision to "make the unseen visible."[24] Amazingly, for the Gates to happen in Central Park took a thirty-year process of extensive approvals, meeting with stakeholders, design development, and evaluating natural and cultural impact.[25] Experiencing the project offered looking at the already beloved landscape of Central Park with all of its complexities and beauty with a renewed sense of spirit, or simply just a little differently.

"The Gates" serves as precedent and prelude to discussing the process of bringing Yann Arthus-Bertrand's Earth From Above exhibit to New York City and the US. Arthus-Bertand is from France, but is a global environmentalist. Experiencing Earth From Above often in a book format, the significance of the exhibit itself is of a greater shared cultural experience. Essentially, his mission is to show the people of the world, the world we live in; and, with it comes an environmental message. The images are honest, beautiful, and carry a heavy reflection of environmental, social and cultural concern; and, yet, they inherently inspire, as witnessed by public response to the traveling exhibit around the world. Taking a look at his work, for example of the Palouse bioregion, reflects a beautifully composed aerial photograph of this distinct dune-like, agricultural region in the intermountain northwest (see Figure 13.4). He puts forth, "…The use of biotechnologies, especially in the production of corn and soya, has led to the creation of varieties that are resistant to parasites, and herbicides that are believed to increase yield. Controversial because little is known about the extent

---

[21] Heizer 1968.

[22] Shaffer 1983.

[23] Long 2014.

[24] Weilacher 1996, 33.

[25] Christo and Jeanne-Claude 1979- 2005.

of their side effects on health and the environment, the cultivation of genetically modified organisms (GMOs) is nonetheless on the increase worldwide..."[26]

Figure 13.4: Farming near Pullman, Washington, United States (46°42' N, 117°12' W), 1997 © Yann Arthus-Bertrand / Altitude.[27] Permission to reprint has been granted by Altitude to Wolfram Hoefer; image printed with permission and fees supported by Rutgers University and Wolfram Hoefer.

The interpretive descriptions accompanying his work reveal what has been happening to our resources, in this case in terms of erosion, monoculture, and agrobusiness while pointing us in new directions. In Ecuador, his image of the City of Guayaquil reminds us of the value of seeing, and thus re-seeing, our cultural and built environments with a new lens or perspective via aerial photography. Accompanying his image, "Volcanoes on the Galápagos Archipelago, Ecuador (0°20' S, 90°35' W)," Arthus-Bertrand's words inform of a global cultural heritage and responsibility (see Figure 13.5):

> The nineteen islands of volcanic origin that constitute the Galápagos archipelago emerged from the Pacific between 3 and 5 million years ago. Despite their lunar landscape, they are exceptionally rich biologically. In particular, they are home to the world's biggest colony of marine iguanas, as well as the giant—or *Galápagos* —tortoise, which gave the archipelago its name. Visitors who sail there are entranced; Charles Darwin, however, was inspired to develop his theory of the evolution of species. The Galápagos were designated as a national park in 1959, and in 1978, UNESCO added the islands to its World Heritage list. However,

---

[26] Arthus-Bertrand 2005, 353.

[27] Arthus-Bertrand 1997.

nothing has prevented the increase in human settlement, the introduction of exotic species, or a boom in tourism (though this has been strictly controlled since 1998) from endangering this natural laboratory of evolution. The archipelago miraculously escaped being polluted by some 600 tons of fuel oil that leaked from the tanker *Jessica*, shipwrecked in January 2001—but other coastlines were not as lucky.[28]

The breathtaking image of the Galápagos Islands is well suited with his team's philosophy, "For the will to protect is indeed strongest towards that which one has learnt to understand and love."[29]

Environmental Art as Catalyst for Global Stewardship & Cultural Cohesion

Earth From Above brings to fore the discussion of the utility of art - as cultural connector, as inspiration, and realization trigger. Giving us a rare bird's eye perspective from a helicopter that is much more personal than satellite imagery, the collection lends a common bond to a large majority of the viewing public that do not normally have the opportunity to view the world in this way. Broadening the public worldview provides a profound shared experience. This is where the potential for positive and responsible change lies. This is also where art and technology meet.

> It must be remembered that what has happened in art is itself a part of a very broad movement in which science has made the major contribution. Through its dynamics of rigorous logic 20th Century scientific understanding has come to conclusions not unlike those of the artists. Scientists recognize that in the most precise ranges of observation the observer and the observed interact. When observed and measured with maximum precision, the environment in both its largest and its smallest realism cannot be considered an independent objective world anymore.

> - Gyorgy Kepes, *Art of the Environment*[30]

Given the current evolution toward more holistic, and potentially collective, viewing of the current world versus discreet parts as previously witnessed by the split of art and science, we are witnessing the potential for art and technology to hold hands again. A reintroduction of art's relationship to the environment beyond the self and in context of a larger shared responsibility and consciousness, may once again position the role of art as the self-regulating device protecting from extreme imbalance more than forty years after Gyorgy Kepes pointed out this significant role of art in the world at a time of extreme environmental alarm of the 1960s.[31] Pointing in the direction of the collective once again, creates an

---

[28] Arthus-Bertrand 2005, 201.

[29] Arthus-Bertrand n.d.

[30] Kepes 1972.

[31] Kepes 1972.

avenue for art to re-emerge as a necessary part of the whole of human existence in light of commerce, technology, and unchecked growth and development dimming the role of beauty and creativity. It is time to bring beauty back into the conversation of earthly health as necessary to wellbeing and no longer a derogatory, superfluous folly.

Once said, the age-old debate of attempting to quantify the subjective awaits. Realizing that this in itself is one inherent aspect of art, the point can be studied with the work of Arthus-Bertrand's Earth From Above. The photography of the collection conjures a relatively objective response to environmental conditions as there is a level of quantifiable truth in photorealism. Artistry is expressed in the choice of subject, composition, color, texture, rhythm, and other fundamental aspects of the work. This brings to the fore universal emotive responses to shared experience, poignantly in this case, of the Earth. Because the subject is the Earth, the link is readily made to engage each viewer, each inhabitant, as members of the same global household, so to speak, as witnessed in the demonstration of Arthus-Bertrand's film, "Home." He is giving a voice to the many landscapes and ecosystems in consortium with the people of and affecting place.

Earth From Above is reminiscent of the effect historically witnessed with the advent of space travel, a well-known shift in perception happened when the Earth was viewed from space for the first time across television media to a mass audience as the "pale blue dot," a classic example of parts to whole thinking. Significantly, this historic, shared experience revealed how high technology can inspire awareness of the Earth's fragility; and, at the same time, it is just that technology that makes the Earth so fragile. The Earth From Above exhibit, reaching over 130 million visitors worldwide since 2000 in 150 cities, holds the potential for such a shift. [32] The global presence of Arthus-Bertrand's work provides viewers the opportunity to "...see the extent of human impact on our landscapes. And not a moment too soon: In the past 50 years – a single lifetime – the Earth has been more radically changed than by all previous generations of humanity...(His work) has shown its commitment to awaken a collective responsibility and conscience. In this awareness 'campaign' the objective is to reach to the most people possible."[33] To the Earth's detriment, our global home, the exhibit has yet to tour the United States. Obstacles to valuing art and culture, a continued emphasis on economic development, global recession, events like 9/11, and denial of environmental responsibility by the business community have all influenced the tour not travelling across the U.S. to date.

The cancelled premiere of the U.S. tour in New York City's Bryant Park promised close to 2 million visitors over an eight-week period. The proposed program elements included 130 four-foot by six-foot photographs, an Inspiration Center with an outdoor theater, the Exhibit for the Blind, an on-site classroom, a large-scale climbable World Map, and a gift shop that sells merchandise that helps finance the continuation of the artist's work. The Inspiration Center serves as interpretive hub for the various stories, processes, and educational information. The exhibition criteria for the Earth From Above exhibit is that it is free and open to the public 24/7 for several weeks. However, the design, materials and

---

[32] Picture Earth, Inc. 2007-2010.
[33] Arthus-Betrand 2012.

construction of the exhibit are not free, and these costs are fundamentally covered by fund-raising efforts of the self-forming, organizing and production group in each city that believes in the cause, is dedicated to the mission, and sees the exponential value of hosting the exhibit in their city. Despite the struggle to bring the exhibition to the United States, a production team of multi-disciplinary, collaborators worked diligently from 2007-2010 to make it happen at this critical time. Picture Earth, Inc., the production, marketing, and operations non-profit organization, formed with a mission "to use art for inspiring and teaching people to take personal responsibility for our world's future and to make better, more sustainable living choices."[34] Emerge Studio, in partnership with Picture Earth, Inc., was a research-based, sustainable design advisory and education component driven by the authors as Principal Investigators from two universities spanning the country as a collective of educators, designers, students, and related experts.[35] Our primary efforts included site selection, analysis and layout; sustainable design and education consultation; visualization; and conceptual, schematic, and design development, fabrication and installation guidance. In association with BKSK Architects of New York City, the intention was for this research to be brought forth into construction documentation.

Rethinking the paradigm of large-scale public art within the context of landscape architecture for the 21st century, while elevating the value of public education, was the foundation in the classroom and our partnership. University student work informed the design process of the exhibit while broadening their scope of design and responsible global citizenship, including exploring the intention and consequences of design, not just the end product. With imagination and innovation at the fore, participants conceived "cradle-to-cradle" solutions for the U.S. tour that, upon the end of its life cycle, "waste" would become "food" for something new. According to MBDC, a product and process design firm founded by William McDonough & Michael Braungart in 1995 and authors of *Cradle to Cradle*, "The Next Industrial Revolution is the emerging transformation of human industry from a system that takes, makes, and wastes to one that celebrates natural, economic, and cultural abundance."[36] This theory in practice adds a scientific layer to the technical process. Thus, designers of the built environment are giving consideration to the life cycle of every material, each engineering decision, and overall construction technique. Given the role of Emerge Studio in the realm of landscape architectural design practice and academia, this model strives to become a benchmark for the sustainable design of large-scale public art and places.

Discussion of Cultural Significance

The questions Emerge Studio has been raising include, "Can art change behavior? Can aesthetic experience impact positive stewardship?" As current, daily technology allows us to be increasingly more disconnected from place, does the exhibit relocate us in the physical realm of the Earth like a cultural positioning

---

[34] Picture Earth, Inc. 2007-2010.
[35] Emerge Studio 2007-2010.
[36] McDonough Braungart Design Chemistry (MBDC) 2013.

system? Does this aid in re-connecting people to actual place and increase sensitivity to monotonous and environmentally lifeless landscapes? The exhibit offers a unique human perspective of viewing the world from the eyes of a bird, along with uplifting and positive reflection, and a sense of celebration. It begs the question, "What is this profound shared experience doing for global, cultural cohesiveness and environmental stewardship?" With a focus not just on the beauty of the photographs themselves, but on the beauty and significance of the civic open spaces as context to the event around the world, the value, protection, and maintenance of these cultural landscapes also comes to the fore. And, the event really starts to question the utility of art while attempting to bring the exhibit to a universal and intergenerational audience while giving the Earth a voice. As precedent, the Ecuadorian Constitution of 2008 serves as the global model of granting inalienable rights to nature.[37]

Figure 13.5: Volcanoes on the Galápagos Archipelago, Ecuador (0°20' S, 90°35' W), 1999 © Yann Arthus-Bertrand / Altitude.[38] Permission to reprint has been granted by Altitude to Elizabeth More Graff; image printed with permission and fees supported by WSU and editors.

The role of education was at the fore of the exhibit planning, not just in the university classroom or proposed on-site event classroom, but also for the visitor. Taking education out of the classroom and art out of the museum, Earth From Above fosters the potential for the passer-by public to begin discussion, taking action, and solving problems. Such spontaneous interaction combined with a sense of inspiration can lead to innovation and collaboration in a wide expanse of life and work, in planning for the built environment, rethinking the way we make things, and simply questioning why and how we do what we do. Addressing American dynamics and the methods, motivations, and culture of business are

---

[37] Revkin 2008.
[38] Arthus-Bertrand 1999.

critical to positive change. Incidentally, the ability to travel the exhibit across the U.S. is largely dependent upon corporate sponsorship for feasibility, and thus, awareness.

We experienced how very compartmentalized American society can be - most significantly, the separation between business and environment. We learned that corporate funding was largely withheld because companies were unwilling to support a "green" mission for fear their customers would scrutinize or hold their practices accountable. An additional barrier was finding public space, in New York City particularly, that was still public. As a result of declining governmental support for acquisition and maintenance of public land, private corporate and public-private ventures are acquiring public spaces; and those remaining in the public realm are often struggling with having enough funds to maintain well-loved, and often over-used, landscapes with hesitation to host events that may cost even more in maintenance once the event is over to recover. Though new park management ventures bring vitality of activity and programming to city spaces, a turn from underutilization and dangerous urban spaces resulting from lack of public funds and management, these spaces once having a primary purpose of meditative refuge for city dwellers, now have a mission as revenue source with limited hours and restricted uses. Consequently, these highly visible spaces that are conducive to far-reaching awareness, are potentially off limits because the usage fees prove prohibitive.

Needed are strategies for evolutionary partnership in support of interconnectedness of business, culture, community, and environment. The capacity of American society for organization, diligence, and innovation can be leveraged in the re-structuring of business initiatives, especially as public-private partnerships emerge, perhaps a bridge for private and corporate business to follow. In public development, art programs serve as models for not only more strategies in the public realm, but for building frameworks in the public-private and private domains. As global citizens and neighbors, it is time to hold business and government accountable. One way we can do this is by speaking out at every chance we have.

This takes a commitment to change and a belief that everyone can make a difference. The changing tide that is amassing critical awareness for the purpose and need for collaboration and nurturing spirit as the current tools for survival has grown since our project and can be utilized. This is a portal to the arts or any creative act, beauty, and love having their rightful place in a healthy society. Thus, despite the postponement of the exhibit, our belief in increased exposure to truth and beauty as opportunity to expand individual and collective consciousness as a catalyst for potential positive change is reinforced. However, new, enhanced, or resurrected strategies and frameworks need to be in place as mechanism for intentional efforts to manifest. Firstly, developing professional-academic, community design collaborations such as Emerge Studio, need to be recognized by academic institutions as scholarly work that has integrity and credibility as it not only leads to expansion of knowledge to a larger audience than journal publication alone, but involves students in inspirational experiential learning, promotes community-centered design, develops applied innovative solutions, and directly works to address some of the world's most pressing issues. Another strategy is for professional academics and practitioners to develop ways of enforcing, in their contracts, ethical responsibility for all involved stakeholders in scope of services as outlined by professional organization's code of ethics; and, reciprocally, for professional organizations to provide universal guidelines for ethical practice beyond a seemingly incidental code. Perhaps this serves as a revisit and evolution of the settlement covenants maintaining a valued and shared worldview as a guide. Consequently, we developed collaborative networks based on a shared vision and learned how essential these networks are to build anything of value. As Bert Mulder in *Between Grace and Fear: The Role of the Arts in a Time of Change* underscores, "You need networks of cooperating organizations that are not competing for territory. A creation society requires a special kind of network."[39] There needs to be mutual agreement, giving and receiving, and commitment to values of well-being and abundance where monetary abundance is only one indicator amongst a system of both qualitative and quantitative criteria. Only this way, can we even begin to consider a constitutional amendment that gives other living communities, besides human, a voice, and thus subsequent funding that public lands, spaces, and projects need.

It has become clear with time, that our project attempt was not a failure, but cause for a deeper understanding of how things work, an increased clarity of interconnected or systems thinking, and evident inspiration that such efforts engender – most notably in student participants - as we evolve and expand our life-work where nature, art and the built environment unite. If and when Earth From Above tours the U.S., it will be evidence that, in the words of Bell Hooks, "We create and sustain environments where we can come back to ourselves, where we can return home, stand on solid ground, and be a true witness."[40] The more we make a place for nature-centered creativity in society, the more opportunity to bear witness.

References

Arthus-Bertrand, Yann. n.d. Yann Arthus-Bertrand: Earth From Above. *Ecorevolution.IT 2.0 Website.* Available at
    http://www.ecorevolution.it/ita/pho/yab.html, accessed April 25, 2014.
Arthus-Bertrand, Yann. 1997. Farming near Pullman, Washington, United States. © Yann Arthus-Bertrand / Altitude.
Arthus-Bertrand, Yann. 1999. Volcanoes on the Galápagos Archipelago, Ecuador (0°20' S, 90°35' W) © Yann Arthus-Bertrand / Altitude.
Arthus-Bertrand, Yann. 2003. Highlands Ranch, Outskirts of Denver, Colorado, United States. © Yann Arthus-Bertrand / Altitude.
Arthus-Bertrand, Yann. 2005. *Earth From Above, Third Edition.* New York: Harry N. Abrams, Inc.
Arthus-Bertrand, Yann. 2009. Home. *National Geographic.* Available at
    http://channel.nationalgeographic.com/channel/videos/home-adventure/,
    accessed March 26, 2014.

---

[39] Mulder 2010, 354.
[40] Hooks 2009, 120.

Arthus-Bertrand, Yann. 2012. Biography. *Yann Arthus-Bertrand Official Website.* Available at http://www.yannarthusbertrand.org/en/biography, accessed April 25, 2014.

Arthus-Bertrand, Yann. 2014. Photos of Home: Highlands Ranch. *La Terre vue du Ciel – Earth From Above Website.* Available at http://www.yannarthusbertrand2.org/index.php?option=com_datsogaller y&Itemid=2&func=detail&catid=37&id=950&lang=en&l=1680, accessed April 25, 2014.

Butzer, Karl. 1990. The Indian Legacy in America. In *The Making of the American Landscape,* edited by Michael P. Conzen. London: Harper Collins.

Christo and Jeanne-Claude. 1979-2005. *The Gates.* Available at http://www.christojeanneclaude.net/projects/the-gates?view=info, accessed March 26, 2014.

Cole, Thomas. 1836-1837. *View on the Catskill-Early Autumn.* The Metropolitan Museum of Art, ARTstor Digital Library.

Dictionary.com. 2014. Wilderness. Definition available at http://dictionary.reference.com/browse/wilderness, accessed March 26, 2014.

Emerge Studio. 2007-2010. Picture Earth, Inc. Available at http://pictureearth.org/about_us/emerge_studio.html, accessed March 26, 2014.

Farrer, Henry (American). 1843-1903. Untitled (landscape), ca. late 19th century Watercolor 22.8 x 28.8 cm (image). The Fine Arts Museums of San Francisco, gift of Davis and Langdale Company, 1981.2.25.

Gatti, Ronald. 1969-1989. The Radburn Association. Available at http://www.radburn.org/, accessed March 26, 2014.

Gifford, Sanford Robinson (1823-1880). *Hunter Mountain, Twilight,* 1866. Oil on canvas, 30 5/8 x 54 1/8in. (77.8 x 137.5cm). Daniel J. Terra Collection, 1999.57. Terra Foundation for American Art, Chicago, IL, U.S.A.

Gruchow, Paul. 1988. *The Necessity of Empty Places.* New York: St. Martin's Press.

Hamer, David. 1990. *New Towns in the New World: Images and Perceptions of the Nineteenth-Century Urban Frontier.* New York: Columbia University Press.

Heizer, Michael. 1968. *Dissipate Black Rock Desert,* Nevada, *#8 of Nine Nevada Depressions.* University of California, San Diego, ARTstor Digital Library.

Hooks, Bell. 2009. *Belonging: A Culture of Place.* New York: Routledge.

Howard, Ebenezer. 1946. *Garden Cities of To-Morrow* (London, 1902), reprinted and edited with a Preface by F. J. Osborn and an Introductory Essay by Lewis Mumford. London: Faber and Faber.

Kepes, Gyorgy. 1972. Art & Ecological Consciousness. In *Art of the Environment,* edited by Gyorgy Kepes. New York: George Braziller, Inc.

Kunstler, James Howard. 1993. *The Geography of Nowhere.* New York: Simon & Schuster.

Lingeman, Richard. 1980. *Small Town America.* New York: G. P. Putnam's Sons.

Long, Richard. n.d. Index. *Richard Long Official Website.* Available at http://www.richardlong.org/index.html, accessed April 25, 2014.

McDonough Braungart Design Chemistry (MBDC). 2013. *Beyond Sustainability: Cradle to Cradle Science, Innovation + Leadership.* McDonough Braungart Design Chemistry LLC.

Mulder, Bert. 2010. The Architecture of Inspiration. In *Between Grace and Fear: The Role of the Arts in a Time of Change,* edited by William Cleveland and Patricia Shifferd, 343-354. Champaign: Common Ground Publishing.

National Park Service. 1916. US Department of the Interior. National Park Service Organic Act of 1916. Available at http://www.nps.gov/grba/parkmgmt/organic-act-of-1916.htm, accessed March 26, 2014.

National Park Service. 2014. US Department of the Interior. *Ansel Adams.* Available at http://www.nps.gov/yose/historyculture/ansel-adams.htm, accessed March 26, 2014.

Picture Earth, Inc. 2007-2010. Available at http://pictureearth.org/, accessed March 26, 2014.

Revkin, Andrew C. 2008. Ecuador Constitution Grants Rights to Nature. *New York Times*, September 29, 2008. Available at http://dotearth.blogs.nytimes.com/2008/09/29/ecuador-constitution-grants-nature-rights/, accessed March 26, 2014.

Shaffer, Diana. 1983. Nancy Holt: Spaces for Reflections or Projections. In *Art in the Land: A Critical Anthology of Environmental Art*, edited by Alan Sonfist, 169-177. New York: E.P. Dutton.

Smith, Page. 1966. *A City Upon a Hill.* Cambridge: MIT Press.

Stilgoe, John. 1982. *Common Landscapes of America 1580-1845.* New Haven: Yale University Press.

US Department of Transportation. 1956. Federal Highway Administration. Dwight D. Eisenhower National System of Interstate and Defense Highways. Available at http://www.fhwa.dot.gov/programadmin/interstate.cfm, accessed March 26, 2014.

Weilacher, Udo. 1996. *Between Landscape Architecture and Land Art.* Berlin: Birkhauser – Verlag fur Architektur.

Chapter 14: Sculpting Sustainability: Art's Interaction with Ecology

Jade Wildy

The Industrial Revolution, which began in the late 1800s in the western world, could be viewed as the beginning of the downward spiral in the health of the natural environment. While many argue over the root cause, be it the result of capitalism, democracy, culture, or the needs of a rapidly growing world population, there is extensive devastation inflicted upon the natural environment each day. Recognition of this fact is evidenced by the growing societal emphasis on green alternatives for energy and in the interest in environmental disasters from popular culture including Al Gore's award winning documentary, *An Inconvenient Truth,* produced in 2006,[1] and in disaster movies like *The Day After Tomorrow,* in 2004.[2]

These movies demonstrate that through entertainment mediums attention can be drawn to the health of the environment to encourage individuals to make alterations in their own lifestyle and become proactive in broader societal change. However, there is a growing field of creative, expressive visual art forms that highlight environmental concerns and seek to actively remedy some of the environmental destruction that has occurred over the years. These expressive forms have arisen from substantial changes to what is considered art, and has culminated in various artists actively working to promote change, several of whom are presented in this paper.

Art that is about ecological and environmental change has arisen from several changes in art history. Historically, art forms were expected to conform to a particular style, and students were preoccupied with repetition of a master's style and technique, but gradually apprentice artists began to experiment on their own. Most people are familiar with the beautiful paintings by artists such as Vincent Van Gogh with his brightly colored expressive paintings; Claude Monet, with his softly colored water lilies; or the oddly angled figures of Pablo Picasso's Cubist paintings. However, what most don't know is that these artists, who began

---

[1] *The Day After Tomorrow* 2004.
[2] *An Inconvenient Truth* 2006.

playing with form and color in art, began a change in what was considered acceptable as "art."

In the 1950s, art underwent a further drastic change during a conceptual revolution. Artists began to experiment with what an artwork meant or represented; the concept behind the work which, in many cases, began to be more important that the work itself. This lack of emphasis on the final art object, combined with other expressive theatrical art forms, developed into new art form that incorporated performative elements or actions into the art practice, which came to be known as performance art.[3] Alan Kaprow, who termed this new art form "Happenings," explains that:

> ...once foreign matter was introduced into the picture in the form of paper [in reference to collage, introduced by Picasso and Braque], it was only a matter of time before everything else foreign to paint and canvas would be allowed to get into the creative act, including real space.[4]

An idea that was embraced by various artists, a "Happening" can be defined as "a work of art involving the interaction of people and things in a given setting or situation."[5] In these works there is emphasis on interactivity either physically or by engaging the viewer to cerebrally connect with the artwork. It is essentially an intimate performance that may not necessarily result in a final product because the *art* is in the *action* not the final object. Art began to address social themes and become more active in promoting societal change. These changes in the history of art: the progressive change in the forms of the object, the inclusion of concept and the lack of emphasis on a final art object essentially set the scene for environmental art. This evolution of art continued during the 1960s and began to include non-tradition art forms and materials, like carving into the Earth or creating vast spirals of rock.

Environmental Art

The words "environmental art" are generally used as an umbrella term to describe an artistic process or artwork where the artist actively engages with the environment. It covers a widely diverse range of interactions, styles, approaches, methodologies and philosophies. The term "environmental art" does not exclusively refer to ecologically conscious artworks as it is also used to describe artworks made in an urban environment in addition to the natural landscape. However, it is more frequently used to refer to natural environments.

The environmental art genre began in the late-1960s. 1968 is marked as the birth year,[6] as this was the year of the landmark exhibition *EARTHWORKS* [sic], held by the Virginia Dwan Gallery in New York, featuring ten prominent environmental artists: Carl Andre, Herbert Bayer, Walter De Maria, Michael

---

[3] Higgins 1976, 268.

[4] Kaprow, 1960, 165-166.

[5] Stangos 1994, 233.

[6] 1968 is recognized as the date of the first prominent exhibition, however several environmental artists were practicing far earlier.

Heizer, Stephen Kaltenbach, Sol LeWitt, Robert Morris, Claes Oldenburg, Dennis Oppenheim and Robert Smithson.[7] Described as "Revel[ing] in their geophilia,"[8] these artists responded to the concept of the exhibition by linking the gallery with the land in various ways, showing a combination of both documentary photographs of works in the environment, sculptures and installations in the gallery.[9] The following year in 1969, an exhibition called *Earth Art* was held at the Andrew Dickson White Museum of Art in Ithaca, New York, and Michael Heizer held a solo exhibition at Virginia Dwan Gallery. These ground breaking exhibitions paved the way for a new art genre to develop and evolve.

Environmental artworks can be both small and intimate or on a grand scale spanning several kilometers. There is no specific stylistic approach and it is not limited to a specific group of artists or region; rather it is a broad genre. There are however several distinctive sub-genres of environmental-based art with which artists can be identified. The most relevant to this paper concerns the natural environment and environmentalism.

A growing trend in environmental art is the form known as "Ecovention;" a contraction of "Ecology" and "Invention," encompassing artworks that repair damage to a natural environment.[10] It can also be known as reclamation art or eco-art.[11] These forms of works are more environmentalist in motivation and represent a growing sub-genre of environmental art, that actively engages with the body of knowledge on environmental concerns.

The term "Ecovention" was first coined by Sue Spaid in her 2002 book of the same name, and describes an artist-initiated project that employs an inventive strategy to physically transform a local ecology.[12] Also referred to as restoration art, the conceptual basis for the artistic process is specifically to repair damage done to the environment. Ecovention has also been historically described as "Landscape Reclamation,"[13] typically referring to industrial sites that were used for artistic purposes, a concept notably embraced in the 1970s with several initiatives and exhibitions being held.[14] The idea of Ecovention presents itself both as large landscaping projects, which drastically transform a local ecology, as well as smaller projects that remedy a particular situation, like habitat loss in a specific area. As a sub-genre of environmental art, these pro-active forms could be referred to as environmentalist art.

Environmentalist Artists

Most relevant to today's environmental concerns are the environmentally motivated sculptures that combine aesthetics and ecological remediation. While there are many artists whose practice falls under this genre there are several artists

---

[7] Boettger 2004, 55-63.
[8] Glueck 1968, 38.
[9] Glueck 1968, 38.
[10] Spaid 2002, 1.
[11] Tufnell 2006, 96.
[12] Spaid 2002, 1.
[13] Tufnell 2006, 96.
[14] Beardsley 1998, 103.

that are prominent in the field and who highlight the restorative interactions between artist and environment.

Jackie Brookner

Jackie Brookner is an environmental artist who began her practice in the 1990s, following in the footsteps of the early big name artists like Robert Smithson and Andy Goldsworthy. In the tradition of the expanding art genre, Brookner collaborates with ecologists to design works that function ecologically as well as aesthetically, representing a conceptual merger between aesthetics, design and ecological function. Essentially, her sculptures represent a merger between science and art to form aesthetic sculptures with a remedial function.

Water is one of the most precious resources on this planet, yet it is subjected to vast pollution and waste. Brookner's biosculptures are a notable example of remedial, small-scale installation and gallery-based works. These works actively cleanse polluted water using moss and porous cement, like her 1995 work, *Prima Linga,* which involved the sculpting of a large cement tongue, on which moss was encouraged to grow.

In *Prima Linga,* mist sprays polluted water over the moss that covers the sculpture, allowing the water to be partially absorbed and run down and be collected in a small pool at the bottom, containing fish and aquatic plants.[15] In effect, the tongue licks the water clean enabling the fish to live in the purified water forming a tiny functioning ecosystem. This ecological aestheticism is mirrored in several of Brookner's works, including *I'm You,* 2000 and *The Gift of Water,* 2001, producing a body of work that is both conceptual aestheticism and ecologically functional.[16]

Integral to these works are the natural processes that enable them to function, and demonstrate the remedial power that art can possess. However, conceptually these works can metaphorically express renewal in the cleansing of the water and the continuation of life in the organisms that survive in the pool.

Patricia Johanson

Patricia Johanson's work also has a remedial yet aesthetic function. However, her practice differs substantially from Brookner's through the sheer size of her works. Conceptually, through her large-scale, remedial works, Johanson seeks to "transform sites to make us aware that we are citizens, not masters of the biosphere..." through carefully designed symbiosis between butterflies, birds and other creatures.

In the 1960s, while studying at Bennington College, she began writing about designing the world as a work of art, suggesting "total environmental design – aesthetic, ecological, psychological and social – such that the person would be placed *inside* the work of art..." [17] Thus, the artist functions as creative intelligence rather than isolated idealist. Essentially, Johanson sought to make

---

[15] Brookner 2013.

[16] Orenstein 2003, 108.

[17] Blum 1989, 337.

works that people could engage with and walk through that had ecological functionality. This culminated in her practice of producing, large landscaped gardens and sculpture parks that could be enjoyed, by the public while also remedying the environmental health of the location.

This merger between the aesthetic art object and ecological restoration is reflected in Johanson's works like *Fair Park Lagoon,* built in 1981. This work is built around Leonhardt Lagoon, (part of the park complex of the Dallas Museum of Natural History), which had become an algal bloom infested, eutrophicated eyesore due to the many years of fertilizer that had washed into the water from the surrounding lawns, killing off the native water life.[18] The lagoon had also gained a negative reputation due to the drowning death of a child some years prior.[19] Johanson was commissioned to design the parks and sculptures surrounding the lagoon, to ecologically transform the area and to restore the waters of the lagoon itself.

As the surrounding area became healthier and as the fertilizer was no longer washing into the water, the lagoon itself became healthier and various bird and animal species began to return to the site. Thus, through utilizing strong design elements that incorporated several varieties of native flora and fauna, combined with meandering paths and sculptures, the dangerous polluted waters were restored to an enjoyable, biologically diverse park featuring a vibrant lagoon.[20]

Works and projects such as *Fair Park Lagoon* demonstrate Johanson's belief that, through art, environmental devastation can be healed.[21] Through the production of sculptural gardens that have aesthetic elements that are engaging to the viewer while reviving the surrounding ecosystem, Johanson's works bring ecology into the surrounding culture. She establishes an interactivity between humanity and ecology, whereby art can engage with society through an ecologically sound sculpture park, to bring about a change in society's ecological thought. As with Brookner's biosculptures, projects like Johanson's *Fair Park Lagoon* marry enjoyable aesthetics with environmental restoration that have resulted in improved water, and biodiversity.

Lynne Hull

Another form of environmental remediation focuses not on the landscape, but on the wildlife that have suffered due to human interventions. Lynne Hull began her artistic career as a potter, with the functionality of useable objects in mind. This functionality evolved when she began to use recycled, fabricated materials that were compatible with nature in order to benefit wildlife. She refers to this process as a form of "eco-atonement for human encroachment."[22] This sense of remediation or "eco-atonement" is an increasing trend amongst artists (as demonstrated by Johanson's work). While in wider art sculptures are generally considered from an aesthetic viewpoint for the human viewer's benefit Hull's

---

[18] Spaid 2002, 65-67.
[19] Kelley 2006, 3.
[20] Spaid 2002, 65-67.
[21] Kelley 2006, 3.
[22] Warshall 2000, 93.

practice is set apart from her sculptural counterparts as her works are made directly for the animal kingdom.

With a similar instigation of human produced environmental devastation that results in loss of habitat, seen in many environmental artists practice, Hull's artworks provide safe havens and habitats for wildlife that have lost their natural habitat, through a sculptural form she refers to as "trans-species art."[23] Comprised of scrap, driftwood and other materials, Hull builds trees and rafts from scavenged and recycled materials that animals can safely live and nest in.[24] These are produced in consultation with biologists and zoologists, with the aim of nurturing endangered species,[25] and to encourage the wider community to consider the void left by human environmental degradation.[26] For example, the *Raptor Roosts* she began producing in the 1980s were made in response to the number of large predatory birds that were dying, as urbanization had removed the large trees they nested in, which resulted in these birds trying to use the elevated power poles for nesting. Hull's answer was large, tree-like sculptures, that tower above the ground providing a high, safe vantage point with nest friendly plateaus.

While still sculptural, Hull's works are not aesthetic in the same sense as Brookner's or Johanson's, as they are primarily for the animals to fulfil their living needs, rather than human's visual aesthetic needs. Perhaps they could be argued as prescribing to a different set of aesthetics: one of animal preference for a safe, good-looking tree in which to nest.

Agnes Denes

The environmental problem of deforestation has also been addressed by several other artists including Agnes Denes, though from a difference conceptual angle. While Denes' most prominent work was *Wheatfield: A Confrontation,* created in Lower Manhattan in 1982, which addressed ideas like consumerism and food security, a much later work, *Tree Mountain,* addressed the idea of deforestation and the environment on a multi-generational grand scale.

Denes' practice began in 1964 when she attended Columbia University and the New School of Art. Her early practice began with a movement away from the traditional art forms of painting and drawing and she began to make extensive investigations into the ideas surrounding triangulation, with her research crossing over into science. This research substantially influenced her artistic practice and as a result, Denes could be considered as one of the leading artists who began incorporating scientific concepts into their art. This has become substantially important for environmental art, particularly works that are ecologically motivated.[27]

Tree Mountain is a constructed mountain measuring 420 meters long, 270 meters wide and 28 meters high in an elliptical shape, in the gravel pits at Pinziö, Ylöjärvi, Finland,[28] as part of a generational reclamation project. Conceptually

---

[23] Fullerton 2009.

[24] Hull 2008.

[25] Orenstein 2003, 107.

[26] Song 2009, 4-13.

[27] McEvilley 2004, 158–160.

[28] Kastner and Wallis 1998, 161.

Tree Mountain crosses several boundaries. 100,000 trees were planted in an intricate pattern "...derived from a combination of the golden section and sunflower/pineapple pattern designed by Denes."[29] Each tree has an average lifespan of 400 years allowing the trees to be passed down through the generations to the descendants of the original person invited to plant a tree.[30]

An impressive visual image, the mountain is also representative of nature's chaos and vitality, as the geometric patterning will eventually dissolve with the dropping of seeds and the growth of new saplings that will take the legacy and message of *Tree Mountain* far beyond the lifespan of the original trees and owners. Thus, while this work may not "reforest" the Earth it does provide a long-term conceptual sign, pointing to the environment and reforestation.

The triangulation that Denes has extensively researched underpins this work, which merges science, art, and ecology. *Tree Mountain* was produced on a site that had been heavily affected by mining, in an effort to not only raise awareness but also to restore a degraded site.[31] This effectively connects present generations with future generations. The trees themselves are also significant as Denes explains:

> Trees see so much history, they sway and whisper, hibernate and turn to blossoms. My forests are not landscaping where a tree is put there for contrast among paths and bushes to decorate a park or garden. A serious forest means business, not cutting business or profit, but demanding attention, respect, awe if beautiful and mysterious.[32]

This work is reflective of aesthetic sculpture on a massive scale that combines the constructed elements of a carefully planned artwork with the natural beauty of the landscape in the eventual organic growth. Essentially, with the choice of this location: to reclaim the gravel pits, *Tree Mountain* transcends several of the Ecoventionist forms in landscape reclamation, biosculpture and has an element of generational social comment.

Alan Sonfist

Where *Tree Mountain* is forward looking to multiple future generations, Alan Sonfist's *Time Landscape* is comparatively retrospective, yet still points to the environment and forest clearing. This work started in 1965 and reflects the pre-colonial history of Manhattan, before the city was built. Through the planting of trees and shrubs from Manhattan's pre-colonized state, Sonfist created an environmental artwork that is a combination of a botanical museum, a recreational park for people to enjoy and an ecological reminder of human intervention and domination over the landscape.

A significant element of *Time Landscape* was in the preparation, as Sonfist, like Johanson, researched the natural history of the area, establishing the native

---

[29] Denes 1995.
[30] Denes 2010.
[31] Brady 2007, 296.
[32] Denes 2010.

plants that had originally grown there to produce a park-like area, which pays homage to the past rather than a standard monument or statue. This work also has personal resonance for Sonfist as it reminds him of the forests he played in as a child. Thus, *Time Landscape* has a duel meaning: one referencing an earlier time in Manhattan's history, the other an earlier time in Sonfist's (and perhaps others) personal history.[33]

These two works, *Tree Mountain* and *Time Landscape,* which focus on forests, demonstrate a duality of environmental education. *Tree Mountain* looks forward to future generations and their responsibility for the environment of the future, while *Time Landscape* points back to a pre-history and the environment that existed before.

Helen Mayer and Newton Harrison

Similar to *Time Landscape* in the use of planting, but with a different conceptual focus, Helen Mayer and Newton Harrison produce works that reclaim the landscape through a process of remediation. For the Harrisons, the effect of their practice on the natural environment is of the upmost importance, rather than aesthetic considerations, which is in many ways similar to Hull's "trans-species art" for the animals. Their practice is for the *environment,* rather than *people.*

The Harrisons represent one of the earliest practices of landscape restoration as an artistic practice. Their early practice involved the production of many gallery-based works, of which *Portable Orchid* (1972-73) from the *Survival Piece Series* is a striking example. These works parallel their conceptual concerns through the production of an indoor installation, involving "twelve four foot in diameter hexagonal redwood boxes, three feet deep, planted with assorted citrus trees, topped by hexagonal redwood light boxes,"[34] that were produced in reaction to the industrial development in Orange Country that was threatening the food orchards.[35]

Works such as this represent their early concern with the environment, which later culminated in their large-scale restorative works. These works involve an extensive consultation process in which the Harrisons' confer with a variety of people including journalists, mayors, public officials, artists and farmers to "discover an appropriate solution that optimizes twin components: biodiversity, which depends upon the continuity and connectivity of living organisms; and cultural diversity, which requires framing and distinction between communities."[36]

An example of the Harrisons' long-term reclamation works is their early-1977 project, *Spoils Pile*, which involved diverting approximately 3,000 truckloads of earth and organic material, including mulch of leaves and grass, to transform approximately 20 acres from clay and rock to a viable meadow with trees and berry patches. The site was formally part of a "spoils pile" from a quarry filled with debris generated by the construction of the Niagara Power Plant. The introduction of organic material revitalized the soil and produced

---

[33] Sonfist 2004, 68.

[34] Converted measurements are 1.2 meters in diameter by 0.91 meters tall.

[35] Harrison and Harrison 2010a.

[36] Spaid 2002, 32.

fertile growing foundations that started a process that continued to evolve and grow, eventually giving life to a vibrant flowering meadow. This process is similar in concept to today's popular "no-dig" gardens, which involves laying heaped organic material like straw over poor soil (or even no soil) and top dressing it with viable soil, like compost, that plants can grow in. It adds fertile soil over infertile material to allow organic growth and doesn't require digging in to the garden.

A key element in the Harrison's practice is that they are collaborating with and for nature, rather than benefitting an individual or company. As they state:

> Our work begins when we perceive an anomaly in the environment that is the result of opposing beliefs or contradictory metaphors. Moments when reality no longer appears seamless and the cost of belief has become outrageous offer the opportunity to create new spaces – first in the mind and thereafter in everyday life.[37]

Their works function for the environment rather than human's desire for using the environment to suit their own needs, which also has the beneficial element of communicating environmental consciousness to the general community. While works such as *Spoils Pile* may not have the same level of design and aesthetics as found in the landscape reclamation projects of Johanson or Denes, the natural beauty, that has continued to grow since the project was started, is the main strength of these works.

Joseph Beuys

Contrasting works that change and remedy the environment on a massive scale are those that seek to transform the way people interact with the environment on the individual level. Raising awareness and changing minds is one of the key elements needed for positive environmental improvement, and some artists see art's role as functioning to improve the condition of society for the benefit of both the planet and humanity as a joint, mutually beneficial goal. This is significantly present in the environmental works of Joseph Beuys who, aside from being a progressive performance artist, was instrumental in establishing the German Green Party (Die Grünen) in Germany in 1980.[38]

Beuys referred to his practice as "social sculpture," in which he viewed every person as an artist (by which he meant a creative person) and every artist working towards the goal of positive change for society. His work, *7,000 Oaks (7,000 Eichen)* exemplifies his belief in the role of art, societal remediation and environmentalism as a combined concept. Completed in 1987 for *Documenta 8* in Kassel, it involved the installation of 7,000 oak trees with 7,000 companion pillars of basalt along an urban street. The trees represented the environment, life, growth and renewal, while the pillars represented industrialism, strength and the built environment. Combined, they represented a unity of urban and natural environments.

---

[37] Harrison and Harrison 2010b.
[38] Adams 1992, 26.

These works were culturally inclusive and combined environmentalism with cultural reclamation and inclusion. The conceptual project based on Beuys' *7,000 Oaks* has been continued at other sites including New York City in 1988 (extended in 1996). Beuys originally intended *7,000 Oaks* to become a worldwide phenomenon "...as part of a global mission to effect environmental and social change" as well as a local event that was "a gesture towards urban renewal."[39] Through works such as these, Beuys intended an integrated process of environmental, urban and social renewal.

Hans Haacke

Works like *7,000 Oaks* bear similarity with other works that crossed boundaries between art and other fields. While combining art with the sciences or humanities has become increasingly common in contemporary art, in the 1950s and 60s there were very few artists that addressed non-art concerns. However, like Beuys, Hans Haacke also began experimenting with artworks that brought an environmental message to society and bridged the gap between the art gallery and the environment.

While Haacke is better known for his politically motivated art from 1970 onwards, Haacke's early installations also included works that are aligned with the concepts of environmental art. Conceptually, Haacke's works used non-verbal language to convey or gather information to broach complex, non-visual issues that were related to time and change.[40] Rather than dictating a message, Haacke's works alluded to a concept or idea that required the viewer to think about the message, and essentially strived to produce art that existed and developed in "real-time." Haacke also developed a concept that he referred to as "systems art," which relates to the interconnected nature of various elements.[41]

The elements of Haacke's "systems art" that could be more prominently linked with environmental art, included what he referred to as "physical systems," where the work actively interacts with environmental conditions in works such as *Condensation Cube* (1963-1965), where the atmospheric conditions and ambient temperature caused condensation to form in a plexiglass cube. Essentially, this work became interactive and responsive to the changing environmental conditions within the gallery space including the viewers, whose body heat had the potential to raise the temperature. (The role of the viewer in the gallery became a focus in Haacke's later political art.) These physical systems later linked into "biological systems" which involved various works like *Grass Cube* (1967) and *Grass Grows* (1969), which centered on natural growth and change and involved allowing grass seeds to germinate and grow from a small mound of soil in the exhibition venue.[42] Of these works, Haacke stated:

> [it is to]...make something which experiences, reacts to its environment, changes, is non-stable ... make something sensitive to light and

---

[39] Beuys 2014.
[40] Stangos 1994, 265.
[41] Skrebowski 2008, 54-83.
[42] Stangos 1994, 42.

temperature changes, that is subject to air currents and depends, in its functioning, on the forces of gravity...articulate something natural.[43]

These were extended into biological works that were ecological in focus including *Chickens Hatching* (1969), *Transplanted Moss Supported in an Artificial Climate* (1970), *Directed Growth* (1970-1972), and *Rhine Water Purification Plant* (1972). These works directly focused on the interactions between living plants and animals and the external environment.[44]

Setting historical precedent for contemporary works, early works like these established the ecological and conceptual role that art can play in positive environmental change. Viewable as a conceptual precursor to practices like Brookner's, *Rhine River Purification Plant* involved the purification of water, gathered from the Rhine River, using a system of pools and filtration units in the gallery to reveal and then cleanse the pollutants from the water, to a condition that would allow fish to live in the water. This installation created a physical link between the gallery space and a degraded natural environment and had a positive impact for environmental change, both through the physical cleansing of the water and through the raising of awareness that there was a problem with the water.

Conclusion

There is an increasing interest in environmentalism and maintaining the health of the planet. Through a merger between art and environmental social activism, positive change can be instigated to improve the planet for future generations. This chapter has demonstrated how artists can proactively interact with the natural environment in a wide variety of different ways, and with a multiple of different conceptual angles and practices. However, it is through those works that actively engage with the environment and promote positive change that the role art can play in sculpting a more positive, cleaner, ecologically conscious future is most evident.

Art has evolved from traditional art forms to encompass different outlets of expression, concepts and practices. This evolution has allowed new forms of art to exist like environmental and ecologically engaged art, which can have considerable impact on both the viewer and their attitudes towards nature as well as directly on the natural environment. This chapter demonstrates that there are a wide variety of artworks addressing various figments of environmental damage, using a variety of approaches.

It is increasingly common for art to cross the boundary into non-art territory with artists like Jackie Brookner, Agnes Denes and Lynne Hull researching the science behind the concepts in their work. All of these artists have conducted extensive research into the ways of reaching their individual environmental goals. Brookner's "biosculptures" are based on scientific understandings about water purity and the ways in which plants can be used as natural, non-technological purifiers. Denes' *Tree Mountain* employs mathematical principles to demonstrate

---

[43] Kastner 1998, 32.
[44] Skrebowski 2008, 54-83.

the gradual change of a forest from order to natural disorder as well as providing a multi-generational message about reforestation. Hull's "trans-species art" is underpinned by knowledge about the habitat requirements of animals. These artists demonstrate how art can actively employ the use of biologists, ecologists and other science-based fields in order to allow artworks to have both an aesthetic and remedial function in repairing environmental damage.

Through art produced by artists with environmental vision, artworks can contribute to positive environmental change. As the Harrisons' landscape reclamation projects demonstrate, some artworks can be direct and active. The Harrisons' projects involve them assessing a degraded site due to processes such as mining or landfill, and making changes to it, like bringing in soil and replanting the area effectively reclaiming the damaged site. In contrast, other artists address environmental issues passively through presenting a changed environment, either for the future like *Tree Mountain* or the past like *Time Landscape.* Both of these projects involved planting: *Tree Mountain* utilized trees that have an average lifespan of 400 years allowing it to last through multiple generations carrying the message of reforestation. *Time Landscape* recalled a past, pre-settlement landscape to remind people of the untouched natural environment.

Where most environmental projects act directly on the environment of a specific area of concern, this chapter has also highlighted artists that have focused directly on humanity and their role in a positive environmental future. Joseph Beuys saw art as a figment of wider community and social change as with his project *7000 Oaks*. This work not only carried an environmental message through the planting of trees, but also one of society's strength and resilience, through the accompanying basalt pillar. The role of viewer was also a focus of Hans Haacke's early "systems art" and the interaction between artwork, gallery and the degraded environment highlighted in projects like *Rhine Water Purification Plant.* This work demonstrates some of the early interest in ecological systems within the gallery that artists were beginning to experiment with.

Artworks that have been both directly and indirectly inspired by these experimentations with ecology and repair of damaged environments are continuing to be produced in increasing numbers. With the aid of technology that is growing in refinement and ability, artworks can attain greater levels of sophistication as well as be produced on the foundations of a more in-depth knowledge of our planet and ecosystems. Ergo, artists are now able to use a much greater variety of technology and scientific understanding to produce works that are proactive in attaining a healthy natural environment.

Ecological aestheticism, which combines visually striking artistic practice with environmental remediation, provides art with a functional springboard to promote a brighter, more environmentally friendly future. The productions of such works, like those highlighted in this chapter have the benefit of not only immediately improving areas of environmental concern, but of encouraging new generations to improve and promote environmental conscience. It is through the practices of these artists that sustainability can be sculpted.

References

Adams, David. 1992. Joseph Beuys: Pioneer of a Radical Ecology. *Art Journal* 51 (2): 26.

An Inconvenient Truth. 2006. *The Internet Movie Database.* Available at http://www.imdb.com/title/tt0497116/, accessed March 27, 2014.

Beardsley, John. 1998. *Earthworks and Beyond: Contemporary Art in the Landscape*, 3rd ed. New York: Abbeville Press.

Beuys, Joseph. 2014. Joseph Beuys: 7000 Oaks *Dia Art Foundation.* Available at http://www.diacenter.org/sites/page/51/1364, accessed March 27, 2014.

Blum, Andrea.1989. From the Other Side: Public Artists on Public Art. *Art Journal* 48 (4): 336-346.

Boettger, Suzaan. 2004. Behind the Earth Movers. *Art in America* 92 (4): 55-63.

Brady, Emily. 2007. Aesthetic Regard for Nature in Environmental and Land Art. *Ethics, Place & Environment* 10 (3): 287-300.

Brookner, Jackie. 2013. Biosculptures. Artist. *JackieBrookner.com*, http://jackiebrookner.com/project/biosculptures/ accessed April 8, 2014.

Denes, Agnes. 1995. Agnes Denes: Tree Mountain - A Living Time Capsule. *Arts and Science Collaborations, Inc.* Available at http://www.asci.org/news/featured/denes/denes.html, accessed March 27, 2014.

Denes, Agnes. 1996. *Tree Mountain - A Living Time Capsule.* 11,000 Trees, 420x270x38 meters, Ylöjärvi, Finland 1992-1996.

Denes, Agnes. 2010. What It Means to Plant a Forest. *Greenmuseum.org.* Available at http://greenmuseum.org/content/artist_content/ct_id-198__artist_id-63.html, accessed March 27, 2014.

Fullerton, Anne. 2009. Eco-art. *Reportage-enviro.com.* Press Release, August 5. Available http://www.reportage-enviro.com/2009/08/eco-art/, accessed March 27, 2014.

Glueck, Grace. 1968. Moving Mother Earth. *The New York Times:* Art Notes. October 6.

Harrison, Helen Mayer and Newton Harrison. 2010a. Survival Pieces, *The Harrison Studio*, Available at http://theharrisonstudio.net/?page_id=826, accessed April 8 2014.

Harrison, Helen Mayer and Newton Harrison. 2010b. The Harrison Studio Artpark. Artist Website. *The Harrison Studio*, http://theharrisonstudio.net/?page_id=131, accessed April 8, 2014.

Higgins, Dick. 1976. The *Origin of Happening*. American Speech 51 (3/4): 268-271

Hull, Lynne. 2014. Artist Statement – Lynne Hull. *Eco-Art.* Available at http://eco-art.org/?page_id=748, accessed April 8 2014.

Kaprow, Allan. 1960. *Assemblages, Environments and Happenings*. New York. Harry N. Abrams.

Kastner, Jeffrey and Brian Wallis. 1998. *Land and Environmental Art: Themes and Movements*. London: Phaidon Press.

Kelley, Caffyn. 2006. *Art and Survival: Patricia Johanson's Environmental Projects*. Salt Spring Island, BC: Islands Institute.

Orenstein, Gloria Feman. 2003. The Greening Of Gaia. *Ethics & the Environment* 8 (1): 103-111.

Popper, Frank. 1975. *Art: Action and Participation.* London, UK: Studio Vista.

Skrebowski, Luke. 2008. All Systems Go: Recovering Hans Haacke's Systems Art. *Grey Room* 30 (Winter): 54-83.

Sonfist, Alan. 2004. *Nature, the End of Art: Environmental Landscapes.* Florence, Italy: Gli Ori Publishers.

Song, Young Imm Kang. 2009. Community Participatory Ecological. *Art and Education.* 28 (February): 4-13.

Spaid, Sue, and Contemporary Arts Center (Cincinnati, Ohio). 2002. *Ecovention: Current Art to Transform Ecologies.* Cincinnati, OH: Greenmuseum.org,Contemporary Arts Center, Ecoartspace

Stangos, Nikos. 1994. *Concepts of Modern Art: From Fauvism to Postmodernism.* 3rd ed. New York, NY: Thames and Hudson

The Day After Tomorrow. 2004. *The Internet Movie Database.* Available at http://www.imdb.com/title/tt0319262/, accessed March 27, 2014.

Tufnell, Ben. 2006. *Land Art.* London; New York, NY: Tate Publishing, distributed in the U.S. by Harry N. Abrams.

Warshall, Peter. 2000. Eco-Art. *Whole Earth,* No.101: 92.

Wildy, Jade. 2011. *Shades of Green: Changes in the Paradigm of Environmental Art Since the 1960s,* Master of Art History Thesis, University of Adelaide, South Australia. WorldCat Dissertation database, OCLC 767779747.

Part V: Rights

Editors' Note: In April 2010 Bolivia hosted the World People's Conference on Climate Change and the Rights of Mother Earth in Cochabamba. The conference, attended by nearly 30,000 campaigners, activists and members of civil society groups, adopted a Universal Declaration on the Rights of Mother Earth. Article 2 of the declaration reads:

> Mother Earth and all beings of which she is composed have the following inherent rights:
> a. the right to life and to exist;
> b. the right to be respected;
> c. the right to regenerate its bio-capacity and to continue its vital cycles and processes free from human disruptions;
> d. the right to maintain its identity and integrity as a distinct, self-regulating and interrelated being;
> e. the right to water as a source of life;
> f. the right to clean air;
> g. the right to integral health;
> h. the right to be free from contamination, pollution and toxic or radioactive waste;
> i. the right to not have its genetic structure modified or disrupted in a manner that threatens it integrity or vital and healthy functioning;
> j. the right to full and prompt restoration the violation of the rights recognized in this Declaration caused by human activities

Universal Declaration of the Rights of Mother Earth (2010)[1]

---

[1] Universal Declaration of the Rights of Mother Earth 2010), article 2. Available at http://www.rightsofmotherearth.com/declaration/, accessed June 6, 2014.

Chapter 15: Know Your Rights: Earth Jurisprudence and Environmental Politics in the Americas

David Humphreys

Introduction

In October 2012 a group of environmental activists entered the West Burton gas-fired power station in England and occupied one of the site's chimneys for a week. The owners, EDF Energy (a subsidiary of Électricité de France), responded by suing the activists for £5 million in damages resulting from lost production and increased security costs.

From one perspective the lawsuit made sense: EDF had legally acquired private property rights to the site, and it operated the power station within British law. The company was therefore fully entitled to seek compensation given that the trespassers had prevented the company from operating the site leading to significant financial losses. However, the protesters took a very different view: EDF is a major greenhouse gas emitter and its activities contribute to anthropogenic climate change. They were, they believed, fully justified in protesting as they were operating in the service of a greater good, namely the rights of future generations to inherit an habitable planet.

Following a public backlash against EDF the company quietly dropped the lawsuit. However, the story is important in illustrating the contradictions between the different types of rights that inform environmental politics. This chapter begins by examining four clusters of right: the rights of states, human rights, property rights (or, more accurately, the legally sanctioned rights of actors to property, such as territory, products or patents) and corporate rights (namely the rights of business enterprises under national and international commercial law). The chapter then examines the emerging jurisprudence of rights of nature with particular reference to Ecuador, Bolivia and the United States. Throughout the paper contradictions between these five clusters of rights are analyzed.

Rights and Environmental Politics

The study of rights lies at the heart of academic disciplines such as the law and political philosophy and is central to the study and practice of environmental politics. Rights may be defined as legal or moral freedoms or entitlements that an actor is entitled to expect from other member of a moral community. This paper focuses primarily on the rights of nature. First, however, it briefly examines the four sets of rights that have hitherto dominated the study and practice of environmental politics, starting with the rights of states.

Rights of States

The Charter of the United Nations of 1945 is based on the principle of sovereign equality between states.[1] Shortly after its creation the UN set out to negotiate a declaration on the rights and duties of states. A draft was concluded by the International Law Commission and presented to the General Assembly which in 1949 passed resolution 375(IV) agreeing to circulate the draft for comments and suggestions from member states.[2] Two years later the General Assembly noted that the number of states that had responded was too small "to base thereon any definite decision" and passed resolution 596(VI) postponing consideration of the matter.[3] The draft was not subsequently passed[4] although some lawyers consider it a key text of the UN General Assembly.[5] It provides an indication of the rights that states now enjoy under customary international law including "the right to independence …including the choice of its own form of government" (article 1), "the right to exercise jurisdiction over its territory" (article 2), "the right to equality in law with every other State" (article 5) and "the right of individual or collective self-defence against armed attack" (article 14).[6]

Throughout the 1960s the General Assembly passed several resolutions codifying principles of international law leading in 1970 to resolution 2625(XXV) adopting the Declaration on Principles of International Law concerning Friendly Relations and Co-operation among States in accordance with the Charter of the United Nations. This amplifies some of the rights in the 1949 draft including "Each state enjoys the rights inherent in full sovereignty."[7] The declaration makes clear that states are equal, rights and duties are equal and that states have the duty to respect the rights of other states.

The principle of sovereignty has since been clarified for international environmental issues. The 1972 Stockholm Declaration of the Human Environment asserts that states have "the sovereign right to exploit their own resources pursuant to their own environmental policies, and the responsibility to

---

[1] United Nations 1945, article 2.1.

[2] United Nations 1949.

[3] United Nations 1952.

[4] The draft was criticised from a legal point of view as it was unclear whether it was codifying existing international law on the rights and duties of states or providing a guide for its future development. See Kelsen 1950.

[5] Rauschning et al. 1997.

[6] International Law Commission 1949.

[7] United Nations, 1970.

ensure that activities within their jurisdiction or control do not cause damage to the environment of other States or of areas beyond the limits of national jurisdiction."[8] This principle was repeated in the Rio Declaration on Environment and Development of 1992 with one important difference: whereas the Stockholm Declaration mentioned "environmental policies" the Rio Declaration mentioned "environmental and developmental policies."[9] The principle marries the right of states to exploit natural resources with the duty to avoid transboundary environmental harm.

Human Rights

The second set of rights invoked in environmental politics is human rights. In the seventeenth and eighteenth centuries natural rights theorists such as John Locke, Thomas Hobbes and Thomas Jefferson argued for the existence of natural rights, which are universal, inalienable and cannot therefore be taken away, such as the rights to worship, to free speech and to own property. This work has influenced the development of liberal political thought and contributed to contemporary notions of human rights.

The cornerstone piece of international human rights law is the Universal Declaration of Human Rights adopted by General Assembly resolution 217(III) in 1948. The principle asserts that humans have the rights to "life, liberty and security of person" (article 3), "to own property alone or in association with others" (article 17), to "food, housing, clothing and medical care and necessary social services" and to education (article 26).[10] These so-called first generation human rights focusing primarily on the individual have since been supplemented by "second generation," or collective, rights. In 1986 the General Assembly affirmed the right to development as an "inalienable human right."[11] The General Assembly recognized the right to water and sanitation in 2010.[12]

There has been no equivalent declaration on rights to the environment. Richard Hiskes presents a human rights argument that all citizens, both present and future, have environmental entitlements to clean air, water and soil. He argues that for these rights to be realized states that ascribe to them should promote a universal consensus on their applicability.[13] Such a consensus is emerging: by 2005 some 50 countries had established a constitutional right to a clean environment with many recognizing that future generations have the right to inherit a clean environment.[14] In 2007 the General Assembly adopted the United Nations Declaration on the Rights of Indigenous Peoples (UNDRIP)

---

[8] United Nations 1972, principle 21.

[9] United Nations 1992a, principle 2.

[10] United Nations 1949. Other key international declarations on human rights are the *United Nations International Covenant on Economic, Social and Cultural Rights* (opened for signature 16 December 1966; entered into force 3 January 1976) and the *United Nations International Covenant on Civil and Political Rights* (opened for signature 16 December 1966; entered into force 23 March 1976).

[11] United Nations 1986.

[12] United Nations 2010.

[13] Heskes 2009.

[14] Hayward 2005, 22.

which recognizes the principle of free, prior and informed consent whereby indigenous peoples have the right to participate in and be fully informed about decisions that affect their traditional lands.[15]

Two regional legal instruments have endorsed a human right to a clean environment. In 1981 the Organisation of African Unity adopted the African Charter on Human and People's Rights (Banjul Charter), which states "all peoples shall have the right to a general satisfactory environment favourable to their development." [16] In 1988 the San Salvador Protocol to the American Convention on Human Rights in the Area of Economic, Social and Cultural Rights recognized that "Everyone shall have the right to live in a healthy environment" and states "shall promote the protection, preservation, and improvement of the environment."[17]

Property Rights

As noted above, the right to own property appears in the Universal Declaration of Human Rights. The right to property ownership applies not only to individuals but also to communities, businesses and organizations. A property right may be defined as an entitlement to own and use a prescribed area or piece of property under the law. The modern state thus plays a central role in establishing property rights. This notion of property rights owes much to political philosophers such as Hobbes and Locke who argued that there were no legal property rights prior to the modern state, which alone decides the rules of property ownership. [18] The private property rights granted by the state often conflict with traditional and collective ideas of property. In particular, the idea that property did not exist before the creation of the modern state clashes with the notion of customary rights to land claimed by indigenous and traditional communities who have lived on and farmed land for generations, with communally-owned and managed land passed down from generation to generation within families and communities.

Corporate Rights

Some of the main beneficiaries of legal private property rights created by the state are agricultural business corporations which, often in alliance with host governments, control much of the world's most fertile agricultural land, often through "land grabs" resulting in the displacement of traditional communities.[19] Business corporations and other investors have also been granted rights under international law, including the right to sue. The World Trade Organization (WTO) agreements have adopted the idea of the business corporation as an entity with a legal personality that was first introduced in the United States through the 1886 court case, *Santa Clara County v. Southern Pacific Railroad,* which ruled that business corporations have rights of protection that are equal to those of

---

[15] United Nations 2007, articles 10, 11.2, 19, 28.1, 29.2 and 32.2. See also Ward 2011.

[16] Banjul Charter 1981, article 24. For an analysis of the Charter see Umozurike 1983.

[17] Organization of American States 1988, article 11.

[18] Hobbes, 2002 [1651]; Tuck 2002.

[19] Pearce 2012; Linklater 2013.

natural persons.[20] The decision, it can be argued, was instrumental in turning the US away from a society governing by the rights of people to one governed by business. [21] Jurisprudence on corporate rights now finds expression in international law and the WTO agreements that promote trade and investment liberalization, such as the Agreement on Trade-Related Investment Measures (TRIMS) and the Agreement on Trade-Related Intellectual Property Rights (TRIPS).

The WTO has no agreement dealing specifically with the environment and has been criticized for promoting a rules-based economy that places the rights of businesses over and above those of people and environments.[22] For proponents of corporate rights the WTO has the advantage of providing a relatively harmonized and predictable international business climate. The WTO has enforcement and compliance mechanisms that require states to implement international trade and trade-related law on pain of sanctions.

Both complementarities and tensions may exist between these four sets of right. One example of a complementarity concerns the rights of states and human rights: the 1970 *Declaration of Principles of International Law*, as well as codifying rights of states, mentions the importance of maintaining human rights, noting that "all peoples have the right freely to determine, without external interference, their political status and to pursue their economic, social and cultural development, and every State has the duty to respect this right."[23] States have thus voluntarily adopted human rights which all states then have a duty to respect.

However, tensions between different rights may exist. In particular, states may grant rights to corporations in international law that subsequently undermine government environmental protection policy. For example, when the Mexican government decided to close a waste disposal facility owned by Metalclad Corporation after a geological survey revealed that the site would contaminate water supplies Metalclad sued the Mexican government arguing that the closure of the site represented an expropriation of its assets. The decision from the International Centre for Settlement of Investment Disputes (ICSID) ruled in favor of Metalclad, awarding the business US$15.6 million. [24] (The ICSID is an autonomous institution. However, the legal convention that brought it into existence was formulated by the World Bank. It entered into legal effect in 1966.) The government of Mexico appealed the decision, only to find that it had surrendered to corporations rights that undermined its own autonomy.

The advent of the rights of nature movement has added a fifth ingredient to the mix of rights that inform environmental politics. The next section traces the origins of the rights of nature discourse and analyzes the growing, albeit still tenuous, political acceptance of these rights.

---

[20] Horwitz 1985; Korten 1995, 59; Drutman and Cray 2004, 63.

[21] Monks 2008.

[22] Drutman and Cray 2004.

[23] United Nations 1970.

[24] Weiler 2001, 702,

Rights of Nature in South America

On April 22, 2009 the Bolivian president Evo Morales addressed the United Nations General Assembly and said that 60 years after the UN had adopted the Universal Declaration of Human Rights "Mother Earth is now, finally, having her rights recognised."[25] The General Assembly subsequently passed a resolution designating 22 April as International Mother Earth Day.[26] The agreement of this resolution has its origins in traditional Andean beliefs which have been politically recognized in two countries: Ecuador and Bolivia. In 2008 Ecuador became the first country in the world to include rights of the Earth in its constitution, article 71 of which declares that

> Nature, or Pacha Mama, where life is reproduced and occurs, has the right to integral respect for its existence and for the maintenance and regeneration of its life cycles, structures, functions and evolutionary processes. All persons, communities, peoples and nations can call upon public authorities to enforce the rights of nature.[27]

The constitution allows any individual or group to take action through the courts to uphold nature's rights. Indigenous peoples played an important role in the drafting of the constitution and were represented in the drafting process by the Confederation of Indigenous Nationalities of Ecuador (CONAIE). Their involvement paved the way for the inclusion of rights of nature in the constitution. In 2011 the first court case to uphold the rights of nature was brought, namely *Wheeler v. Director de la Procuraduria General Del Estado de Loja.* The court ruled that the dumping of road debris into the Vilcamba River violated nature's rights and found the local provincial council liable. The council was ordered to remove the debris in order to restore the right of the river to flow.[28]

In 2009 Bolivia passed a new constitution stipulating that Bolivians have a duty to "protect and defend an adequate environment for the development of living beings."[29] The following year the Bolivian legislature passed the Law of the Rights of Mother Earth which recognized seven rights of Mother Earth: the rights to life and to exist; not to have cellular structure modified or genetically altered; to pure water; to clear air; to balance; to continue vital cycles and processes free of human existence; and not to be polluted.

The idea of rights of nature expressed in the constitutions of Ecuador and Bolivia reflects a particular South American worldview. For example, Pacha Mama, or Mother Earth, is an Andean goddess, the giver of life, who to Andean indigenous peoples has rights irrespective of human desires and wants. The term Pacha Mama appears in article 71 of the 2008 constitution of Ecuador (above). It does not appear in the text of Bolivian constitution (although it is mentioned in the introduction to the published edition[30]). The idea of Pacha Mama underlies

---

[25] United Nations 2009a.

[26] United Nations 2009b.

[27] Constitution of the Republic of Ecuador 2008, Article 71.

[28] Daly 2012, 64.

[29] Constitution of the Plurinational State of Bolivia 2009, Article 108.16.

[30] Constitution of the Plurinational State of Bolivia 2009, 1.

Bolivia's 2010 Law of the Rights of Mother Earth, although this law used the Spanish for Mother Earth (Madre Tierra).

The notion of *buen vivir* is central to the 2008 constitution of Ecuador. The term has no single translation into English but is usually translated as "living well" or "good living" and is not to be confused with a higher standard of living defined in economic terms. Good living includes a spiritual component, cultural identity, community (of which the natural world is part) and harmony between people and nature. *Buen vivir* does not mean that humans should be prevented from using nature, but it does redefine human use of nature. Humans are not separate from nature but have an interdependent, complementary and indivisible relationship with it. The idea of *buen vivir* articulates a collective notion of community and citizenship that embraces all life and not solely humans with collective rights prevailing over individual rights. As Villalba argues, "Community does not imply a lack of individuality, since individuality is expressed through complementarity with other beings in the group." [31] Ecuadorians have a collective and inclusive notion of citizenship[32] and this is reflected in the constitution with a chapter on the collective rights of "communities, peoples and nationalities."[33] The Bolivian constitution mentions "collective well-being" [34] and grants indigenous peoples the right to the "collective titling of rights and resources."[35]

The emphasis on collective titling and collective ownership of land runs counter to the notion of individual ownership of property in Western societies derived from work of 17th political philosopher John Locke who viewed property as land with which man (sic) has mixed his labor.[36] This has been used to justify enclosure of land on the basis that if a person is prepared to till land then it may be claimed as private property. As recently as the 1980s deforestation was one route for an aspiring property owner in Brazil or Ecuador to stake a legal claim to land.[37]

Those who recognize rights of nature seek to promote a world view whereby human rights are dependent on, and cannot be realized without, the recognition and defense of the rights of Mother Earth. The relationship between rights of nature and human rights is thus seen not as one of equivalence but one whereby rights of nature trump those of the humans with the latter proscribed by the former. As Morales has argued "our rights [the rights of humans] end where we begin to provoke the extermination or elimination of nature."[38]

---

[31] Villalba 2013, 1430.

[32] See, for example, Stober , chapter 17, this volume; Dellert, chapter 18, this volume.

[33] Constitution of the Republic of Ecuador 2008, articles 56-60.

[34] Constitution of the Plurinational State of Bolivia 2009, article 35.I

[35] Constitution of the Plurinational State of Bolivia 2009, article 30.II.6.

[36] Locke 1997. Note, however, that Locke also stipulated the "sufficiency restriction": one must take only what one needs, and should leave enough for others.

[37] Myers 1989.

[38] Morales 2011, 124.

Rights of Nature in the United States

The idea that nature has rights is recognized in many traditional cultures throughout the Americas.[39] It has also gained a tentative status in the US judiciary through Christopher Stone's landmark paper of 1972, "Should trees have standing?"

Standing (*locus standi*) is the ability of a party to demonstrate harm from an action in order to support the party's involvement in a court case. Stone argues that trees, and the whole natural environment in general, should be afforded legal rights. He insists that it is unfair for trees to be denied legal protection because they cannot speak and concludes that guardians, those who wished to defend the rights of trees, should be allowed to bring legal action against those whose actions would harm trees.[40] Stone's paper led to a dissenting opinion in the US Supreme Court. In *Sierra Club v. Morton* the Sierra Club had opposed on ecological grounds the development of a valley in the Sequoia National Forest. The court ruled that the Sierra Club had no standing in the case as neither the club nor its members would suffer injury from the proposed development.[41] However, Justice William Douglas dissented, citing Stone's paper to argue that standing should be conferred upon natural objects so that guardians can sue for their preservation.[42] The provision that any person may take action to defend the rights of nature in the constitution of Ecuador can be seen as consistent with Stone's arguments.[43]

While the US federal government does not recognize rights of nature there has been some recognition at the sub-federal level. In Tamaqua Borough in 2006 an ordinance was issued that recognized natural ecosystems within the borough as "legal persons" for the purpose of preventing sewage sludge dumping on wild land. Significantly, the ordinance asserts not only that "the Borough must take affirmative steps to subordinate the powers of those corporations to the will of the majority within the Borough of Tamaqua" but declares that corporations that apply sludge to the land "shall not be "persons" under the United States or Pennsylvania Constitutions, or under the laws of the United States, Pennsylvania, or Tamaqua Borough, and so shall not have the rights of persons under those constitutions and laws."[44] The ordinance represents the first time that a public body in the United States has granted personhood to nature and stipulated that corporations causing environmental degradation will lose the rights of personhood.

In November 2010 the city of Pittsburgh issued an ordinance that banned natural gas drilling and fracking, elevating community rights and the rights of nature over and above those of corporate personhood. The ordinance was passed after state lawmakers prevented Pittsburgh from taking action to protect the city. The language used in the ordinance, which remains extant at the time of writing

---

[39] Gill 1987; Weaver 1996.

[40] Stone 1972.

[41] Baude 1973.

[42] Hogan 2007.

[43] Constitution of the Republic of Ecuador 2008, article 71.

[44] Tamaqua Borough Sewage Sludge Ordinance, cited in Community Environmental Legal Defense Fund 2006.

(July 2014), appears to draw from Stone's 1972 paper and the Ecuadorian constitution:

> Natural communities and ecosystems, including, but not limited to, wetlands, streams, rivers, aquifers, and other water systems, possess inalienable and fundamental rights to exist and flourish within the City of Pittsburgh. Residents of the City shall possess legal standing to enforce those rights on behalf of those natural communities and ecosystems.[45]

The city of Pittsburgh thus establishes legal standing for any citizen to protect the local environment, including those that have no interest in the threatened ecosystem.

The cases of Tamaqua Borough and Pittsburgh illustrate that although the federal government has not recognized rights of nature the idea is being adopted in other political power centers in the United States. There is a similarity here with the Kyoto Protocol of 1997, signed by the Clinton administration but renounced by the administration of Bush junior. Since then a number of cities and other local public authorities have committed themselves to meeting the greenhouse gas emissions reduction target that the United States agreed at Kyoto.[46]

Having considered the adoption of rights of nature by the governments of two countries in South America and by certain sub-federal level actors in the United States, the next section will examine the tensions that may exist between the rights of nature and the other sets of rights introduced at the start of this paper.

When Rights Collide

Based primarily on the experiences of Bolivia and Ecuador this section will address the question of where rights of nature stand in relation to the rights of states, human rights, property right and corporate rights. It will ask the question: when different rights collide, which rights tend to prevail.

Upholding one set of rights does not preclude the promotion or advocacy of others. For example, many of those who advocate the primacy of rights of nature argue that upholding the rights of communities and indigenous peoples is necessary for upholding nature's rights. However, to many environmental and human rights activists, upholding the rights of nature and of indigenous communities necessarily challenges corporate rights. For example, to Atossa Soltani, the founder of the campaigning group Amazon Watch, "The Rights of Nature movement is the antidote to reigning in the unbridled power of corporations whose drive for short-term profits is pushing humanity and countless species to extinction."[47] Anuradha Mittal argues that corporations should be denied legal personhood and made "accountable to communities and ecosystems

---

[45] Pittsburgh Pennsylvania Code of Ordinances 2013.
[46] Resnik, Civin and Frueh 2008.
[47] Soltani 2011, 38.

where they extract wealth." [48] Throughout the activist community there is a widespread view that people and communities have lost control of their lands to external actors such as business corporations and that meaningful implementation of both human rights and the rights of nature requires curtailing the rights of corporate to own land where this results in environmental degradation.

The view that corporate rights should be limited has found support in the political establishments of Ecuador and Bolivia. However, the reasons for this are less to do with upholding the rights of nature and more with a concern to assert the rights of the state to exercise sovereignty over its territory and natural resources. In 2007 Bolivia became the first country to withdraw from the International Centre for Settlement of Investment Disputes (ICSID), the body that ruled against the government of Mexico in the Metalclad case. [49] Ecuador withdrew in 2010.

The governments of Ecuador and Bolivia have thus asserted their right to manage their own environments without submitting to external adjudication. Indeed the adoption of rights of nature in Ecuador and Bolivia has been accompanied by more strident assertions of state control over natural resources. The 2008 constitution of Ecuador claims food sovereignty and economic sovereignty and asserts that "The State shall exercise sovereignty over biodiversity, whose administration and management shall be conducted on the basis of responsibility between generations." [50] The 2009 constitution of Bolivia asserts that

> natural assets are of public importance and of strategic character for the sustainable development of the country. Their conservation and use for the benefit of the population will be the responsibility and exclusive authority of the State, and the sovereignty over natural resources may not be compromised. [51]

But while asserting sovereignty, the constitutions of Ecuador and Bolivia also uphold the rights of citizens to participation. [52] There is thus a tension between the state and the executive branch on the one hand, and people, citizens' groups and participatory democracy on the other. [53] In the case of Bolivia this dualism is exemplified in the phrase "natural resources are the property of the Bolivian people and will be managed by the State." [54]

While promoting rights of nature the governments of both countries also promote economic development. The constitution of Bolivia makes clear the country defends its right to fossil fuel-driven development by asserting that its "hydrocarbons ...are the inalienable and unlimited property of the Bolivian people. The State, on behalf of and in representation of the Bolivian people, is the owner of the entire hydrocarbon production of the country and is the only one

---

[48] Mittal 2011, 41.

[49] Gaillard 2007.

[50] Constitution of the Republic of Ecuador 2008, article 400.

[51] Constitution of the Plurinational State of Bolivia 2009, article 346.

[52] For example, Constitution of the Republic of Ecuador 2008, articles 61 to 65.

[53] Schilling-Vacaflor 2011, 16.

[54] Constitution of the Plurinational State of Bolivia 2009, article 311.II.2

authorised to sell it."[55] In 2014 Ecuador, speaking also for Bolivia and Argentina, emphasized at a session of the UN General Assembly Open Working Group on Sustainable Development Goals the "central role of the state" in coordinating environmental and social policies.[56]

In both countries the practicalities of both upholding rights of nature and promoting a state-driven economistic model of development that remains reliant on external investment has led the government into conflict with citizens seeking to protect local environments. In Bolivia there have been protests against hydrocarbon and mineral mining projects from indigenous peoples invoking the need to respect Pacha Mama. The government has ignored these protests.[57] Similarly, since the adoption of the 2008 constitution in Ecuador there have been indigenous protests against oil extraction and mineral mining leading to arrests.[58] President Correa has responded that mining will create new employment opportunities.[59]

In a 2012 court case, *Kichwa Indigenous People of Sarayaku v. Ecuador*, the Inter-American Court of Human Rights ruled that the government of Ecuador should have consulted with the Kichwa people in line with the principle of free, prior and informed consent before commencing oil drilling on customary lands and called on the government to comply with the ruling in future.[60] It should be noted that the Ecuadorian constitution does not mention the principle of free, prior and informed consent,[61] an omission which, it can be argued, undermines the ability of community groups to defend the rights of nature. Given Ecuador's position as the first country to adopt rights of nature, the defeat of its government in the Inter-American Court of Human Rights on an environmental issue was an international embarrassment. According to one legal analyst the implementation of rights of nature and other environmental law in Ecuador has lagged behind the commitments due to a lack of political will and, despite government resistance, continued corporate control of environmental decision-making.[62] In December 2013 the Global Alliance for the Rights of Nature, a global civil society network, announced that the Ecuadorian Ministry of the Environment had ordered the dissolution of Fundación Pachamama, one of the Ecuadorian groups that had campaigned for the inclusion of rights of nature in Ecuador's constitution.[63] Fundación Pachamama has been a vocal critic of the Ecuadorian government's decision to allow oil mining on Amazonian indigenous peoples' land.

---

[55] Constitution of the Plurinational State of Bolivia, article 359.

[56] *Earth Negotiations Bulletin* 2014, 7. (The words in quotation marks represent a quote from the *Earth Negotiations Bulletin* and not necessarily the actual words used by the Ecuadorian delegate.)

[57] Stevenson 2013, 22.

[58] For a history of these protests, which predate the election of Correa and the adoption of the 2008 constitution, see Anguelovski 2007.

[59] Becker 2011, 58.

[60] Inter-American Court of Human Rights 2012, 35, 49-50.

[61] Although it does mention the softer formulation "free prior informed consultation, within a reasonable period of time" on proposals for prospecting of producing non-renewable resources on indigenous lands. Constitution of the Republic of Ecuador, article 57.7.

[62] Kimerling 2013, 62.

[63] Email from Global Alliance for the Rights of Nature, December 6, 2013.

A further case illustrating the tension between rights of nature and the sovereign rights of states over their natural resources in Ecuador concerns the Yasuni-ITT scheme. In 2007 Ecuador offered to desist indefinitely from deforestation in order to exploit the Ishpingo-Tambococha-Tiputini (ITT) oilfield in the Yasuni national reserve if the international community was prepared to compensate it for so doing. A United Nations Development Program trust fund was established to receive donations for the protection of the Yasuni reserve. Ecuador's proposal was consistent with the notion of common but differentiated responsibilities, a legal principle included in the UN Framework Convention on Climate Change which holds that responsibilities for addressing climate change should be shared among states, with those states that have polluted most bearing the greater responsibility.[64] As Correa commented in 2007 when making the proposal, "Ecuador doesn't ask for charity but does ask that the international community share in the sacrifice and compensates us with at least half of what our country would receive, in recognition of the environmental benefits that would be generated by keeping this oil underground."[65]

In August 2013 Correa announced that Ecuador was abandoning the scheme after just $13 m. of the $3.6 bn. being sought was deposited.[66] It can be argued that Ecuador is fully entitled to exploit its oil if other countries are not prepared to share the costs of conservation. Against this it may be argued that Ecuador should desist from mining the oil if it is serious about respecting the rights of nature. The decision of the government to abandon the scheme suggests that while Ecuador is prepared to stand some economic costs to protect nature it is also prepared to use the country's resources as a political bargaining chip with other countries and to assert its sovereign right to exploit these resources should such bargaining fail.

The fate of the Yasuni-ITT scheme illustrates the dominant approach to environmental conservation, namely payment for environmental services (PES). The underlying rationale of the PES approach is that property owners should be financially compensated for maintaining the environmental services of their property, as without such compensation owners are free to use their property for other, less environmentally friendly, uses. This approach underpins many policies to halt tropical deforestation. On this view, if forest owners are compensated for the environmental services that standing forests provide, such as carbon sequestration and biodiversity conservation, they are less likely to fell their forests for, say timber or conversion to cattle pasture. Payment for environmental services may take place through markets (for example, for carbon sink functions or watershed services) or may be negotiated bilaterally or multilaterally between buyers and sellers. Ecuador's Yasuni-ITT proposal was consistent with the latter approach. However, the idea of rights of nature runs counter to the idea of nature as property that can be exploited, bought and sold. The PES approach is a logical and rational one in a world where private property rights prevail. However, in a world where the rights of nature were dominant, all other rights would be subsumed under nature's rights and acceptable only if they did not undermine or erode nature's rights. A fundamentally different type of law would prevail, human societies would be governed and regulated very differently and the

---

[64] United Nations 1992b, preamble, articles 3.1 and 4.1.

[65] Environmental News Service 2007.

[66] Watts 2013, 13.

human-nature relationship would be placed on a very different legal basis than today.

Conclusion

Through examining the tensions between the idea of rights of nature and four other sets of rights this paper has provided some indication of why upholding the rights of Pacha Mama has proved so difficult in practice in Ecuador and Bolivia. In neither country should the adoption of rights of nature be seen as absolute, unfettered or consistent. In both there are unresolved questions between conserving nature to benefit the present and future generations, and between resource use to promote national development or localized community-driven development. It seems likely that some of the political support that rights of nature has enjoyed in the these two countries is less about respect for nature or for the rights of future generations per se but as a political tool to assert right of access to the environment by favored local or national actors, with foreign businesses admitted on terms set by the national government. That said, there is no doubt that many of those who support the idea do so out of firm moral conviction and principled belief. The challenge for those who support rights of nature is how to promote both a wider uptake of the idea among political leaders and civil society and a clearer long term vision of how upholding nature's rights may be operationalized in practice.

References

Anguelovski, Isabelle. 2007. Indigenous resistance to oil extraction in southern Ecuador: The need for state recognition of autonomy demands for cultural sustainability. *International Journal of Environmental, Cultural, Economic and Social Sustainability* 3(5): 95-105.

Banjul Charter. 1981. *African (Banjul) Charter) on Human and Peoples' Right*. Available at http://www.africancourt.org/en/images/documents/Sources%20of%20La w/Banjul%20Charta/charteang.pdf, accessed September 14, 2013.

Baude, Patrick L. 1973. Sierra Club v. Morton: Standing Trees in a Thicket of Justiciability. *Indiana Law Journal* 48(2): 197-215.

Becker, Marc. 2011. Correa, Indigenous Movements, and the Writing of a New Constitution in Ecuador. *Latin American Perspectives* 38(1): 47-62.

Community Environmental Legal Defense Fund. 2006. Tamaqua Borough Corporate Waste and Local Control Ordinance. Available at http://www.celdf.org/article.php?id=439, accessed September 25, 2013.

Constitution of the Plurinational State of Bolivia. 2009. Translated by Luis Francisco Valle V. Printed in San Bernadino, California. Spanish version available at http://pdba.georgetown.edu/constitutions/bolivia/bolivia09.html, accessed October 11, 2013.

Constitution of the Republic of Ecuador. 2008. English translation from Political Database of the Americas, Georgetown University. Available at http://pdba.georgetown.edu/Constitutions/Ecuador/english08.html, accessed October 11, 2013. Spanish version available at http://pdba.georgetown.edu/Constitutions/Ecuador/ecuador08.html, accessed October 11, 2013.

Daly, Erin. 2012. The Ecuadorian Exemplar: The first ever vindications of constitutional rights of nature. *Review of European Community and International Environmental law (RECIEL)* 21(1): 63-66.

Dellert, Christine. 2010. Protecting the 'Other': Ecological Citizenship Under Ecuador's Constitution. *International Journal of Environmental, Cultural, Economic and Social Sustainability* 6(6): 117-128.

Drutman, Lee, and Charlie Cray. 2004. *The People's Business: Controlling Corporations and Restoring Democracy*. San Francisco: Berrett-Koehler.

Earth Negotiations Bulletin. 2014. Summary of the Seventh Session of the UN General Assembly Open Working Group on Sustainable Development Goals: 6-10 January 2014. *Earth Negotiations Bulletin* 32(7), 18pp.

Environmental News Service. 2007. Ecuador seeks compensation to leave Amazon oil undisturbed. Available at http://www.ens-newswire.com/ens/apr2007/2007-04-24-04.asp, accessed October 28, 2013.

Gaillard, Emmanuel. 2007. The denunciation of the ICSID Convention. *New York Law Journal* 237(122). Available at http://www.shearman.com/files/Publication/a4ce24f1-83de-445d-a50a-b82baf2f89fc/Presentation/PublicationAttachment/ce3cbe9a-ca49-4eaa-b3f5-d0a26ba0680c/IA_NYLJ%20Denunciation%20ICSID%20Convention_0 40308_17.pdf, accessed October 25, 2013.

Gill, Sam D. 1987. *Mother Earth: An American Story*. Chicago: University of Chicago Press.

Hayward, Tim. 2005. *Constitutional Environmental Rights*. Oxford: Oxford University Press.

Heskes, Richard P. 2009. *The Human Right to a Green Future: Environmental Rights and Intergenerational Justice*. Cambridge: Cambridge University Press.

Hobbes, Thomas. 2002 [1651]. *Leviathan*. London: Penguin.

Hogan, Marguerite. 2007. Standing for Nonhuman Animals: developing a Guardianship Model from the Dissents in *Sierra Club v. Morton. California Law Review* 95(2): 513-534.

Horwitz, Morton J. 1985. *Santa Clara* Revisited: The development of corporate theory. *West Virginia Law Review* 88: 173-224.

Inter-American Court of Human Rights. 2012. *Case of the Kichwa Indigenous People of Sarayaku v. Ecuador*, Judgement of June 27, 2012. Available at http://www.escrnet.org/sites/default/files/Court%20Decision%20_ English _ .pdf, accessed October 28, 2013.

International Law Commission. 1949. Draft Declaration on Rights and Duties of States 1949. Available at http://untreaty.un.org/ilc/texts/instruments/english/draft%20articles/2_1_ 1949.pdf, accessed September 12, 2013.

Kelsen, Hans. 1950. The Draft Declaration on Rights and Duties of States: Critical Remarks. *American Journal of International Law* 44(2): 259-76.

Kimerling, Judith. 2013. Oil, Contact, and Conservation in the Amazon: Indigenous, Huaorani, Chevron, and Yasuni. *Colorado Journal of International Environmental Law and Policy* 24: 43-115.

Korten, David C. 1995. *When Corporations Rule the World.* London: Earthscan.

Linklater, Andro. 2013. *Owning the Earth: The Transforming History of Land Ownership.* London: Bloomsbury.

Locke, John. 1997. *Political Essays,* edited by Mark Goldie. Cambridge: Cambridge University Press.

Mittal, Anuradha. 2011. The nature of farming and farming with nature. In *The Rights of Nature: The Case for a Universal Declaration of the Rights of Mother Earth,* edited by Council of Canadians, Fundación Pachamama and Global Exchange, 40-1. San Francisco: Council of Canadians, Fundación Pachamama and Global Exchange.

Monks, Robert L. 2008. *Corpocracy,* Hoboken NJ: John Wiley and Sons.

Morales, Evo. 2011. The Century of the rights of Mother Earth. In *The Rights of Nature: The Case for a Universal Declaration of the Rights of Mother Earth,* edited by Council of Canadians, Fundación Pachamama and Global Exchange, 122-125. San Francisco: Council of Canadians, Fundación Pachamama and Global Exchange.

Myers, Norman. 1989. *Deforestation rates in tropical countries and their climatic implications.* London: Friends of the Earth.

Organization of American States. 1988. *Additional Protocol to the American Convention on Human Rights in the Area of Economic, Social and Cultural Rights.* Available at http://www.cidh.oas.org/Basicos/English/basic5.Prot.Sn%20Salv.htm, accessed September 14, 2013.

Pearce, Fred. 2012. *The Landgrabbers: The new fight over who owns the Earth.* London: Transworld.

Pittsburgh Pennsylvania Code of Ordinances. 2013. Code of Ordinances City of Pittsburgh Pennsylvania. Codified through Ordinance No. 12-2013, April 19, 2013. Available at http://library.municode.com/index.aspx?clientID=13525&stateID=38&statename=Pennsylvania, accessed February 4, 2014.

Rauschning, Dietrich, Katja Wiesbrock, and Martin Lailach (eds). 1997. *Key Resolutions of the United Nations General Assembly, 1946-1996.* Cambridge: Cambridge University Press.

Resnik, Judith, Joshua Civin, and Joseph B. Frueh. 2008. Ratifying Kyoto at the Local Level: Sovereigntism. Federalism and Translocal Organizations of Government Actors (TOGAs). *Arizona Law Review* 50: 709-786.

Schilling-Vacaflor, Almut. 2011. Bolivia's New Constitution: Towards Participatory Democracy and Political Pluralism. *European Review of Latin American and Caribbean Studies* 90: 3-22.

Sheehan, Linda. 2012. Establishing Earth-based governance for the rights of the environment. In *Future Perfect: Rio +20 – United Nations Conference on Sustainable Development,* edited by Michele Witthaus, Cherie Rowlands and Jacqui Griffiths, pp.18-21. Available at http://digital.tudor-rose.co.uk/future-perfect/#6, accessed January 31, 2014.

Soltani, Atossa. 2011. The Heart of the World: U'Wa perspectives on Mother Earth. In *The Rights of Nature: The Case for a Universal Declaration of the Rights of Mother Earth,* edited by Council of Canadians, Fundación Pachamama and Global Exchange. pp.36-8. San Francisco: Council of Canadians, Fundación Pachamama and Global Exchange.

Stevenson, Hayley. 2013. Representing Green Radicalism: the limits of state-based representation in global climate governance. *Review of International Studies* August: 1-25.

Stober, Spencer, 2010. Ecuador: Mother Nature's Utopia. *International Journal of Environmental, Cultural, Economic and Social Sustainability* 6(2): 229-239.

Stone, Christopher D. 1972. Should Trees Have Standing? Towards Legal Rights for Natural Objects. *Southern California Law Review* 45: 450-501.

Tuck, Richard. 2002. *Hobbes: A Very Short Introduction.* Oxford: Oxford University Press.

Umozurike, U.O. 1983. The African Charter on Human and Peoples' Rights. *American Journal of International Law* 77(4): 902-912.

United Nations. 1945. *Charter of the United Nations,* San Francisco: United Nations Available at http://www.un.org/en/documents/charter/index.shtml, accessed August 1, 2013.

United Nations. 1949. UN General Assembly Resolution 375(IV). Available at http://daccess-dds-ny.un.org/doc/RESOLUTION/GEN/NR0/051/94/IMG/NR005194.pdf?OpenElement, accessed August 1, 2013.

United Nations. 1952. UN General Assembly Resolution 596(VI). Available at http://daccess-dds-ny.un.org/doc/RESOLUTION/GEN/NR0/068/51/IMG/NR006851.pdf?OpenElement, accessed August 1, 2013.

United Nations. 1970. Declaration on Principles of International Law concerning Friendly Relations and Co-operation among States in accordance with the Charter of the United Nations. Available at http://legal.un.org/avl/ha/dpilfrcscun/dpilfrcscun.html, accessed July 4, 2014.

United Nations. 1972. Stockholm Declaration on the Human Environment. Available at http://www.unep.org/Documents.Multilingual/Default.asp?documentid=97&articleid=1503, accessed September 14, 2013.

United Nations. 1986. A/RES/41/128 Declaration on the Right to Development. Available at: http://www.un.org/documents/ga/res/41/a41r128.htm, October 9, 2013.

United Nations. 1992a. Rio Declaration on Environment and Development. Available at http://www.unep.org/Documents.Multilingual/Default.asp?documentid=78&articleid=1163, accessed September 14, 2013.

United Nations. 1992b. UN Framework Convention on Climate Change. Available at http://unfccc.int/files/essential_background/background_publications_ht mlpdf/application/pdf/conveng.pdf, accessed October 28, 2013.

United Nations. 2007. United Nations Declaration of the Rights of Indigenous Peoples. Available at http://www.un.org/esa/socdev/unpfii/documents/DRIPS_en.pdf, accessed September 21, 2013.

United Nations. 2009a. General Assembly GA/10823, Sixty-third General Assembly, General Assembly proclaims 22 April 'International Mother Earth Day', adopting by consensus Bolivian-led resolution, 22 April, press release.

United Nations. 2009b. A/RES/63/278 Resolution adopted by the General Assembly on 22 April 2009, 63/278. International Mother Earth Day. Available at http://www.un.org/en/ga/search/view_doc.asp?symbol=A/RES/63/278& Lang=E, accessed September 17, 2013.

United Nations. 2010. A/RES/64/292 Resolution adopted by the General Assembly on 28 July 2010, 64/292, The human right to water and sanitation. Available at http://www.un.org/en/ga/search/view_doc.asp?symbol=A/RES/64/292, accessed September 15, 2013.

United Nations. 2011. A/66/302. Harmony with Nature: Report of the Secretary-General, 15 August. Available at http://www.un.org/ga/search/view_doc.asp?symbol=A/66/302, accessed January 6, 2014.

Villalba, Unai. 2013. *Buen Vivir* vs Development: A paradigm shift in the Andes? *Third World Quarterly* 34(8): 1427-1442.

Ward, Tara. 2011. The Right to Free, Prior and Informed Consent: Indigenous Peoples' participation Rights within International Law. *Northwestern Journal of International Human Rights* 10(2): 54-84.

Watts, Jonathan. 2013. Oil drilling to start in pristine Amazon as eco fund fails. *Guardian*, 17 August, p.13.

Weaver, Jace (ed.) 1996. *Defending Mother Earth: Native American Perspectives on Environmental Justice*. Maryknoll NY: Orbis Books.

Weiler, T. 2001. *Metalclad v. Mexico:* A Play in Three Parts. *Journal of World Investment* 2(4): 685-711.

Chapter 16: Women, Sustainability, and Biodiversity: Vandana Shiva's Arguments for Earth Democracy

Jennifer E. Michaels

Introduction

The Indian physicist Vandana Shiva has long been a respected, if sometimes controversial, leader in promoting environmental sustainability, biodiversity, and the empowerment of women and has written numerous books and articles focusing on these issues. Through the clarity of her writing and her outspokenness she succeeds in bringing these concerns to the attention of wide audiences. This chapter discusses her comprehensive criticism of unsustainable practices brought about in her opinion by monocultures, industrialization and globalization, her strong arguments for biodiversity, sustainability, and social justice, her activism, and the solutions she proposes for a sustainable future. While her specific focus is India, she also examines the impacts of unsustainable practices globally in such places as Africa, China, South America, the United States and many other countries. In her recent work, her voice has become ever more urgent since she fears that such practices threaten the survival of the human race. The choices we make, she stresses, will "decide whether or not we survive as a species."[1] In her opinion, "we can either keep sleepwalking to extinction or wake up to the potential of the planet and ourselves."[2]

Vandana Shiva's Path to Ecological Activism

Shiva was born in 1952 in Dehra Dun in the foothills of the Himalayas. Her father was a forest conservator, and her mother, an early feminist, was a farmer. Her

---

[1] Shiva 2008a, 142.
[2] Shiva 2008a, 144.

parents adopted the name Shiva to erase their caste identity. After receiving her bachelor's degree in physics in India, she studied in Canada, receiving an M.A. in the philosophy of science from the University of Guelph in Ontario and a PhD with a dissertation on quantum theory at the University of Western Ontario. On her return to India she trained as a nuclear physicist and recalls that some of the most exciting times were when she worked at the Bhabha Atomic Center in Mumbai. She gave up this career, however, after her sister Mira, a physician, pointed out that she understood little about nuclear hazards. This was, Shiva reflects, a "lesson in humility that precipitated my shift toward sciences that defend life and away from those that annihilate life."[3] She had expected "to spend a lifetime solving puzzles in quantum theory. Instead, I have spent the past two decades solving puzzles in agriculture."[4] Shiva was inspired by Chipko, a grassroots movement of rural Indian women resisting the destruction of forests, and she has been active in numerous conservation, animal, and human rights movements in India and worldwide.

When Shiva was thirty she founded the Research Foundation for Science, Technology and Ecology, at first in her mother's cowshed in Dehra Dun, but now located in New Delhi and which Shiva still directs. Following her parents' beliefs Shiva opposes caste and religious discrimination, and in her research organization Muslims, Hindus and Christians work together.[5] Besides raising awareness of sustainability and biodiversity through her voluminous writing, lecturing and activism, Shiva started building Navdanya (nine seeds) in the early 1990s, an organization that preserves indigenous seed and encourages a return to traditional sustainable organic farming methods.[6] Shiva, a strong ecofeminist, has received numerous prizes and honors, including in 1993 the prestigious Right Livelihood Award (seen as the Alternative Nobel Prize) that recognizes "outstanding vision and work on behalf of our planet and its people." This prize honored her for her work on sustainability and "for placing women and ecology at the heart of modern development discourse."[7] She has served as an advisor to governments in India and abroad, continues to be a leader in grassroots movements and is in demand as a contributor to works on sustainability and as a lecturer worldwide. She has been active in opposing corporate control and intellectual property rights to living organisms such as seeds. In *Speaking of Earth: Environmental Speeches that Moved the World,* Alon Tal notes that Shiva received more nominations for inclusion in his book from more countries than anyone else.[8] The speech he included was "Sharing and Exchange, the Basis of Our Humanity and Our Ecological Survival Has Been Redefined as a Crime."[9]

Mahatma Gandhi has strongly inspired Shiva's work. She underscores that for Gandhi, whose philosophy and actions she greatly admires, "nonviolence is

---

[3] Shiva 2005a, 41-42.

[4] Shiva 2005a, 92.

[5] Barsamian 2002.

[6] Navdanya 2014.

[7] The Right Livelihood Award: Vandana Shiva 1993. The Chipko movement also received this award in 1987.

[8] Tal 2006, 218.

[9] Shiva 2006a, 221-231.

not just the absence of violence. It is an active engagement in compassion."[10] According to Gandhi, she observes, "the earth has enough for everyone's needs, but not for some people's greed,"[11] and he advised: "Recall the face of the least privileged person you know ... and ask if your action will harm or benefit him/her."[12] Gandhi accused Western industrialization and colonialism of causing poverty, dispossession, and destruction of livelihoods in India. The *charkha*, the spinning wheel, the symbol of India's independence, represented for him a technology that "conserves resources, people's livelihoods and people's control over their livelihoods."[13] Shiva draws on Gandhi's thought in her work to help marginalized and dispossessed people, especially in rural areas, and to preserve resources and livelihoods. Inspired by Gandhi's commitment to nonviolence, Shiva adopted strategies such as *satyagraha*, which she defines as "the duty to exercise non-cooperation and civil disobedience," to protest laws that abuse human freedom. She was also influenced by Gandhi's notion of *swaraj*, "self-rule in terms of the freedom of all within society, including the last person." [14] Following Gandhi she envisions self-rule not through a centralized state, but as decentralized democratic local communities. [15] Like Gandhi she believes that *swadeshi*, "the spirit of regeneration," is "central to the creation of peace and freedom" and is based on "building on the resources, skills, and institutions of a community, and when necessary, transforming them." [16] She emphasizes: "Gandhi's legacy carries the seeds for the freedoms of humans and all species. Gandhi's legacy is humanity's hope."[17]

Biodiversity and Traditional Sustainable Practices

Shiva defines biodiversity as "a web of relationships which ensures balance and sustainability."[18] Essential to biodiversity, in her view, is reverence for the Earth and for nature. Shiva observes that many indigenous cultures consider the Earth to be "terra mater," the sacred mother, Gaia, and she stresses the importance of "*Vasudhaiv Kutumbakam* (the whole world is one family)," which means living in harmony with nature by valuing and respecting all diverse forms of life.[19] For Shiva, diversity means not only protecting biodiversity, but also accepting people of different ethnicities, social backgrounds and religions. In her opinion, diversity "creates harmony; and harmony, in turn, creates beauty, balance, bounty, and peace in nature and society, in agriculture and culture, in science and politics."[20] Sustainability, she emphasizes, requires "the protection of all species and all people, and the recognition that diverse species and diverse people play an

---

[10] Shiva 2005b, 116.
[11] Shiva 2006a, 231.
[12] Mies and Shiva 1993, 86.
[13] Shiva 1995a, 197-98.
[14] Shiva 2008b, 203-204.
[15] Shiva 2005a, 131.
[16] Shiva 1997, 125.
[17] Shiva 2005b, 95.
[18] Mies and Shiva 1993, 171.
[19] Shiva 2006b, 199.
[20] Shiva 2005a, 73.

essential role in maintaining ecosystems and ecological processes." [21] Shiva dismisses the term "sustainable development" as a contradiction since its goal is to ensure a reliable supply of raw material for industry and it prioritizes short-term profit. She concurs with a Native American elder who observed: "Only when you have felled the last tree, caught the last fish and polluted the last river, will you realize that you can't eat money." [22]

Shiva stresses that all people belong to the Earth family and are responsible for the planet's wellbeing: "We share this planet, our home, with millions of species. Justice and sustainability both demand that we do not use more resources than we need." [23] She points out that over centuries human societies respected nature's limits and its cycles and rhythms, and this ecologically sustainable paradigm should be followed today: "In feeding the earthworms, we feed ourselves. In feeding cows, we feed the soil, and in providing food for the soil we provide food for humans." [24] Indigenous peoples such as those in the Amazon, the Andes or the Himalayas, who have long recognized the need for conserving resources, are for her examples of centuries-old sustainable living cultures. [25] Shiva argues that traditional farming methods conserve biodiversity and enrich the soils: "Seeds come from the farm, soil fertility comes from the farm and pest control is built into the crop mixtures." [26] Sustainable agriculture, based on recycling soil nutrients, uses mixed cropping such as planting legumes and pulses interspersed with cereals to fix nitrogen in the soil and maintain the water cycle. [27] Indigenous cattle provide organic fertilizer, and many plants provide nutrition. In India, a major region of biodiversity, indigenous knowledge of medicine and agriculture contains centuries of people's innovation. Over time farmers developed numerous crops and crop varieties appropriate to different regions and adapted to different climates. [28] India alone developed thousands of rice varieties bred to survive floods and droughts and to resist pests, [29] and farmers in the Andes developed over three thousand varieties of potatoes. [30] Rural communities conserved forests, important gene reservoirs, and used them sustainably as sources of "food, fodder, fuel, fiber, medicine, and security from floods and drought" and as reliable supplies of water for drinking and irrigation. [31]

Custodians of Biodiversity: The Role and Dignity of Women in Traditional Sustainable Farming

Shiva believes strongly that for centuries women have shaped economies of care, wellbeing and happiness and she links food security to women's food producing

---

[21] Shiva 2006a, 229.
[22] Quoted in Shiva 1992, 192-93.
[23] Shiva 2005b, 50.
[24] Shiva 2006a, 229-30.
[25] Shiva 2005b, 51.
[26] Shiva 1995a, 200.
[27] Shiva 2000a, 113.
[28] Shiva 2000b, 144.
[29] Shiva 1997, 89.
[30] Shiva 2000a, 79.
[31] Shiva 2000a, 1.

abilities.[32] They are responsible for producing more than half of the world's food and provide more than 80 percent of food needs "in food-insecure households and regions."[33] Women have maintained seed continuity over millennia despite war, flood and famine and, as selectors and preservers of seeds, their knowledge and skills have shaped crop improvement.[34] Like women in other cultures, rural women in India did much of the farming and became "the custodians of biodiversity," and both their work and their knowledge, Shiva stresses, are crucial to biodiversity.[35] In their small-scale agriculture women conserve and renew natural resources. They work, for instance, as "soil builders" rather than "soil predators."[36] In protecting biodiversity, Shiva believes, Indian women are shaped by the ancient Indian worldview in which nature is viewed as Prakriti, as "a living and creative process, the feminine principle from which all life arises."[37] In her opinion, the feminine principle is nonviolent and sustains life by maintaining and valuing nature's interconnectedness and diversity.[38] Traditionally women played and continue to play an essential role not only in farming, but also in indigenous forest management.[39] They possess a reservoir of knowledge about seeds, soils and forests and share their expertise with each other. As they collect fodder in the forests, for example, they learn from each other in "informal forestry colleges."[40]

Industrialized Farming: Globalization and "the Corporate Control of Life"[41]

Shiva is outspoken in her criticism of industrialized farming and its resulting monocultures that require ever increasing inputs of herbicides and pesticides. Many environmentalists have long expressed concern about the impact of monocultures on biodiversity and sustainability. In her pioneering 1962 book *Silent Spring* Rachel Carson was one of the first to make a wide readership aware that monocultures "set the stage for explosive increases in specific insect populations" and that "single-crop farming ... is agriculture as an engineer might conceive it to be." By disrupting nature, people disturb "the built-in checks and balances by which nature holds the species within bounds."[42] Since Carson wrote her book, monocultures have increased immensely. Like Carson, the former senator and US vice-president Al Gore writes to make the general public aware of scientific evidence about the global environmental crisis and thereby hopes to encourage change in environmental policies. He argues that dependence on

---

[32] Shiva 2013, 15. This book grew out of Shiva's Sydney Peace Prize lecture in 2010 and was first published in 2012 by Women Unlimited, New Delhi.
[33] Shiva 2010, xi. First published in India in 1988. Shiva wrote a new introduction for the 2010 edition.
[34] Shiva 1995b, 61.
[35] Mies and Shiva 1993, 167-68.
[36] Shiva 2010, 108.
[37] Shiva 2010, xxxiii.
[38] Shiva 2010, 14.
[39] Shiva 2010, 61.
[40] Shiva 1993, 310.
[41] See Shiva 2011.
[42] Carson 1964, 20.

monocultures threatens food security. People have, for example, ignored the lessons of the Irish famines of the mid-1800s when the potato blight wiped out Ireland's main crop.[43] Like Shiva, other environmentalists are concerned that monocultures lead to "the global homogenization of culture" and dismantle "local traditions and economies."[44] Such homogenization causes "a breakdown of both biological and cultural diversity, erosion of our food security, an increase in conflict and violence, and devastation of the global biosphere."[45]

For Shiva, monocultures are not only unsustainable, but they are also "a declaration of war against nature's diverse species."[46] Instead of "celebrating and conserving life in all its diversity" the sanctity of life has been "substituted by the sanctity of science and development."[47] In her opinion market-driven reductionist science views some plants such as soy and cotton as commodities and eradicates other, often nutritious, plants as weeds. Industrial farming has driven seventy-five per cent of agricultural diversity to extinction.[48] It reduces living organisms to mechanical systems, in which sacred cows become milk machines and living forests commercial wood, and it destroys the "organic processes and rhythms and regenerative capacities" of nature.[49] It is based "on the metaphor of 'man's empire over inferior creatures' rather than the metaphor of 'the democracy of all life.'"[50]

Throughout her work Shiva stresses that monocultures destroy biodiversity, livelihoods, and spread chemical pollution. Tropical forests, for instance, "the creators of the world's climate," are cut down or submerged by large dams, destroying the delicate balance of their lifecycles and disrupting rivers and other water sources, often resulting in desertification of the land. In colonial times, the British viewed Indian forests as timber mines and exploited them for military purposes such as shipbuilding, which led to deforestation.[51] More recently, monocultures of such non-native trees as eucalyptus, a tree suited for pulpwood, have been planted. This tree, however, destroys the water cycle, particularly in semi-arid regions, because of its high water needs, and its failure to produce humus depletes the soil.[52] Robbing forests of their biodiversity through clear-cutting or commercial tree monocultures leaves behind impoverished communities and depleted ecosystems "which can no longer produce biomass or water."[53]

Similarly, in contrast to small-scale sustainable shrimp farming, practiced for centuries, large commercial shrimp aquacultures, known as a "rape and run" industry and often subsidized by the World Bank, have a devastating impact on

---

[43] Gore 1992, 77.

[44] Mander, 1996, 5.

[45] Norberg-Hodge 2002, 13.

[46] Shiva 1997, 102.

[47] Shiva 1993, 303.

[48] Shiva 2013, 9.

[49] Mies and Shiva 1993, 25.

[50] Shiva 1995b, 50.

[51] Shiva 1993, 304, 306-07.

[52] Shiva 1993, 313.

[53] Mies and Shiva 1993, 71.

their surroundings.[54] Mangroves are cleared to make room for factory farms that require enormous amounts of fish to feed the shrimp, thereby damaging local fishing industries. Waste from the farms pollutes the sea and damages coastal agriculture. The farms cause salinization of the surrounding land and drinking water and destroy nearby trees and crops.[55] Clearing mangroves for shrimp farms or other development not only destroys habitat for fish, but also endangers people. The Indian Ocean tsunami in December 2004 was especially destructive where mangroves had been cut down: "Wherever mangroves survived, people survived." The tsunami should remind us, Shiva observes, "that we are all interconnected through the earth."[56]

Shiva is especially critical of the change from traditional sustainable farming to unsustainable agricultural monocultures that, in her opinion, began with the Green Revolution and increased under globalization. As the glowing tributes on the death of Norman Borlaug, "the father of the Green Revolution," demonstrate, many credit the Green Revolution with preventing famines. Shiva believes, however, that the Green Revolution actually caused hunger. By integrating Third World farmers "into the global markets of fertilizers, pesticides and seeds," it disintegrated "their organic links with their soils and communities" and replaced the regenerative organic nutrient cycle with chemicals.[57] While rice yields were often high, the new seeds took nutrients out of the soil and required irrigation, causing in some regions "large-scale mining of groundwater."[58] Industrial farming created waterlogged or salinized soils, desertification, and diseased crops.[59] She points out, for example, that the new rice hybrids produce less straw for cattle feed and supplant indigenous seeds with a small number of engineered strains. The required herbicides and pesticides destroy native flora and fauna, pollute the soil and contribute to invasive super weeds, resistant to herbicides, and insect pests, resistant to pesticides.[60] When, however, the economic yield of monocultures is measured, the costs of herbicides, pesticides, and environmental degradation are not calculated. The focus is on short-term profits and how crop yields contribute to GNP with no consideration given to the long-term consequences in the future.

In Shiva's view there is a "corporate hijacking of our seed and food."[61] Transnational corporations such as Monsanto view seed as a resource "to be engineered, patented and owned for corporate profit."[62] These genetically modified seeds do not reproduce themselves and "essentially become seed morgues," forcing farmers to buy seed from seed companies each year.[63] Shiva fears that "every vital, living resource of the planet that maintains the fragile web of life is in the process of being privatized, commodified, and appropriated by

---

[54] Shiva 2000a, 45.
[55] Shiva 2000c, 96.
[56] Shiva 2006b, 206.
[57] Mies and Shiva 1993, 113.
[58] Shiva 1995a, 203.
[59] Mies and Shiva 1993, 28.
[60] Shiva 2000a, 16.
[61] Shiva 2000a, 4.
[62] Mies and Shiva 1993, 28.
[63] Shiva 2000a, 82.

corporations."[64] Transnational corporations increasingly engage in what she terms biopiracy: "Patent claims over biodiversity and indigenous knowledge that are based on the innovation, creativity, and genius of the peoples of the Third World are acts of biopiracy."[65] Patents on life forms such as seeds, allowed through international patent laws and enabled through the World Trade Organization (WTO) and the General Agreement on Tariffs and Trade (GATT), not only prevent the free exchange of seeds, but also steal indigenous knowledge. There are patents on seeds such as basmati rice, varieties of which were developed in India. Another example of biopiracy is the neem, a native Indian tree used for centuries as a biopesticide and for its medicinal and anti-bacterial properties. This was "metaknowledge" in the public domain until Western companies took out patents on it.[66] Biopiracy, she stresses, is a "double theft because first it fosters theft of creativity and innovation, and secondly, it deprives people of ... the full use of indigenous biological resources" previously available for everyone to use.[67]

Shiva has harsh words for globalization, which she calls "environmental apartheid"[68] and "maldevelopment."[69] In her view, globalization is a Western male-dominated paradigm imposed on Third World countries. It creates freedom for corporations, but not for citizens. It has created the "McDonaldization of world food,"[70] deeming locally produced fresh food backward and "stale food clothed in aluminum and plastic" modern.[71] As a consequence of globalization, millions of people have been uprooted worldwide and their links to the soil and their livelihoods sacrificed for such economic development as large dams, factories, and monocultures, thus transforming "organic communities into groups of uprooted and alienated individuals."[72] It has promoted unsustainable agricultural practices, which are, in her view, "programmes of hunger generation" because fertile land is diverted for monocultures.[73] Small farmers are displaced, and the biological diversity, which provided much of the food, is replaced by cash crops, ill-suited to local conditions.

Shiva blames transnational corporations and international institutions for exacerbating environmental problems.[74] The World Bank, for example, finances the relocation of pollution-intensive industry to the Third World and displaces millions of tribal people by financing such projects as dams. In her opinion, GATT and the WTO foster exploitation, greed and profit maximization.[75] She also criticizes the corporate use of Orwellian doublespeak – the metamorphosis, for example, of former pesticide companies into "Life Sciences" which she calls

---

[64] Shiva 2005a, 13.
[65] Shiva, 2005c, 102.
[66] Shiva 1997, 71.
[67] Shiva 2005c, 104.
[68] Shiva 2000b, 135.
[69] Mies and Shiva 1993, 79.
[70] Shiva 2000a, 70.
[71] Shiva 2010, xx.
[72] Mies and Shiva 1993, 99.
[73] Mies and Shiva 1993, 78.
[74] Shiva 1994, 102-106.
[75] Shiva 2000b, 130.

"Death Sciences," because they are "in reality peddlers of death."[76] A glance at a website such as Monsanto's illustrates how it, like other transnational corporations, has co-opted the language of sustainability to legitimize and market its products.

Shiva is equally critical of the "white" revolution. In her estimation, globalization is changing India from a country "where the sacred cow was worshipped to an economy where cows are slaughtered for export" and is trying to convert a predominantly vegetarian society into one with a beef-eating Western-style diet.[77] It replaces indigenous cattle, bred for hardiness, milk production and draft power, with less hardy and disease-prone animals bred for milk or meat. In Shiva's opinion, the industry does not regard cattle as living creatures, but as raw material, and it transforms herbivores into carnivores through feed that has caused bovine spongiform encephalitis (BSE). Shiva believes that "sacred cows are a metaphor of ecological civilization," whereas "mad cows are a metaphor of an anti-ecological, industrial civilization."[78] Such unethical treatment is contrary to "the Indian culture of reverence for farm animals."[79]

Climate Change and its "Pseudo-Solutions"

Recently Shiva has turned her attention to global warming, stressing that our very large ecological footprint drives both species extinction and climate change.[80] The climate crisis threatens the survival of the human species, because there is, as she notes "no place to hide."[81] Shiva discusses severe climate changes in the Himalayas, which she calls the Third Pole: rainfall has become erratic; there is less snow at higher altitudes; perennial streams are drying; and glaciers are retreating. For example, in the last three decades, the Gangotri glacier, the source of the Ganges, has receded rapidly, and the resulting loss of seasonal glacial melt threatens water supplies for 500 million people and 37 per cent of India's irrigated land.[82] Yet the fossil-fuel economy, responsible for the greenhouse gases, continues on its path of "development." In India, she points out, ever more factories are built destroying both fertile land and adding to greenhouse emissions.[83] The increase in Indian car production, which she calls an "ecological catastrophe by design," necessitates consumption of more oil and the building of infrastructure, "substituting sacred rivers with highways" and replacing non-polluting means of transportation such as bicycles, walking, and animals with highly polluting vehicles.[84] Producing fertilizers for monocultures adds to greenhouse gases, and monocultures, especially tree plantations, disrupt water cycles. She points out that corporations such as Walmart not only destroy India's

---

[76] Shiva 2000c, 104, 114.
[77] Shiva 2006b, 193-4.
[78] Shiva 2000a, 73-74.
[79] Shiva 2006b, 194.
[80] Shiva 2013, 8.
[81] Shiva 2008a, 3.
[82] Shiva 2013, 107, 104.
[83] See Shiva 2013, 233-256.
[84] See Shiva 2008a, 49-76.

traditional small producers and small retailers and waste large amounts of food, at a time when many are hungry, but also increase the distance that food is transported, thereby contributing to ever more emissions.[85]

Shiva stresses the need to challenge fundamental assumptions and norms in society that have caused the problems of climate change, but she also stresses that "our dependence on fossil fuels has fossilized our thinking."[86] Current efforts to mitigate the impact of climate change show such fossilized thinking and she terms them unsustainable "pseudo-solutions" to climate chaos. She argues, for example, that carbon trading "defends the rights of corporations to pollute the atmosphere and destabilize the climate."[87] It rewards corporations for polluting and allows them to continue robbing people of their rights to clean air. Nuclear energy, another proposed solution, is neither clean nor sustainable. As she emphasizes: "Nuclear winter is not an alternative to global warming."[88] Biofuels, she believes, are also not effective sources of renewable energy. Producing them requires more fossil fuel energy than biofuels can generate, and, as new lands are cleared for biofuel plantations, the planet increasingly loses its rainforests, such as in Indonesia and Brazil. Monocultures designated for biofuels lead to increased use of fertilizers and pesticides and put stress on already weakened water systems. Biofuels become a threat to food security because prices of staples such as corn, for example, increase and become unaffordable for poor people, and land is taken away from other crops thus placing pressure on food supplies. Corn, soy and other crops feed cars, while people starve.[89]

The Impact of Unsustainable Practices on People

For Shiva corporate globalization is not only a war against the Earth, but also "against people, against democracy and against freedom."[90] Shiva, a strong advocate not only of biological, but also of cultural diversity, believes that sustainability is closely linked with social justice, peace, and democracy.[91] In her work, she focuses in particular on the plight of those who live on the margins of Indian society such as small farmers who depend on biodiversity for survival. They are harmed by environmental degradation, dispossessed by monocultures, and their rights in their local communities are undermined by globalization.[92] Monocultures, industry, and large-scale dams rob people of what she calls the commons, for example land used for grazing animals, water, and clean air. Shiva criticizes Indian oligarchs for promoting unsustainable development and degrading the environment,[93] and she is particularly outspoken against the unethical practices of transnational corporations that aggressively market their transgenic seeds to poor farmers who become overwhelmed with debt when crops

---

[85] See Shiva 2013, 212-220.
[86] Shiva 2008a, 130.
[87] Shiva 2008a, 20.
[88] Shiva 2008a, 27.
[89] See Shiva 2008a, 77-94.
[90] Shiva 2013, 5.
[91] Shiva 1997, 126.
[92] Shiva 2000b, 137.
[93] See Shiva 2013, 233-56.

fail. Seed, previously a renewable resource that indigenous cultures shared with the world, now has to be bought, and genetically modified seeds and the necessary chemicals are expensive.[94] In Warangal in Andhra Pradesh, for example, where farmers used to grow millets, pulses and paddy, land was converted to cotton that requires intensive irrigation and is vulnerable to pest attacks. Because of failed crops and the resulting debt this area suffered an epidemic of farmer suicides.[95]

Poisons in our Food is an important work by Shiva, Mira Shiva, and Vaibhav Singh, that express concerns about the increasing use of agricultural chemicals that not only pollute land and water, but also harm human health. Pesticide residues appear in food, making it no longer healthy, but hazardous, as well as in the human body such as in human breast milk. They call pesticides "weapons of mass destruction."[96] Small Indian farmers, who often overuse such chemicals hoping to maximize profits from their crops, are not trained in their safe application and disposal and are often the victims of acute poisoning from them. The authors point out that safety instructions on the chemical containers are often difficult to follow and are, moreover, written in languages unfamiliar to the farmers. Why, they ask, should safety instructions on pesticide containers sold for example in Maharashtra not be written in Marathi? Farmers are not the only victims of pesticides. Since much of India is agrarian, the country can turn "into the capital of cancers, chronic diseases, and birth defects."[97] As they point out, fetuses, infants and children are particularly vulnerable if they are "exposed to pesticides during the critical stage of their development" and they "may incur such harm that the entire generation of future India may be adversely affected."[98] They cite studies that examine connections between pesticides and a variety of serious health problems, including reproductive problems, such as reduced sperm count and stillbirths – people become, in effect, like genetically modified seeds that cannot reproduce successfully. In their opinion, agrochemicals not only lead to species extinction, but also threaten the sustainability of healthy human beings in the present, and harm the health of future generations.

In Shiva's view, the destruction of biodiversity is closely linked to the erosion of respect for cultural diversity. Such monocultures of the mind lead to "monocultures of religion," which promote intolerance for cultural and religious diversity and foster religious fundamentalism. Economic insecurity caused by globalization gives rise, she believes, to right wing fundamentalism, extremism, and terrorism that spawn "vicious cycles of violence, injustice, and fear."[99] By destroying local communities globalization "ruptures the delicate fabric of equity, democracy, and pluralism" and sows "the seeds of inequality, exclusion, fundamentalism, and violence."[100] As globalization spreads, she argues, extremist violence and religious intolerance follow. Terrorists, who "are made, not born," are symptoms of unjust societies. Creating sustainability and justice is the only

---

[94] Shiva 2005c, 100.
[95] Shiva 2006a, 222.
[96] Shiva, Shiva and Singh 2013, 128.
[97] Shiva, Shiva and Singh 2013, 19.
[98] Shiva, Shiva and Singh 2013, 49.
[99] See Shiva 2005a, 54-55.
[100] Shiva 2005a, 78.

effective strategy in her view for controlling the emergence of both pests and terrorism. The Western war mentality, she notes, will fail to prevent young people from becoming terrorists, but will instead "create more resilient superterrorists, just as pesticides, herbicides, and genetic engineering have created superpests and superweeds."[101]

The main victims of globalization, Shiva believes, are poor rural women. The increasing scarcity of natural resources makes it harder for women to fulfill their traditional gender roles of feeding their families. As forests are cut down, women have to spend longer searching for fodder and fuel; as wells are poisoned by herbicides and pesticides and water sources disrupted, they have to walk farther for clean water. Although women are major producers of food, in the global market economy their work is invisible and they have become increasingly marginalized and made dispensable. Like other feminists she highlights the critical role women's expertise has played in the survival of humankind.[102] Western male-dominated science and technology have, however, devalued women's knowledge about sustainable agriculture and ignored their wealth of expertise. As women's work has been devalued, so has their status as human beings been devalued. To underscore her argument, Shiva points out the difference between the traditional "bride price" and the growing dowry system. The "bride price," which the bridegroom's family typically gave to the bride's family, recognized the value of women's work and was intended to compensate the bride's family for losing her productive work because of her marriage. In contrast, a dowry, which the bride's family has to pay, does not recognize women's work and in fact "devalues women by defining them as a burden."[103] Increasing incidents of rape, female feticide and dowry deaths in India illustrate in her view "how capitalist patriarchy and traditions of religious patriarchy are converging to unleash new levels of violence against women." In Shiva's view, "the displacement of women from productive work, the destruction of economies of sustenance, and the growth of the culture of consumerism and commodification, have all combined to devalue women in society." With the availability of modern technology to determine the sex of a fetus, some families think it cheaper to abort a female fetus than to later pay for a dowry.[104]

Earth Democracy: Toward a Sustainable Future

To combat problems caused by unsustainable practices that have damaged soil, water and air, Shiva urges a return "to a women-centered, nature-friendly agriculture."[105] Biodiverse, organic small farms provide more nutrition per acre than monocultures and do not deplete the soil and water. To combat Western male-oriented reductionist science and globalization that, in her view, have turned the planet into a supermarket where everything is for sale, she urges the rejection of male hegemony over knowledge and a return to feminist perspectives that locate "production and consumption within the context of regeneration" and view

---

[101] Shiva 2005a, 73.
[102] See, for example, work by Merchant 1980 and Harding 1986.
[103] Shiva 2005b, 135.
[104] Shiva 2005b, 134-138.
[105] Shiva 2005a, 93.

the Earth as "an active subject, not merely as a resource to be manipulated and appropriated."[106] In Hindu terms, this means a return to *Shakti*, the feminine creative principle, from which all life arises and requires a reverence for the Earth and its varied life forms. Although rural women have, in Shiva's view, been the main victims of globalization, they have also been at the forefront of ecological struggles to preserve forests, lands, and water. For example, poor women in Ecuador struggled to save mangroves as breeding grounds for fish and shrimp.[107] Shiva pays particular tribute to the Chipko movement of peasant women in the Himalayas who embraced trees to prevent logging. These women, who like Shiva followed Gandhi's principle of nonviolence, became role models for her and her university in ecology.

Besides raising awareness of sustainability and biodiversity through her writing, Shiva started building Navdanya (nine seeds) in the early 1990s, an organization that preserves indigenous seed and encourages a return to traditional sustainable organic farming methods that she calls nonviolent agriculture. Shiva distinguishes, however, between local organic farming intended for local consumption and pseudo-organic farming where produce is grown on large farms and shipped long distances.[108] Shiva views seed as a symbol of freedom. Navdanya's slogan is: "Native seed – indigenous agriculture – local markets." It builds community seed banks, strengthens farmers' seed supplies, and searches for sustainable agricultural options for different regions.[109] Navdanya's numerous and increasing members conserve biodiversity, use organic farming methods, and share seeds, refusing to cooperate with patent laws that make saving seeds a crime, and Navdanya continues to mobilize Indian farmers to reject transgenic seed.[110] Through outreach it educates farmers about the advantages of using native seeds and biodiverse farming methods. The Navdanya network has already set up over a hundred freely accessible community seed banks across India, has trained thousands of farmers across the country through large camps or attendance at its schools, and has run teacher training and research on farms.[111] Seed becomes for them a symbol of hope: "We spread seeds of freedom instead of seeds of slavery and seeds of suicide."[112] Like Gandhi's spinning wheel, seed, in which "cultural diversity converges with biological diversity," is for Shiva a symbol of freedom and protest against globalization.[113] In keeping with Navdanya's focus on outreach and education, Shiva and Satish Kumar co-founded Bija Vidyapeeth (the school of the seed) in 2002 with goals similar to those of its sister college (Schumacher College in the UK) to foster sustainability, education in Earth democracy and communal living. Its productive eight-acre farm has rejuvenated soil left desertified by eucalyptus monoculture and produces more than 600 varieties of plants, including 250 rice varieties. To fight for

---

[106] Mies and Shiva 1993, 33-34.
[107] Mies and Shiva 1993, 3.
[108] Shiva 2008a, 125.
[109] Shiva 2000b, 148.
[110] Shiva 2000a, 3.
[111] Navdanya 2014.
[112] Shiva 2008a, 119.
[113] Shiva 1997, 126.

sustainability, the school promotes holistic solutions based on biodiversity, justice and democracy.

Shiva's outrage at the patenting of life forms has made her an outspoken critic of the World Trade Organization, the World Bank, GATT and the Agreement on Trade-Related Aspects of Intellectual Property Rights (TRIPS) and she has been active in grassroots campaigns opposing corporate patents on indigenous knowledge. [114] Shiva's research foundation took part in the 1994 challenge to a patent held by the transnational corporation W. R. Grace on properties of the neem tree, and after a ten-year legal struggle the European Patent Office withdrew the patent. This was the first successful challenge against a transnational corporation's patent on genetic resources of the South. Among other successes were challenges to RiceTec's patent claims to basmati rice and Monsanto's claim for wheat. In June 2002 a data bank to preserve traditional knowledge about plants and animals and the therapeutic effects of natural substances was begun as a strategy to prevent such biopiracy by transnational corporations. [115]

Protecting Earth democracy means, in Shiva's view, being an activist. She has long been involved in many demonstrations, such as grassroots protests against large-scale shrimp farms. She supports the growing ecological movements in India that work to implement a holistic development process to ensure "justice with sustainability and equity with ecological stability." [116] Following Gandhi's example of the salt march, Shiva participates in yatras (marches). A march to raise awareness about farmer suicides started in Nagpur in Gandhi's ashram and went to Bangalore. [117] On March 5, 1999, about 2,000 groups in India joined with Navdanya to launch the Bija (seed) Satyagraha to defend people's rights to biodiversity, to protest unjust seed laws, and to reject "the colonization of life through patents." [118] She reflects: "I think that we need satyagraha everywhere." [119]

Shiva strongly advocates the rejection of the globalized Western reductionist paradigm and the creation of resilient people-centered local communities and nonviolent sustainable practices: "The age of soil symbolizes the age of Gaia, of the flowering of diversity and democracy, of justice, sustainability, and peace." [120] It is crucial for what she calls "Earth democracy" to empower people at the local level rather than allow decisions to be imposed on them by people far away such as scientists in university laboratories, businessmen in corporate boardrooms, or institutions like the WTO who have no understanding of local conditions or traditions. Local communities help protect local ecosystems and local cultures, produce locally grown fresh foods, reduce food miles and greenhouse emissions, thereby "enriching our lives while lowering our consumption without impoverishing others." [121]

---

[114] Shiva 1994, 102-106.
[115] Shiva and Brand 2005, 52-54.
[116] Bandyopadhyay and Shiva 1995, 313-14.
[117] Shiva 2008b, 205.
[118] Shiva 2005c, 105-06.
[119] Shiva 2008b, 203.
[120] Shiva 2008a 7.
[121] Shiva 2008a, 46.

Despite the numerous problems she highlights and the bleak picture she paints Shiva nevertheless remains optimistic that people can bring about positive change and she envisions a future based "on inclusion, not exclusion; on nonviolence, not violence; on reclaiming the commons, not their enclosure; on freely sharing the earth's resources, not monopolizing and privatizing them."[122] She stresses that people need to move beyond an oil economy and reinvent society: "We need to do it fast and we need to do it creatively. We can."[123] She is convinced that people have "the potential to participate creatively in building alternatives to the systems designed for total control and limitless profits" and that "we have just begun to tap our potential for transformation and liberation. This is not the end of history, but another beginning."[124]

Criticisms of Shiva's Thinking

Some scholars have taken issue with some aspects of Shiva's voluminous writings while at the same time praising her leadership in raising important issues. Some criticize her for ignoring the negative aspects of nature. They point out correctly that nature "is not always harmonious or comforting" and contains "pathogens, parasites and processes that do not always lead to such an idyllic vision."[125] Shiva is faulted for romanticizing pre-colonial life, a time when "women not only did not own property, they were considered to *be* property."[126] Others fault her for downplaying caste, class and patriarchy in Vedic and pre-Vedic cultures and argue that her focus on the *Shakti* tradition, the feminine principle expressed in Hindu terms, makes it difficult for her to speak to Dalits, tribals, Muslims, Sikhs, and Christians.[127] Shiva is criticized for idealizing Indian subsistence agriculture. By so doing she conjures up a past "where people lived in perfect harmony with nature, and women were highly respected," a past that may not have existed.[128] Another criticism is that Shiva's examples relate to poor rural women primarily from Northwest India, but her generalizations "conflate all Third World women into one category." Agarwal observes that by not differentiating between women of different classes, castes, and races, Shiva essentializes women and she argues that Shiva's emphasis on the feminine principle relates to Hindu discourse alone and does not apply to all Indians. By attributing existing problems to colonialism and Western science, Shiva ignores pre-existing inequalities, such as in the Mughal period, which was "considerably class/caste-stratified."[129] As Agarwal notes, Shiva's studies of poor rural women do not apply to all women. Women in wealthier rural households, for example, have deep wells and do not have to spend many hours each day procuring drinkable water, and women in urban environments face a variety of different economic and environmental problems. Shiva, however, does not claim to

---

[122] Shiva 2005b, 4.
[123] Shiva 2008a, 1.
[124] Shiva 2005b, 184-6.
[125] Subramaniam et al. 2002, 204.
[126] Menon 1999, 4.
[127] Dietrich 1999, 78-79.
[128] Braidotti et al. 1997, 58.
[129] Agarwal 1999, 102-104.

represent the situation of all women. Her focus is on those she considers most affected by monocultures and globalization. Although she is criticized for her use of Hindu discourse, Shiva reaches out successfully beyond the Hindu context, as her Right Livelihood Award demonstrates. The practical steps she has initiated, such as the founding of Navdanya and the Bija Vidyapeeth, and her Gandhian methods of protest are effective models for a wide variety of communities and locales and for people of different faiths and class.

Conclusion

Shiva has shaped debate about hybrid seeds and commercial agriculture and she has been and continues to be a tireless crusader for sustainability, biodiversity and economic and social justice. As Tal argues, "No individual has been as important in raising questions about agriculture, the environment, and the social impacts of free trade and transnational corporate abuse as Vandana Shiva."[130] She challenges the Western model of development and argues for a return to local sustainable farming as a solution to the present crisis and a recovery of indigenous, especially female knowledge. Her promotion of sustainable farming, her seed banks, organic farm and school, and her leadership in grassroots movements are all models that others can and have already emulated. For Shiva "ecological resilience is born of reverence for all species and protection of biodiversity. In caring for the Earth family, we secure our own future." [131] "Our own future" includes our responsibility to care for the Earth for the sake of future generations. As the Native American Oren Lyons expresses it: "Take care how you place your moccasins upon the earth, step with care, for the faces of the future generations are looking up from the earth waiting for their turn for life."[132]

References

Agarwal, Bina. 1999. The Gender and Environment Debate: Lessons from India. In *Gender and Politics in India*, edited by Nivedita Menon, 96-142. Oxford: Oxford University Press.

Bandyopadhyay, Jayanta, and Vandana Shiva. 1995. Development, Poverty and the Growth of the Green Movement in India. In *A Survey of Ecological Economics*, edited by Rajaram Krishnan, Jonathan M. Harris, and Neva R. Goodwin, 311-314. Washington: Island Press.

Barsamian, David. 2002. Monocultures of the Mind: An Interview with Vandana Shiva. Available at http://www.thirdworldtraveler.com/Vandana_Shiva/Monocultures_Mind .html, accessed February 5, 2014.

Braidotti, Rosi, Ewa Charkiewicz, Sabine Hausler, and Saskia Wieringa. 1997. Women, the Environment and Sustainable Development. In *The Women, Gender & Development Reader*, edited by Nalini Visvanathan, Lynn

---

[130] Tal 2006, 218.
[131] Shiva 2006a, 206.
[132] Quoted in Mies and Shiva 1993, 88.

Duggan, Laurie Nisonoff, and Nan Wiegersma, 54-61. London: Zed Books.

Carson, Rachel. 1964. *Silent Spring*. New York: Fawcett Crest.

Dietrich, Gabriele. 1999. Women, Ecology and Culture. In *Gender and Politics in India*, edited by Nivedita Menon, 72-95. Oxford: Oxford University Press.

Gore, Al. 1992. *Earth in the Balance: Ecology and the Human Spirit*. Boston, New York, London: Houghton Mifflin.

Harding, Sandra. 1986. *The Science Question in Feminism*. Ithaca: Cornell University Press.

Mander, Jerry. 1996. Facing the Rising Tide. In *The Case Against the Global Economy and for a Turn Toward the Local*, edited by Jerry Mander and Edward Goldsmith, 3- 19. San Francisco: Sierra Club Books.

Menon, Nivedita, 1999. Introduction. In *Gender and Politics in India*, edited by Nivedita Menon, 1-36. Oxford: Oxford University Press.

Merchant, Carolyn. 1980. *The Death of Nature: Women, Ecology, and the Scientific Revolution*. New York: Harper and Row.

Mies, Maria, and Vandana Shiva. 1993. *Ecofeminism*. London: Zed Books.

Navdanya. 2014. Nine Seeds. Webpage available at http://www.navdanya.org, accessed February 10, 2014.

Norberg-Hodge, Helena. 2002. Global Monoculture: The Worldwide Destruction of Diversity. In *Fatal Harvest: The Tragedy of Industrial Agriculture*, edited by Andrew Kimbrell, 13-15. Washington: Island Press.

Shiva, Vandana. 1992. Recovering the Real Meaning of Sustainability. In *The Environment in Question: Ethics and Global Issues,* edited by David E. Cooper and Joy A. Palmer, 187-193. London and New York: Routledge.

Shiva, Vandana. 1993. Colonialism and the Evolution of Masculinist Forestry. In *The "Racial" Economy of Science: Toward a Democratic Future*, edited by Sandra Harding, 303-314. Bloomington: Indiana University Press.

Shiva, Vandana. 1994. International Institutions Practicing Environmental Double Standards. In *50 Years is Enough: The Case Against the World Bank and the International Monetary Fund*, edited by Kevin Danaher, 102-106. Boston: South End Press.

Shiva, Vandana. 1995a. Biotechnological Development and the Conservation of Biodiversity. In *Biopolitics: A Feminist and Ecological Reader on Biotechnology*, edited by Vandana Shiva and Ingunn Moser, 193-213. London and New Jersey: Zed Books.

Shiva, Vandana. 1995b. Democratizing Biology: Reinventing Biology from a Feminist, Ecological and Third World Perspective. In *Reinventing Biology: Respect for Life and the Creation of Knowledge*, edited by Lynda Birke and Ruth Hubbard, 50-71. Bloomington: Indiana University Press.

Shiva, Vandana.1997. *Biopiracy: The Plunder of Nature and Knowledge*. Boston: South End Press.

Shiva, Vandana. 2000a. *Stolen Harvest: The Hijacking of the Global Food Supply*. Cambridge MA: South End Press.

Shiva, Vandana. 2000b. Ecological Balance in an Era of Globalization. In *Principled World Politics: The Challenge of Normative International*

Relations, edited by Paul Wapner and Lester Edwin J. Ruiz, 130-149. Lanham: Rowman and Littlefield.

Shiva, Vandana. 2000c. War Against Nature and the People of the South. In *Views from the South: The Effects of Globalization and the WTO on Third World Countries*, edited by Sarah Anderson, 91-125. Chicago: Food First Books.

Shiva, Vandana, 2005a. *India Divided: Diversity and Democracy Under Attack*. New York: Seven Stories Press.

Shiva, Vandana. 2005b. *Earth Democracy: Justice, Sustainability, and Peace*. Brooklyn, NY: South End Press.

Shiva, Vandana. 2005c. Global Trade and Intellectual Property: Threats to Indigenous Resources. In *Rights and Liberties in the Biotech Age: Why We Need a Genetic Bill of Rights*, edited by Sheldon Krimsky and Peter Shorett, 98-106. Lanham: Rowman and Littlefield.

Shiva, Vandana. 2006a. Sharing and Exchange, the Basis of Our Humanity and Our Ecological Survival Has Been Redefined as a Crime. In *Speaking of Earth: Environmental Speeches that Moved the World*, edited by Alon Tal, 221- 231. New Brunswick: Rutgers University Press.

Shiva, Vandana. 2006b. The Implications of Agricultural Globalization in India. In *Animals, Ethics and Trade: The Challenge of Animal Sentience*, edited by Jacky Turner and Joyce D'Silva, 193-207. London: Earthscan.

Shiva, Vandana. 2008a. *Soil Not Oil: Environmental Justice in an Age of Climate Crisis*. Brooklyn, NY: South End Press.

Shiva, Vandana. 2008b. Fighting Indiscriminate Globalization. In *India Revisited: Conversations on Contemporary India*, edited by Ramin Jahanbegloo, 200-209. Oxford: Oxford University Press.

Shiva, Vandana. 2010. *Staying Alive: Women, Ecology and Development*. Brooklyn, NY: South End Press.

Shiva, Vandana 2011. *The Corporate Control of Life/Die Kontrolle von Konzernen über das Leben*. Ostfildern: Hatje Cantz Verlag.

Shiva, Vandana. 2013. *Making Peace with the Earth*. London: Pluto Press.

Shiva, Vandana, and Ruth Brand. 2005. The Fight Against Patents on the Neem Tree. In *Limits to Privatization: How to Avoid too Much of a Good Thing*, edited by Ernst Ulrich von Weizsäcker, Oran R. Young, and Matthias Finger, 51-54. London: Earthscan.

Shiva, Vandana, Mira Shiva, and Vaibhav Singh. 2013. *Poisons in our Food: Links Between Pesticides and Diseases*. Dehra Dun: Natraj.

Subramaniam, Banu, James Bever, and Peggy Schultz. 2002. Global Circulations: Nature, Culture and the Possibility of Sustainable Development. In *Feminist Post-Development Thought: Rethinking Modernity, Post-Colonialism and Representation*, edited by Kriemild Saunders, 199-211. London: Zed Books.

Tal, Alon, ed. 2006. *Speaking of Earth: Environmental Speeches that Moved the World*. New Brunswick: Rutgers University Press.

The Right Livelihood Award: Vandana Shiva. 1993. Available at http://www.rightlivelihood.org/v-shiva.html, accessed February 10, 2014.

Chapter 17: Ecuador: Towards Mother Nature's Utopia

Spencer S. Stober

Introduction

"La naturaleza o Pacha Mama, donde se reproduce y realiza la vida, tiene derecho a que se respete integralmente su existencia y el mantenimiento y regeneración de sus ciclos vitales, estructura, funciones y procesos evolutivos." [1] These revolutionary words introduce a section on the rights of nature in Ecuador's 2008 constitution, and their significance is not lost with the following English translation: "Nature, or Pacha Mama, where life is reproduced and occurs, has the right to integral respect for its existence and for the maintenance and regeneration of its life cycles, structure, functions and evolutionary processes." [2] The Community Environmental Legal Defense Fund (CELDF) [3] assisted the Ecuadorian Constituent Assembly in preparing this language which passed with an overwhelming majority of votes cast by Ecuadorians in a national referendum on September 28, 2008. On that day, Thomas Linzey, Executive Director of the CELDF, announced that "Ecuador is now the first country in the world to codify a new system of environmental protection based on rights."[4] The Carter Center monitored the voting process which followed a period in Ecuador when popular protests were common, with eight presidents from 1997 to 2007 when President Rafael Correa assumed office and called for a popular referendum process that ultimately led to a new constitution in 2008. The Carter Center reported that "new institutions outlined in the approved constitution will have as much legitimacy and relevance as bestowed upon them by the main social and political actors."[5]

---

[1] República de Ecuador, Constitución de 2008, Article 71.

[2] Constitution of the Republic of Ecuador. 2008, Article 71. Note that Pacha Mama is sometimes written in English as Pachamama.

[3] CELDF 2014.

[4] Linzey 2008.

[5] The Carter Center 2008.

This chapter considers the cultural and sociopolitical factors that influenced Ecuador's recent move to recognize the rights of nature in its 2008 constitution. Factors influencing this unprecedented event are better understood when examined sequentially. First, there is an important historical context that enabled indigenous populations to garner a political voice that was eventually codified in Ecuador's 1998 constitution. Second, the recognition of constitutional rights for nature (ecosystems) in Ecuador's 2008 constitution is an event that was predicated on the granting of rights for indigenous people in the 1998 constitution. Their culture and sociopolitical influence contributed to this ecocentric step by Ecuador. Third, it is necessary to discuss practical aspects of environmental provisions in constitutions, and fourth, to consider the rights of nature language in Ecuador's 2008 constitution. Finally, questions concerning the implications of this bold step are raised, but Ecuador may be working toward a perfect place for Mother Nature to thrive.

Gaining a Political Voice

This section considers the historical context that enabled indigenous populations to achieve a political voice and rights set forth in Ecuador's 1998 constitution. Their efforts were then instrumental in Ecuador's move to recognize the rights of nature in its 2008 constitution. The historical context can best be described by taking a look at the people and their history. Eduardo Galeano, in *Open Veins of Latin America*, calls an important historical fact to our attention: "Latin America's underdevelopment arises from external development, and continues to feed it."[6] An exodus of resources from Latin America has and continues to fuel the developed world's appetite, and throughout this process, new and different cultures entered the sociopolitical structure of the continent. Ecuador is a microcosm of Latin America with its varied geography and a diverse population that is described as follows by John D. Martz.

> Its [Ecuador's] inhabitants include proud descendants of white European colonizers, Indian heirs to the great Incan civilization, blacks carrying the legacy of African origins, primitive nomadic hunters and fishermen, and the racial mixture produced by generations of miscegenation and intermarriage. Political ideology runs the gamut from the unbounded radicalism of the extreme Marxist left to the reactionary falangist Hispanidad which worships at the grail of sixteenth-century Spain. Between these extremes you can find conservatives, liberals, populists, Christian democrats, communists, and socialists.[7]

Given their colonial history, it took a concerted action for indigenous peoples to gain a political voice in Ecuador. To that end, the Confederation of Indigenous Nationalities of Ecuador (CONAIE) became a powerful force in securing rights for indigenous people, and ultimately, in the development of a constitution to

---

[6] Galeano 1997, 285.
[7] Martz 1972, 2.

sustain those rights.[8] Robert Andolina reports being a participant observer during this social movement and he provided the following summary. CONAIE, in 1986, along with human rights organizations, environmentalists, and evangelical federations, pushed for collective control over land, infrastructure, natural resources, government programs, and education. Indigenous groups were seeking autonomy and, at the same time, access to, and participation in, social and political life. Their goal was an administratively decentralized and culturally heterogeneous "plurinational state." *Proyecto Político* was a constitutional reform platform put forward by CONAIE in 1994 to promote their vision of a plurinational state by focusing on basic needs, local markets, and inter-cultural dialogue and respect. This democratic forum movement then expanded with support from class-based peasant unions and international organizations including the Catholic Church. The Democratic Forum became the *Coordinadora de Movimientos Sociales*, and along with CONAIE and citizens, the movement was called *Pachakutik* and won national congressional seats and positions in local government. Social protests organized by the Patriotic Front which included CONAIE, labor unions, teachers and students, on February 5, 1997, upset President Bucaram's so-called "delegative democracy" for failing to act on populist promises made during his campaign, and congressional President Fabián Alarcón assumed the interim presidency.[9] The constitutional reforms that followed were grass-roots driven. Andolina concludes that the results were unprecedented.

> [The 1998 Ecuadorian Constitution]…contains rights that are unprecedented in their collective character and in the pertinence to non-Western cultural beliefs and practices: communal land, indigenous (and Afro-Ecuadorean) territorial 'circumscriptions,' development with identity managed by indigenous people, education in indigenous languages, indigenous judicial and health practices, representation in all government bodies, participation in resource use decisions, environmental preservation in indigenous lands and collective intellectual property rights.[10]

Many of the sociopolitical factors that led to these unprecedented changes in the 1998 constitution remain active today, and with the indigenous population's voice affirmed in the 1998 constitution, the stage was set for Ecuador's indigenous populations to reclaim their relationship with nature.

Reclaiming a Relationship with Nature

The granting of constitutional rights to ecosystems in the 2008 constitution was an event predicated on the granting of rights for indigenous people in the 1998 constitution. Their voices and a distinctly non-western view of nature informed constitutional language that recognized Pachamama as having rights. This section

---

[8] Andolina 2003, 732.
[9] Andolina 2003, 721-22, 727, 729-31.
[10] Andolina 2003, 748.

calls to attention three factors that influenced this process: an intimate relationship with nature, a "collective" view of citizenship, and a dynamic political system.

The first factor for consideration is the intimate relationship with nature that characterizes indigenous populations in Latin America. The late Chico Mendes (a Brazilian rubber tapper and environmentalist best known as Chico) was very influential in the environmental movement in Latin America. His story illustrates the intimate relationship that indigenous populations have with nature—it is their life and their livelihood. [11] Chico's legend lives on and illustrates how this relationship with nature permeates Latin American culture and influences the way nature is managed to this very day. Chico organized the Amazonian *serigueiros* (tappers of latex from the majestic *seringueiras* trees). Also a tapper, Chico had an intimate relationship with the forest, but not the urban world until later in life. According to legend, Chico did not read a newspaper until the age of 18, and during his formative years, he was influenced by Euclides Tavora, an army officer hiding out in the jungle following a failed populist uprising. Chico credits Tavora for introducing him to socialist and populist ideas. Chico organized the rubber tappers' first *empate* (a standoff) in 1976 to resist forest clear-cutting. The movement was characterized by non-violent civil disobedience and grew in numbers as indigenous communities joined. Chico introduced the concept of "extractive reserves" as a strategic goal because he believed that the forests could sustain both humans and wildlife without being destroyed. For example, the jungles with *seringueiras* trees are like "rubber factories" and should remain a source of livelihood to tappers and a home to countless creatures. Extractive reserves are an important environmental management practice today. Chico spoke these words shortly before his assassination in 1989: "As long as there are Indians and *seringueiros* in the Amazonian jungle, there is hope for saving it." [12] If Chico were alive today, he would be pleased with the environmental provisions in Ecuador's 2008 Constitution (specifically Article 74 to be discussed later). This is one example of how the indigenous movement has made a difference.

A second factor for consideration is how indigenous Ecuadorians view citizenship. According to Luis Macas, former President of CONAIE, the Western concept of a civil society and its focus on "individual citizenship" is in conflict with the Ecuadorian indigenous movement that focuses on "collective citizenship"—indigenous Ecuadorians view themselves as a collective with a history and vision for the future. [13] Land ownership is not a priority, but the ability to collectively interact with nature in a way that supports their livelihood while preserving nature is a top priority. As we have seen earlier, Chico supported the "extractive reserves" approach to managing nature, and these words illustrate his view on land ownership: "We do not want to be landowners. The *seringueiros* want the rain forest to be controlled by the union [rubber tappers] so that those whose livelihood relies on extraction can continue to do so." [14] Chico's declaration characterizes the indigenous relationship with nature not only in his homeland of Brazil, but more broadly across Latin America.

---

[11] Tal 2006.

[12] Tal 2006, 116.

[13] Mato 2008, 424.

[14] Tal 2006, 116.

A third factor, Ecuador's dynamic political system, may have allowed this traditional voice to be heard. The term "dynamic" is preferred here because it is non-judgmental. That said, Susan Alberts suggests that Ecuador is one of Latin America's most "troubled" democracies, with persistent conflict between executive and legislative branches. She proposes that the 2008 constitution may alleviate this situation with a mechanism in the constitution for the legislature to remove the president and vice versa, but she raises a concern that maintaining a practice of "majority run-off for presidential elections" may perpetuate an already fragmented party system. To win, a candidate must have at least 40 percent of the votes with at least a 10 percent spread between the winner and the next contender, and if this is not the case, then a run-off election must take place between the top two contenders. This system makes it possible to win with a small portion of the votes, and according to Alberts, this encourages parties to "maintain their separate identities" and leads to party fragmentation, a factor that Alberts argues precludes the successful institutionalization of democracy in Ecuador. [15] But describing Ecuador as a "troubled" democracy discounts the possibility that party fragmentation created the opportunity for a concerted action by indigenous populations to gain a political voice in Ecuador. In fact, it is possible that the "majority run-off" policy enabled indigenous voices in Ecuador. Recall that the CONAIE became a powerful force in securing rights for indigenous people, and ultimately, in the development of the 1998 constitution to sustain those rights. CONAIE, along with human rights organizations, pushed for collective control over land and natural resources. Indigenous groups were able to reclaim their political voice in an administratively decentralized and culturally heterogeneous "plurinational state." [16] This was a dynamic period in Ecuador's history and set the stage for the codification of a decidedly non-western view of nature in its 2008 Constitution.

Environmental Provisions in Constitutions

Ecuador's move to recognize the "rights of nature" in its 2008 constitution is groundbreaking, because up until this event, countries with environmental provisions in their constitutions were recognizing the "rights of citizens" to a clean environment. To fully appreciate the significance of Ecuador's momentous action, it is necessary to first discuss environmental provisions in their traditional form.

Tim Hayward, in his book on *Constitutional Environmental Rights*, makes a strong case that every individual has a right to an environment that supports their health and well-being, and this right is of sufficient import that it should be protected at the highest level—the constitution. [17] He reminds us that the application of environmental provisions within a constitution can trump laws and administrative decisions within a country, and because environmental provisions within a constitution are a pronouncement of values, they may also promote the development of laws to protect the environment. Hayward argues that the right to

[15] Alberts 2008, 849-69.
[16] Andolina 2003, 727.
[17] Hayward 2005.

an environment that supports our health and well-being is a fundamental human right, and as such, environmental protection must transcend political associations. Thus, countries are increasingly under international pressure to develop policies respective of this fundamental human right.[18] Hayward reviewed the literature and estimates that about 50 nations have establish a constitutional right for persons to a clean environment, and this number grows to more than 100 nations if general language supporting a clean environment is included.[19] The inclusion of environmental provisions in constitutions is now common practice. Whether or not environmental provisions can protect the human right to a clean environment, or as in Ecuador's case, nature's right to exist, maintain and regenerate her vital cycles, is an important question for consideration.

Emmanuel E. Okon examined environmental provisions in Nigeria's constitution and offered three suggestions that may strengthen constitutional efforts to protect the environment. First, environmental provisions should explicitly create fundamental rights for individuals, communities, and states. Second, if courts recognize the symbiotic relationship between human life and the environment, then they may adopt a more harmonious interpretation of the provisions. Third, there needs to be a more liberal application of whether or not one has standing before the court when an environment has been degraded and one is seeking justice.[20] These suggestions assist in our discussion of Ecuador's move to mitigate environmental degradation by extending rights to nature. Okon's first suggestion is that environmental provisions should explicitly create fundamental rights for individuals, communities and states. This was already the case in Article 86 of Ecuador's 1998 Constitution:

> El Estado protegerá el derecho de la población a vivir en un medio ambiente sano y ecológicamente equilibrado, que garantice un desarrollo sustentable. Velará para que este derecho no sea afectado y garantizará la preservación de la naturaleza. Se declaran de interés público y se regularán conforme a la ley...[21]

A translation to English is as follows: "The State will protect the population's rights to live in a clean and ecologically balanced environment that guarantees sustainable development..."[22] The 1998 constitution goes on to declare that the preservation of nature is in the public interest and will be regulated according to the law.[23] Thus, under their 1998 Constitution, Ecuadorians do have a right to a healthy environment, and the government is prepared to protect that right. But Ecuador went even further when it recognized the rights of nature in its 2008 Constitution. Okon was addressing human rights, not rights for nature; but in theory, if the rights of nature are clearly articulated (Okon's first suggestion), then courts may more effectively interpret those rights and decide whether or not

---

[18] Hayward 2005, 1-7.

[19] Hayward 2005, 22.

[20] Okon 2003.

[21] República de Ecuador, Constitución de 1998, Article 86.

[22] This Spanish to English translation was by Jessica Umbenhauer, Alvernia University, US, in June of 2009.

[23] República de Ecuador, Constitución de 1998, Article 86.

nature has standing when an individual or group is seeking justice for harm to ecosystems (Okon's second and third suggestions). The courts will ultimately interpret the rights of nature language in Ecuador's new constitution, but it is useful here to briefly examine the language for its potential as a tool to protect the rights of nature.

The Rights of Nature

Ecuador's 2008 Constitution contains language that recognizes the rights of nature to exist, guarantees its protection and restoration, while allowing for humans to benefit from the environment to a limited extent. To this end, Ecuadorians recognized "La naturaleza o Pacha Mama" (Nature or Mother Earth) as worthy of respect and having rights. The import of this action warrants discussion of Articles 71 to 74 in Chapter Seven, *Rights of Nature*, contained within in Ecuador's 2008 Constitution. An English translation will be used here, but the original Spanish version is available in the Political Database of the Americas.[24]

> *Article 71.* Nature, or Pacha Mama, where life is reproduced and occurs, has the right to integral respect for its existence and for the maintenance and regeneration of its life cycles, structure, functions and evolutionary processes. All persons, communities, peoples and nations can call upon public authorities to enforce the rights of nature. To enforce and interpret these rights, the principles set forth in the Constitution shall be observed, as appropriate. The State shall give incentives to natural persons and legal entities and to communities to protect nature and to promote respect for all the elements comprising an ecosystem.[25]

Article 71 protects Pachamama from human interventions that disrespect her right to exist and maintain herself (ecosystems). Pachamama is nature for the indigenous people of Ecuador; they are at one with her in ways that those of us living in urban centers may fail to understand. Indigenous populations across Latin America are striving to maintain their cultural identities, and their relationship with nature is fundamental to their existence. The Pachamama Alliance is a non-profit organization working to support fundamental environmental, social and cultural rights in Ecuador, and the Alliance assisted the CELDF in drafting this language.[26] Indigenous populations in Ecuador recognize the importance of "strategies of resistance."[27] Struggling to pay U.S. creditors, the World Bank and the International Monetary Fund, external actors such as oil companies have for decades encouraged Ecuador to exploit its Amazon forests in order to service its external debt.[28] Thus, the 2008 Constitution is a "strategy of resistance." Ecuadorians now have the right to demand that the state comply with

---

[24] República de Ecuador, Constitución de 2008.
[25] Constitution of the Republic of Ecuador. 2008, Article 71.
[26] The Pachamama Alliance 2014a.
[27] Hayward 2005, 231.
[28] Smith 2009, 15.

the rights of nature, and that the state must encourage (and thus allow) individuals and groups to take action and protect nature and her ecosystems.

This brings us to Articles 72 and 73. These articles address nature's right to restoration and protection.

> *Article 72*. Nature has the right to be restored. This restoration shall be apart from the obligation of the State and natural persons or legal entities to compensate individuals and communities that depend on affected natural systems. In those cases of severe or permanent environmental impact, including those caused by the exploitation of nonrenewable natural resources, the State shall establish the most effective mechanisms to achieve the restoration and shall adopt adequate measures to eliminate or mitigate harmful environmental consequences.[29]

Article 72 recognizes nature's right to be made whole again, and without obligation on the part of the state or the people to compensate those depending on the affected systems. This language may pave the way for reclaiming of Ecuadorian control over natural resources without compensation to rogue corporations, and in cases of severe environmental degradation, the state will step in with appropriate methods to mitigate the damage.

> *Article 73*. The State shall apply preventive [precautionary] and restrictive measures on activities that might lead to the extinction of species, the destruction of ecosystems and the permanent alteration of natural cycles. The introduction of organisms and organic and inorganic material that might definitively alter the nation's genetic assets is forbidden.[30]

Article 73 calls upon what some refer to as the environmental precautionary principle. This term became popular in the 1970s during the social democratic planning process in the former West Germany.[31] If we apply this principle to nature, then we are required, even when the science is unable to affirm beyond a shadow of a doubt that nature is in need, to take steps and protect her ecosystems for future generations. The first challenge when applying the precautionary principle is how to make it operational, and one way is to establish an institutional mechanism.[32] By recognizing the rights of nature in its constitution, Ecuador has created an institutional mechanism to apply the environmental precautionary principle. It is operational because Article 73 requires the state to exercise caution and to protect ecosystems, including genomes (genetic composition).

A second challenge when applying the precautionary principle is that it works through a continuum, with the "lighter greens" at one pole, and "deeper greens" at the other. The lighter greens might call for cost-benefit analyses,

---

[29] Constitution of the Republic of Ecuador 2008, Article 72.
[30] Constitution of the Republic of Ecuador 2008, Article 73.
[31] Jordan and O'Riordan 1999, 19.
[32] Jordan and O'Riordan 1999, 30.

feasibility studies and scientific explanations, but the deeper greens may demand more fundamental changes in society and its institutions.[33] Articles 71 through 73 are on the deep green side, recognizing nature's rights and demanding restoration and protection, but Article 74 is a lighter shade of green.

> *Article 74.* Persons, communities, peoples, and nations shall have the right to benefit from the environment and the natural wealth enabling them to enjoy the good way of living. Environmental services shall not be subject to appropriation; their production, delivery, use and development shall be regulated by the State. [34]

Article 74 recognizes the right of persons and communities to access nature to live well, but they may not misuse her. Some deep greens might object to Article 74 on the grounds that nature has no obligation to support us, and modern humans can only desire to be at one with nature (we need her, but she does not need us). Some middle greens might support Article 74 because they strive to nurture a relationship with nature, and this requires an intimate knowledge of her vital systems (through cultural understanding and material explanation). Some lighter greens might raise concerns regarding Article 74 because they consider themselves to be good stewards of nature, but consider it appropriate at times to subordinate her rights to their rights (nature is here for us). That said, and assuming the constitution is fully implemented, it appears that Articles 71 to 73 will make it difficult for the light greens to subordinate nature to any great extent.

The give and take that is necessary to interpret and apply these environmental provisions may be influenced by cultural and sociopolitical factors. We have seen that the indigenous cultures gained a political voice in Ecuador and this enabled them to claim rights for Pachamama or nature in Ecuador's 2008 constitution. Recall that the environmental precautionary principle may require us to take action in advance of scientific understanding. Does the granting of rights to nature make sense in terms of modern science? Steven Rose is a forward thinking biologist, and some of his views are relevant to this discussion. He suggests that when it comes to understanding life, there is one world with many legitimate descriptions. In living systems, the causes are multiple, and phenomena are complex and interconnected. Organisms and the environment interact and evolve together through space and time. Living systems are self-creating and maintained by self-regulating processes. Organisms are open systems and dynamic stability is maintained with the input of energy. Thus, life constructs its own future.[35] These words have a lot in common with Ecuador's view of nature as articulated in Article 71 of their 2008 constitution—nature is where life is reproduced and fulfilled, and she exits through the maintenance and regeneration of her vital cycles, structures, functions and evolutionary processes. These processes can be demonstrated in material terms and, because Article 73 identifies nature as comprised of ecosystems, it should also be possible for ecologists to provide expert testimony for those seeking justice.

---

[33] Jordan and O'Riordan 1999, 30.
[34] Constitution of the Republic of Ecuador 2008, Article 74.
[35] Rose 1998, 302-309.

Environmental provisions protecting the human right to a clean environment are likely to be applied differently than provisions protecting the rights of nature. Consider this "human rights" perspective. I have a right to a clean environment, and if my human or corporate neighbor pollutes that environment, then I have legal recourse. In contrast, consider this "nature has rights" perspective. My neighbors and I live within nature and she (nature) has intrinsic value, and if my human or corporate neighbor degrades her ecosystems by polluting her air, then we are obligated to provide her with legal recourse. Both perspectives may work to protect the environment, but when it comes to the rights of nature perspective, the language is decidedly ecocentric. Legal challenges to constitutional interpretations regarding the rights of nature are likely, and it is not clear that constitutional rights for nature are even necessary to adequately protect our environment. But one thing is for sure, the language in Articles 71 to 74 is a clear indication that Ecuadorians understand their relationship with nature, and that they are willing to put her ecosystems front and center for protection and restoration.

One can imagine well-endowed environmental organizations representing nature to the highest court, but one can also imagine strong objections by well-endowed corporations seeking Ecuadorian resources. Assigning nature rights in a constitution is an important first step, but to provide her with access to those rights through legal representation before the courts in a system of justice that is capable of protecting her rights will be an ongoing challenge. Social and political actors are responsible for a constitution's legitimacy.[36] The same voices that put nature front and center in the constitution must continue to be heard.

Conclusions and Future Implications

The term "Utopia" was coined in 1516 by Sir Thomas More as a title for his book about an ideal place in the world.[37] Why might Ecuador be an ideal place in the world for nature to thrive? For the answer, this paper examined the concept of environmental provisions as they apply to constitutions, with particular emphasis on the significance of the "rights of nature" language in Ecuador's 2008 constitution. It has been suggested that the granting of constitutional rights to ecosystems in Ecuador's 2008 constitution was an event predicated on the granting of rights for indigenous people in the 1998 constitution. To that end, important cultural and sociopolitical factors influenced events leading up to Ecuador's momentous decision to grant rights to nature. Indigenous Ecuadorians have a long tradition of living with nature in a collective manner that sustained nature's riches for generations. Recently, a dynamic sociopolitical climate in Ecuador encouraged democratic processes that allowed indigenous communities to gain a voice and reclaim their relationship with Pachamama.

Nature may now have rights in Ecuador, but it will take years before it can be determined how this unprecedented action will translate into environmental policies and practice, and whether or not nature will have standing before the courts for a fair hearing. Professor Kelly Swing directs the Tiputini Biodiversity

---

[36] The Carter Center 2008, 16.
[37] Merriam-Webster 2014.

Station (University of San Francisco de Quito, Ecuador). He is concerned that the world's indifference to President Rafael Correa's request (in 2007) for more than three billion dollars to make oil extraction off-limits on 1,200 square kilometers of land near Ecuador's eastern border may push the Country toward more oil extraction.[38] Swing is also concerned that a recent failed attempt to charge a local government for damages to the Vilcabamba River is evidence that Ecuador's environmental legal system is system is strained and unable to act.[39] It remains to be seen whether the Ecuadorian government can continue to support the rights of nature when faced with a lack of global support. Recently the Pachamama Alliance reported that the Ecuadorian government shut down its foundation in December of 2013, and the Alliance's recent appeal to reopen was denied.[40]

Time may tell whether or not Ecuador's resolve for the rights of nature is at risk. Perhaps Ecuador is "on the cusp of a new oil boom" as reported by David Robinson in the *Guardian*.[41] Nevertheless, many Ecuadorians still view nature as having rights, and their actions may be contagious. The Global Alliance for the Rights of Nature is a worldwide movement to defend and protect the rights of nature. The alliance is following Bolivia's recent efforts to pass legislation to acknowledge nature's rights.[42] It remains to be seen to what extent Ecuador is Mother Nature's utopia, but whatever the outcome, Ecuadorians deserve the highest of commendations for their efforts to protect the rights of nature.

References

Alberts, Susan. 2008. Why Play by the Rules? Constitutionalism and Democratic Institutionalization in Ecuador and Uruguay. *Democratization* 15 (5): 849-69.

Andolina, Robert. 2003. The Sovereign and Its Shadow: Constituent Assembly and Indigenous Movement in Ecuador. *Journal of Latin American Studies* 35 (4): 721-50.

CELDF. 2014. *Community Environmental Legal Defense Fund*. Available at http://www.celdf.org, accessed February 24, 2014.

Constitution of the Republic of Ecuador. 2008. English translation from Political Database of the Americas, Georgetown University. Available at http://pdba.georgetown.edu/Constitutions/Ecuador/english08.html, accessed February 18, 2014.

Galeano, Eduardo. 1997. *Open Veins of Latin America: Five Centuries of the Pillage of a Continent*. New York, NY: Monthly Review Press.

Global Alliance for the Rights of Nature. 2014. Bolivia: Fighting the Climate Wars. Available at http://therightsofnature.org/bolivia-fighting-climate-wars/, accessed on February 24, 2014.

Hayward, Tim. 2005. *Constitutional Environmental Rights*. Oxford, London: Oxford University Press, Inc.

---

[38] Swing 2011, 267.
[39] Swing and Sempertegui 2012, 40.
[40] The Pachamama Alliance 2014b.
[41] Robinson 2014.
[42] Global Alliance for the Rights of Nature 2014.

Jordan, Andrew, and Timothy O'Riordan. 1999. The Precautionary Principle in Contemporary Environmental Policy and Politics. In *Protecting Public Health & the Environment: Implementing the Precautionary Principle*, edited by Carolyn Raffensperger and Joel A. Tickner, 15-35. Washington, DC: Island Press.

Linzey, Thomas. 2008. *Ecuador Approves New Constitution: Voters Approve Rights of Nature*. Press Release, September 28. Available at http://www.celdf.org/article.php?id=302, accessed February 24, 2014.

Martz, John D. 1972. *Ecuador: Conflicting Political Culture and the Quest for Progress*. Boston, MA: Allyn and Bacon, Inc.

Mato, Daniel. 2008. Transnational Relations, Culture, Communication and Social Change. *Social Identities* 14 (3): 415-35.

Merriam-Webster 2014. *Utopia.* In the online dictionary available at http://m-w.com/dictionary/utopia, accessed February 24, 2014.

Okon, Emmanuel E. 2003. The Environmental Perspective in the 1999 Nigerian Constitution. *Environmental Law Review* 5 (4): 256-78.

República de Ecuador Constitución de 1998. Political Database of the Americas, Georgertown University. Available at http://pdba.georgetown.edu/Constitutions/Ecuador/ecuador98.html, accessed February 24, 2014.

República de Ecuador Constitución de 2008. Political Database of the Americas, Georgetown University. Available at http://pdba.georgetown.edu/Constitutions/Ecuador/ecuador08.html#moz TocId822446, accessed February 24, 2014.

Robinson, David. 2014. We're on the Cusp of a New Oil Boom in the Ecuadorian Amazon. *Theguardian.com*. Available at http://www.theguardian.com/environment/2014/feb/19/new-oil-boom-ecuador-amazon, accessed February 24, 2014.

Rose, Steven. 1998. *Lifelines: Biology Beyond Determinism*. New York, NY: Oxford University Press, Inc.

Smith, Gar. 2009. In Ecuador, Trees Now Have Rights. *Earth Island Journal*, Winter: 15.

Swing, Kelly. 2011. Day of Reckoning for Ecuador's Biodiversity. *Nature* 469: 267.

Swing, Kelly and Luis Sempertegui. 2012. Problems Enforcing Ecuador Ecology Law. *Nature* 491: 40.

Tal, Alon. 2006. *Speaking of Earth: Environmental Speeches That Moved the World*. New Brunswick, NJ: Rutgers University Press.

The Carter Center. 2008. *Final Report on Ecuador's Approbatory Constitutional Referendum of September 28, 2008*. Press Release, October 25. Available at http://www.cartercenter.org/resources/pdfs/news/peace_publications/election_reports/Ecuador_referendum_report08_en.pdf, accessed February 24, 2014.

The Pachamama Alliance. 2014a. About Pachamama. The Available at http://www.pachamama.org/about, accessed February 24, 2014.

The Pachamama Alliance. 2014b. Fundacion Pachamama: Appeal for Re-open Denied. Available at http://www.pachamama.org/news/fundacion-pachamama-appeal-for-re-open-denied, accessed February 24, 2014.

Chapter 18: Citizen or Commodity: Nature's Rights in Ecuador's Constitution

Christine Dellert

The 444 articles in Ecuador's new constitution, approved by more than 60 percent of voters in late-2008, provide for many sweeping changes in this South American nation, among them social security benefits for stay-at-home mothers and free education through university.[1] Yet the constitution's extension of inalienable rights to nature is the most progressive and pioneering shift in ecological politics worldwide. Ecuador is the first nation to make such a broad commitment. Countries have long focused on constitutional environmental protection by limiting damages and ecological degradation that could directly harm humans. This is the case in more than 100 national constitutions, many of which provide a human right to clean water or an otherwise healthy environment but offer little or nothing to safeguard the natural world from threats that would not otherwise impact human wellbeing.[2] Growing awareness of the importance of ecological sustainability for global communities (human and non-human alike) is forcing us to act in unprecedented ways. Ecuador's new provision, which grants legal standing to nature, rejects past anthropocentric attempts to protect the non-human world through constitutional action. Its "rights to nature" chapter opens the door for a reframing of citizenship, a concept closely tied to constitutional rights in political literature and practice.

The concept of citizenship has evolved in response to changing social, economic, and political environments.[3] So how does Ecuador's constitution affect and advance our theorizing of citizenship? How does it challenge dominant social paradigms that commodify the natural world? Can we accommodate nature in existing Western or neoliberal development models, or do we need an entirely new conceptualization?

---

[1] Smith 2009.
[2] Hayward 2000; Shelton 2006; Bruckerhoff 2008; Soveroski 2007; Depledge and Carlarne 2008.
[3] Mason 2009.

This chapter rejects much of the normative literature on citizenship dominated by anthropocentric liberal and civic republican frameworks and instead turns to the work of scholars and indigenous thinkers who incorporate the natural world into their political, social and cultural communities. Despite the existing array of literature that binds citizenship with environmental concern, including Andrew Dobson's ecological citizen model, much of it fails to fully engage the more-than-human community in its conceptualization. The following sections offer a reframing of citizenship that encompasses the rights of non-humans and expands our concern for the natural world – a way to protect the "other" rather than put a price tag on it. In this reframing, I explore two key characteristics: shifting emphasis from individual agency to the collective community; and extending human obligations to rights-bearing natural subjects. Both of these characteristics are reflected in Ecuador's new constitution and may ultimately result from indigenous groups' rising prevalence in national politics and social consciousness.

The World's First Eco-constitution

In the months following Ecuadorians' ratification of a constitution that grants legally enforceable rights to nature, some scholars and journalists thought the country an unlikely place for the world's first "eco-constitution."[4] After all, oil and mining comprised more than a quarter of Ecuador's GDP in 2008, and approximately one-third of the country's spending budget comes from petroleum products.[5] The country has opened its pristine Amazon forests to foreign oil companies to service its massive debt to U.S. creditors, the International Monetary Fund and the World Bank.[6] In nearly 30 years, Amazon Watch estimates that ChevronTexaco's drilling damaged 2.5 million acres of rainforest and desecrated the landscape with 600 toxic waste pits, while leaving some 30,000 people from indigenous communities with polluted rivers and streams, a cancer rate that's 130 percent of the national norm, and a rate of childhood leukemia that's four times higher where Texaco has operated than in other parts of Ecuador.[7] Although Texaco (now Chervon) has ceased to operate in the country since the 1990s, crude oil spills and poisoned waterways caused by state-run Petroecuador still devastate the ecosystem and are destroying the livelihood of the people who live in Ecuador's Lago Agrio region.[8] From its unfortunate ecological history, though, Ecuador has uniquely positioned itself in contemporary environmental discourse. It is home to the "trial of the century," where an Ecuadorian court in 2013 ordered Chevron to pay US$ 9.5 billion in

---

[4] Smith 2009. Use of the word "eco-constitution" defines a pioneering document that extends inalienable rights to nature for the first time in human history.

[5] Whittemore 2011.

[6] Smith 2009.

[7] Smith 2009; Kimerling 1994.

[8] For a history of environmental damages in Ecuador from oil extraction, particularly in the region of Lago Agrio, see Kimerling 1994; and Santacruz 2009.

environmental damages.[9] The sum was among the largest-ever levied in an environmental lawsuit, and because it was "one of the first legal cases to address human-rights violations by international petroleum firms in a low-income country" it has "the potential to set precedent for similar disputes around the world."[10]

The lawsuit reflects the growing involvement of indigenous communities in struggles with hegemonic political and economic systems, particularly in the Andes.[11] As human rights have played a crucial role in the convergence of sustainability and development issues, indigenous rights are more recently gaining recognition across Latin America, with Ecuador home to one of the most powerful indigenous movements – the Confederation of Indigenous Nationalities of Ecuador (CONAIE).[12] The movement has a history of successful nationwide inclusive mobilization, incorporation into the political arena, and institutionalization of indigenous rights.[13] In 1998, CONAIE succeeded with a constitutional assembly that validated further indigenous peoples' rights, among them including indigenous languages on the list of the nation's official languages. With a common agenda and identity, CONAIE influenced the 1998 constitution while the group's impact was effected through existing political institutions:

> Political institutions, procedures and rules embody and symbolize certain understandings and practices of nations and citizenship, as they channel authority and meaning about the "relation of state to individuals and the nature of leadership, power, and identity." Thus, institutions not only process social movement demands, but may also be processed by those demands as social movements "transform ideas into institutional purpose."[14]

CONAIE's successful entrance into congressional and local elections, with its electoral arm Pachakutik, facilitated political legitimization of long-ignored cultural priorities, and together communities pushed for a radical transformation involving CONAIE's ideological principles for a plurinational state: "A bottom-up, participatory system rooted in consensus, cultural and social diversity, protection of the environment and human rights, and development that meets basic needs and generates self-sufficiency."[15] The movement won its fight for the inclusion of a plurinational state, along with many other progressive social and environmental measures, in Ecuador's constitutional assembly of September 2007, and in a referendum approved by voters the following year.

---

[9] Jones 2014. The verdict from Ecuador's highest court had stood at $19 billion, but the court dismissed punitive damages imposed in 2011 after Chevron refused to apologize for polluting, see Alvaro and Gilbert 2013.

[10] Valdivia 2007, 42. For further analysis of the decades-long dispute and a 2014 ruling by a US judge in favor of Chevron that claims conspiracy and criminal conduct, see Krauss 2014.

[11] For an ethno history of indigenous influence in politics of Andean countries, see de la Cadena 2010.

[12] Clark and Stevenson 2003; Andolina 2003.

[13] Gerlach 2003.

[14] Andolina 2003, 722.

[15] Andolina 2003, 732.

Among the hundreds of articles proposed, at least seven key issues directly dealt with Ecuadorians' relationship with the natural world: "how to pass from an extractive development model to a sustainable development model; how to introduce rights for Nature into this new constitution; how to incorporate alternative national accounts that consider and monitor Nature; how to limit corporate power; how the resource of water is treated; how renewable resources are managed (involving activities such as fishing, hunting and timber); as well as non-renewable resources (oil extraction and mining)."[16] Each of these areas embodies the objective of national growth with the indigenous ideal of *sumak kausai*, which translates to "harmonious life" in Quechua, as a way of valuing and living among and as part of the natural world that sustains us all.[17] Not surprisingly, reports on the new constitution attribute a long process of social-movement engagements by Ecuador's indigenous organizations and communities as influencing constitutional assembly members to draft and eventually accept these proposals. A North American Congress on Latin America (NACLA) report attributed success to indigenous organizations, such as CONAIE and Ecuarunari, as well as nonprofit organizations like Acción Ecológica.[18] The report, however, failed to mention the Pennsylvania-based Community Environmental Legal Defense Fund (CELDF) and the San Francisco-based Pachamama Alliance, which worked with the 130-member assembly to draft the language that establishes inalienable rights for nature. The result of these groups' work is a constitutional chapter that fuses indigenous values with substantive and procedural rights, while at the same time signaling a shift beyond the instrumentalism dominating anthropocentric-framed "protection" of the natural world thus far. Dr. Mario Melo, a lawyer specializing in Environmental Law and Human Rights and an adviser to Fundación Pachamama-Ecuador, said: "The new constitution reflects the traditions of indigenous peoples living in Ecuador, who see nature as a mother and call her by a proper name, Pachamama."[19] Loosely translated as "Mother Earth" or "Mother World" in Aymara or Quechua, "Pachamama" is invoked in the first article of Ecuador's new constitutional "rights to nature" chapter, "where life is reproduced and exists, [it] has the right to exist, persist, maintain and regenerate its vital cycles, structure, functions and its processes in evolution."[20]

---

[16] The Pachamama Alliance 2008.

[17] Galeano 2008, 19.

[18] Dosh and Kligerman, 2009.

[19] Mychalejko 2008.

[20] Chapter 7, Article 71. English translation available from the Community Environmental Legal Defense Fund at http://www.celdf.org, accessed March 7, 2014. Alternative English translation of the Constitution at http://pdba.georgetown.edu/Constitutions/Ecuador/english08.html, accessed March 7, 2014. Several key language differences exist between CELDF's translation and the version available through Georgetown's Political Database of the Americas, specifically those articles pertaining to enforcement of these rights. While CELDF's translation states, "any person, people, community or nationality, may demand the observance of the rights," the Georgetown translations says they "can call upon public authorities to enforce the rights of nature." Additionally, while CELDF's translation states, "The State will motivate natural and juridical persons as well as collectives to protect nature," the Georgetown translation states, "The State will give incentives to natural persons and legal entities and to

More recent accounts of the inclusion of the "rights to nature" amendments, however, suggest the work of a few calculated bureaucrats rather than a broad-based social movement for non-human equity. Originally a call for animal rights, the "rights to nature" language "did not start from the grassroots, but rather came from above, namely from members of the Constitutional Assembly," a CELDF representative recalled. [21] Indeed, key assembly members, including former energy minister and president of the Constitutional Assembly Alberto Acosta, "strategically included [the rights for nature] in a comprehensive package of rights" and drafted the provisions outside of the formal assembly. [22] Still, its inclusion in the constitution implies the natural world's close ties with indigenous worldview and demand for political recognition. Indeed, several groups within the assembly, indigenous delegates among them, "were talking about the rights of nature as strengthening collective rights, hence it was a natural discussion for them." [23] The amendments, then, were not the "act of a group of enlightened people," Acosta said, but rather "the continuation of a process of building alternatives" to decades of inequities for indigenous groups suffering from Ecuador's resource-intensive development model. [24] Without doubt, the new constitution demands sweeping recognition of the natural world's non-anthropocentric value while calling for greater political inclusion of indigenous communities. At the same time, it importantly extends peoples' power and responsibility to protect nature, stating, "Any person, people, community or nationality, may demand the observance of the rights of the natural environment before the public bodies." [25]

Beyond Nature's Rights

Ecuador's constitution endorses a biocentric political commitment and establishes nature as a subject of justice and rights, as opposed to a commodity "to be justly distributed." [26] Its "rights to nature" chapter resonates with Aldo Leopold's famous claim that humans are members of a biotic community where non-human members can be the recipients of justice. [27] Decades later, Ecuador's measure also

communities to protect nature and to promote respect for all the elements comprising an ecosystem." These language differences are worth further exploration due in part to their differing dates of publication. The CELDF version was published in 2008; while the Georgetown translation was last updated in January 2011.

[21] Tanasescu 2013, 855.

[22] Tanasescu 2013, 854.

[23] Tanasescu 2013, 853.

[24] Constitutional Assembly president Alberto Acosta, cited in Tanasescu 2013, 853.

[25] Chapter 7, Article 71. English translation available from the Community Environmental Legal Defense Fund at http://celdf.org/rights-of-nature-ecuador-articles-of-the-constitution, accessed April 6, 2014. Alternative English translation from http://pdba.georgetown.edu/Constitutions/Ecuador/english08.html, accessed April 6, 2014, reads, "All persons, communities, peoples and nations can call upon public authorities to enforce the rights of nature."

[26] Smith 2006, 345.

[27] As cited in Smith 2006; Leopold 1966. Others have long argued that non-human species have characteristics of rights-bearing entities, including interests and the capacity for suffering; see Singer 1977. The "inconsistency claim," for example, argues that treating

offers a practical engagement of Christopher Stone's seminal article on granting legal rights to non-human subjects. Like "legal incompetents" – children or the mentally disabled – natural subjects should have their rights protected through assigned representation before judges and the public, where the interests of trees, rivers, birds, and humans would gain equal consideration as rights-bearing subjects under the law.[28]

The passage of Ecuador's "rights to nature" chapter, though, has caused some in the West to romanticize or reduce its impact to a kind of idealized "nature's utopia," where a resurgence of traditional values could prompt an entire country's population to live in ecological harmony. Certainly, research and scholarly work by non-natives and native peoples emphasizes an indigenous ontological position of attachment to the land.[29] But simply saying that Ecuadorian people wanted to grant nature rights solely based upon traditionalist environmental beliefs is a reductionist claim. The indigenous scholar Daniel R. Wildcat admonishes these kinds of stereotypes, calling them "reassuring romantic reveries about noble savages living close to nature."[30] Instead, Ecuador's extraordinary legal defense for the natural world is as much about securing the survival of its indigenous people as it is about securing protection for streams and forests. For indigenous communities, which make up about 35 percent of Ecuador's population, the right to biologically diverse and sustainable ecosystems is "a right to maintain their way of life."[31] Indeed, nature is not an "apolitical" entity.[32] Rather it is "at the heart of the antagonism that continues to exclude 'indigenous beliefs' from conventional politics."[33] Conservationism offers a beneficial intersection for environmentalists and indigenous groups' efforts. In his studies of the Cofán people in eastern Ecuador, anthropology professor Michael L. Cepek found that "their experience of the radical ecological consequences of petroleum-based development taught them a valued way of life is difficult to pursue in an environmentally transformed landscape."[34] People increasingly believed they must fight for ecological integrity of the lands to ensure their future.[35] This is not to suggest that environmental discourse is not a part of deeply held indigenous perspectives. Care for nature is embedded in everyday practices tied to peoples' culture. To de la Cadena:

---

animals as lacking rights is based on an assumption that they are not, or are less, sentient than human beings, despite the reality that there are humans (infants or mentally disabled persons) whose mental capacities are closer to those of animals. Theorists invoking this claim contend that because those persons are granted equal status and rights by law, so too should non-humans. For further discussion, see de-Shalit 2000; Regan 1988; Nozick 1974.

[28] Stone 1972, 17.

[29] Many have researched indigenous peoples' attachment to land through various academic fields, including emotional geography of place (see Kearney and Bradley 2009) and traditional ecological knowledge (see McGregor 2004).

[30] Wildcat 2009, 20.

[31] Bruckerhoff 2008, 621.

[32] de la Cadena 2010, 350.

[33] de la Cadena 2010, 350.

[34] Cepak 2012, 14.

[35] Cepak 2012.

> Respect and care are a fundamental part of life in the Andes; they are not a concept or an explanation. To care and be respectful means to want to be nurtured and nurture other, and this implies not only humans but all world beings...We nurture the seeds, the animals and plants, and they nurture us.[36]

Still, Western conventional science-based conservation and inclusion in national political agenda offer indigenous groups the ability to advance their lands' sustainability and their often-marginalized way of life.[37]

Fearing that future natural resources would be depleted by extractive industries and neoliberal economies, Ecuadorian indigenous political leaders in 2009 introduced a biodiversity law to complement and offer a framework of regulation for its progressive constitutional provisions. The *Ley Orgánica de Biodiversidad* recognizes many indigenous nationalities' long-standing ancestry on the land, as well as their ecological dependence, and argues for plurinational *autoridad* (authority or sovereignty) as it relates to preserving and managing biodiversity.[38] In citing indigenous peoples' intimate knowledge of the natural world, the legal proposal simultaneously rebukes modern political and economic systems that "turn life into merchandise" and transnational corporations that "take control of nature" to commodify resources.[39] It is the right and responsibility of the people (in this sense, indigenous groups) to see that this does not happen. As a result, the rights to nature constitutional provision and the new biodiversity law offer additional recourse for peoples to protect their native lands and species, and in doing so, fight for and ensure their own survival.

Debating Ecological Citizenship

So how do the inclusion of nature's rights and the extension of peoples' responsibility to defend it affect our theories on citizenship? Put simply, citizenship is a relationship in which parties enjoy rights as a part of that relationship while incurring obligations or duties as a result of it.[40] If we perceive citizenship to be a characteristic of the state, then constitutional guarantees grant citizens rights, assign duties, and extend institutional protections. Ecuador's action begs us to ask: How does the natural world now fit into our conception of citizenship? Since T.H. Marshall's seminal 1950 essay on the basic rights of citizenship, the concept has involved a litany of theorizing on political and civil rights. As Marshall later put it: "The modern drive toward social equality is, I believe, the latest phase in the evolution of citizenship which has been in continuous progress for 250 years."[41] Perhaps a postmodern drive in citizenship

---

[36] de la Cadena 2010, 354.
[37] For further discussion of the differences between Ecuadorian indigenous social movements and Western assumptions, see Cepak 2012.
[38] *Ley Orgánica de Biodiversidad* of Ecuador, 2009, translated from Spanish by author, February 2010.
[39] *Ley Orgánica de Biodiversidad* of Ecuador, 2009, translated from Spanish by author, February 2010.
[40] Mason 2009.
[41] Marshall 1992, 7, cited in Paehlke 2004, 3.

theory would be a push toward ecological equality, or at least, as this chapter argues, a more complete engagement with the natural world and a rejection of "human-bound" citizenship theories. Paehlke rightly states that: "Environmentalism can be understood, essentially, as a part of this long process, as a set of political ideas that broadens the inclusiveness further by encompassing future generations and other species."[42] Ecuador provides us with a unique position to call for a more inclusive reframing of citizenship and a deeper recognition of the natural world's value.

But to do that, we must first move beyond the traditional anthropocentric framework. Much of the theorizing on ecological citizenship since the 1990s has focused on whether it is a distinctive type of citizenship, or if the concept can be accommodated within existing models of citizenship from the dominant schools of thought: the rights approach or the personal duty approach.[43] Scholars of citizenship typically offer two contrasting frameworks: the liberal model, which emphasizes citizenship as a public status that ensures the holder civil, political, and social rights, and the civic republican model, which emphasizes the public duties, virtues, and practices of citizenship.[44] These two models are equally worthy of criticism in their normative exclusion of the "other," only focusing on dominant human relationships. Even when faced with an opportunity to protect ecological resources, Jagers explains, "The reason for republican citizens to be ecologically cautious is that there is a reciprocal gain within the community from preserving ecological resources, while the reason for liberal citizens is that they can claim some right or benefit in return," namely the right to a healthy environment.[45] By pinning their hopes on the concept of rights, liberal theorists further claim that an extension of human rights also should cover the right to a healthy environment.[46] But because of its human-centered focus, this sentiment completely ignores protection based upon nature's right or a non-instrumental obligation to preserve it.

Still, there are theorists who have attempted to incorporate the natural world into existing theories of citizenship.[47] Unfortunately, though, they have failed to fully engage the value and rights of nature in their discussions. As an example, Bell notes that the main problem with a liberal approach to environmental citizenship is its conception of nature as property.[48] Melo-Escrihuela writes, "His [Bell's] alternative is that liberal citizens should view nature as a provider of basic needs and as 'a subject about which there is disagreement'…He claims that it is possible to deduce two types of rights [procedural and substantive] from a conception of the environment as a provider of basic needs…According to Bell, both types of rights can be regulated and enshrined in constitutions."[49] In reality, they already have. As noted above, more than 100 constitutions throughout the world pledge the right to a healthy environment, impose a duty on the state and

---

[42] Paehlke 2004, 3.

[43] Melo-Escrihuela 2008.

[44] Gabrielson 2008.

[45] Jagers 2009, 20.

[46] Jagers 2009.

[47] Hailwood 2005; Barry 2002; Dobson and Bell 2006.

[48] Bell 2005, cited in Melo-Escrihuela 2008.

[49] Bell 2005, cited in Melo-Escrihuela 2008, 116.

citizens to prevent ecological harm, or mention the protection of natural resources.[50] But none until Ecuador actually granted nature its own inalienable entitlements, thereby extending it as a protected and rights-bearing entity subject to human duties and obligations. In Latta's review of citizenship literature, the author suggests that "the existing literature tends to treat ecological citizenship as a normative and institutional tool for promoting a greener future" and consequently "has muted the democratic sensibility that citizenship might bring to the politics of nature."[51] Ecuador's constitution offers us an impetus for a new ethical-political orientation within which to theorize ecological citizenship, one that embraces our position as a part of the natural world and which seeks to include non-human interests in our decision-making and defending.

Indeed, this orientation could be akin to Mark J. Smith's "ecologism," whereby we remake our relationship with the natural world to encompass rights and justice, as well as a "politics of obligation."[52] Smith calls for extending virtues of care and compassion to natural subjects by "displacing the human species from the central ethical position it has always held."[53] Similarly, Curry constructs a notion of community that applies to both the natural and the social and humbles human interests within a larger "republic of life."[54] Curry's theory says that all communities require certain practices to sustain them and draws upon "an extended sense of embodied relationships, in lived communities and specific places – a link which recognizing and revaluing the natural dimension would strengthen."[55] Despite his too-heavy (and sometimes conflicting) emphasis on a liberal citizenship model, Christoff also argues that the rights of both animals and future generations should be represented by ecological citizens and formalized in culture, law, and constitutional rights.[56] He conceives ecological citizenship as an "institution for all those affected by ecological problems to participate in environmental decision making," either through the creation of regional parliaments, referenda on specific environmental issues across nation-states, or a flexible electorate that changes according to each problem.[57] As such, his theorizing might offer political and epistemological solutions to Ecuador's unprecedented action.

More than Individualism and Human-bound Obligations

In her 2012 critique of neoliberalism in Ecuador, Sarah A. Radcliffe argues that the nation's constitution wholly challenges liberal citizenship. Instead of providing individual anthropocentric rights established and administered by groups committed to protecting property rights, Ecuador's constitution establishes social rights for all beings and "theoretically aims to guarantee rights in an integrated way, advancing them all simultaneously without prioritizing certain

---

[50] Shelton 2006.

[51] Latta 2007, 379, 381, cited in Gabrielson 2008, 430.

[52] Smith 1998.

[53] Smith 1998, 99.

[54] As cited in Gabrielson 2008, 434.

[55] As cited in Gabrielson 2008; Curry 2000, 1069.

[56] Christoff 1996.

[57] As cited in Melo-Escrihuela 2008, 117.

rights over others, or postponing some in favour of a select few."[58] Add to that its indigenous declaration of *sumak kawsay* (or *sumak kausai* as cited earlier) that calls for responsibilities beyond human relationships, and Ecuador's constitution also offers a far cry from traditional civic republican citizenship.

Therefore, like Dobson, I suggest that a third model of citizenship beyond the traditional liberal and civic republican frameworks – ecological citizenship – is necessary. As part of this new model, I propose two essential characteristics for a reframing of ecological citizenship to better incorporate indigenous political and cultural values. The first of these characteristics is shifting political opinion towards favoring an emphasis on the collective community over the individual. In a 2003 interview, former CONAIE president and long-time indigenous leader and scholar Luis Macas addressed the clash between indigenous cultural conceptions and the dominant social paradigm that focuses on the individual within the idea of citizenship.[59] "This is because the idea of individual citizenship works directly against the ideas of collective citizenship and rights that indigenous people have historically held, especially in regards to such important matters as territory, language and the administration of justice," he says.[60] I am not proposing that we abandon the importance of the individual as nested in his or her community or tribe. But instead, we might apply a concern from Holder and Corntassel's human rights literature to ecological citizenship. They state that our existing framework emphases individual needs and entitlements in a way that inadequately compares the collective nature of groups with non-Western worldviews and priorities.[61] On the one hand, evidence suggests that it is our individual needs and entitlements, operating as instrumental rationality, anthropocentrism and economic expansionism, which confound our global environmental problems.[62] On the other hand, and as argued by Wildcat, an indigenous approach to "mitigating the environmental chaos that a civilized fixation with control has brought," would be to "think in terms of cooperation and coordination with the balance of nature beyond our human selves."[63] A more collectivized conceptualization of citizenship would prompt greater focus on the needs and entitlements of the whole human-nature system, including those of marginalized actors, whether indigenous peoples or non-humans. The overarching paradigm of *sumak kawsay* in Ecuador's constitution goes further to encompass collective rights, in addition to individual rights, as well as to "overcome the liberal 'false dichotomy' between individual and collective rights by interweaving different types of rights in legislative and practical ways."[64]

In a similar vein, Clark and Stevenson suggest that as it becomes more apparent that our actions are entangled with the lives of "'others' distant in time or space," we must ask ourselves where our communities begin and end – and

---

[58] Radcliffe 2012, 243. See also Quinterro Lopez 2008; Consejo Nacional de Planificación [Ecuador] 2009, 24.
[59] Luis Macas, cited in Mato 2008, 425-426.
[60] Mato 2008, 425-426.
[61] Holder and Corntassel 2002.
[62] Naess 1973; Goodwin 1992; Gabrielson 2008.
[63] Wildcat 2009, 53.
[64] Walsh 2009, 180; cited in Radcliffe 2012, 242.

who is deserving of the benefits, as well as the responsibilities.[65] In some discussions, ecological citizenship has been dubbed "citizenship without a state."[66] As environmental problems become more global in scale – climate change, biodiversity loss, and ozone depletion – it will take more than accelerated individualization of responsibility, but rather concerted and comprehensive international action to resolve the damage.[67] Along these same lines, both the traditional rights and personal duty approaches to citizenship have been blamed for fomenting an individualistic understanding of citizenship and politics that encourages the privatization of environmental responsibility (i.e. simply do your part for the environment by abandoning certain commodities and restraining consumption).[68] While these actions should certainly not be discouraged, they perpetuate a concept of "ecological citizenship as self-restraint," where sustainability would perceivably emerge through an "inner revolution."[69] Melo-Escrihuela correctly adds that

> too singular a focus on individuals might lead to the conclusion that citizens can satisfy responsibilities of ecological citizenship simply by fulfilling personal duties. Such a focus also suggests that citizens' lifestyles, especially those whose everyday activities have large environmental and social impacts, are the main cause of environmental problems.[70]

The urgency of responding to the global environmental crisis has led to a renewed focus on ecological citizenship, where emphasis has shifted from rights to duties.[71] Dean observes these changes in citizenship studies, as well, pointing out that:

> Green thinking has impacted on our understanding of citizenship in at least three different ways. First, environmental concerns have entered our understanding of the rights we enjoy as citizens. Second, the enhanced level of global awareness associated with ecological thinking has helped to broaden our understanding of the potential scope of citizenship. Third, emergent ecological concerns have added fuel to a complex debate about the responsibilities that attach to citizenship.[72]

We see these changes in Dobson's widely cited ecological citizen model, whereby, according to Jagers, the reason to protect ecological resources is a responsibility to minimize negative ecological impacts on others.[73] Dobson argues that ecological citizens are morally guided by the desire for justice,

---

[65] Clark and Stevenson 2003, 235.

[66] Paehlke 2004; Dobson 2003.

[67] Maniates 2001; Paehlke 2004.

[68] Melo-Escrihuela 2008.

[69] Latta 2007, 380.

[70] Melo-Escrihuela 2008, 122.

[71] Clark and Stevenson 2003.

[72] Dean 2001, 491.

[73] Jagers 2009.

invoking the mantra that "justice demands that individuals act in a way that is not always in their best interest."[74] His ecological citizenship concept differs widely from the previous citizenship camps in several key ways, one of which is its emphasis on the private sphere. Dobson recognizes that private acts, or environmental behavior, can have public consequences. Drawing on the concept of ecological footprints, Jagers argues that Dobson acknowledges that individuals affect each other "by consuming goods and services with global origins and with ecological and social consequences, both spatially and far into the future, their inter-personal obligations are both global and inter-generational." [75] Jagers describes Dobson's model further:

> ...ecological citizenship is an inter-personal relationship among strangers (like all citizenship) founded on responsibility, compassion and social justice, and most importantly captured in the principle: When I live my life I affect others, and to these others I have obligations (regardless of whether or not I know them).[76]

Yet, for as much as Dobson's theorizing offers ecological citizenship, it can be argued that it does not fully engage the non-human world. Dobson openly admits anthropocentrism and, according to Hailwood, doesn't necessarily look upon nature as non-instrumental. [77] The obligations he writes and speaks of are to fellow humans, and not natural subjects. On this view, if a community pollutes a river and its negative ecological effects impact another community downstream, presumably the polluters' obligations are to the people impacted, not the river or its non-human inhabitants.

This scenario leads to a second essential characteristic in reframing ecological citizenship: humans' obligations owed to the natural world as rights-bearing subjects. Wildcat labels this concept "indigenous realism," which "entails that we, members of humankind, accept our inalienable responsibilities as members of this planet's complex life system, as well as our inalienable rights."[78] In Ecuador, the inclusion of rights for nature in its constitution extends humans' duties as citizens and members of the nation-state to care for non-humans by virtue of granting these subjects the rights to exist, maintain, and reproduce. This obligation is explicitly stated in the new constitution: "The State will motivate natural and juridical persons as well as collectives to protect nature; it will promote respect towards all the elements that form an ecosystem." [79] Indeed, the constitution's inclusion of *sumak kawsay* recognizes a "harmonious relationship" among human beings and between humans and nature that is based as much on

---

[74] Dobson 2003, 205.

[75] Jagers 2009, 20.

[76] Jagers 2009.

[77] Hailwood 2005.

[78] Wildcat 2009, 9.

[79] Chapter 7, Article 3 English translation from the Community Environmental Legal Defense Fund at http://www.celdf.org. Alternate translation available from http://pdba.georgetown.edu/Constitutions/Ecuador/english08.html reads, "The State shall give incentives to natural persons and legal entities and to communities to protect nature and to promote respect for all the elements comprising an ecosystem."

rights as on responsibilities.[80] Despite some scholars' reluctance to recognize the possibility of including non-humans under the umbrella of citizenship, Ecuador's provision arguably forces us to consider it. Skeptical of non-humans' status in citizenship theory, Hailwood writes:

> Nature is not a fellow citizen; therefore one, as a citizen, can have no obligations to it. Now, it is true that the political/citizen sphere of duties and rights is only a subset of morality. Not every moral claim is a claim of citizenship, and not every moral 'patient' is a fellow citizen. It is also important to be circumspect in the invocation of 'community,' including "political community." Environmentally reasonable citizens would not take themselves to be 'expanding the political community' to encompass the non- human world.[81]

But if we are granting rights to nature and owe obligations to it, how can we not include it in the political community? As another example, Mason's resistance to ecological citizenship also falters in the face of Ecuador's constitutional action: "It is only when identity is combined with something else, such as the shared enjoyment of rights, or shared participation in a cooperative scheme, that the language of citizenship gains a foothold."[82] Arguably it is this "shared enjoyment of rights" that brings the natural world into our "moral orbit of citizenship."[83] Fully addressing this issue is a complicated task that would require much more space than provided in this brief chapter, but it is one that will continue to surface as additional states adopt more progressive policies addressing the rights of non-humans.

Since the passage of Ecuador's constitution, other nations have recognized the rights of nature and offered it legal standing. In 2010, Bolivia passed its Law of the Rights of Mother Earth, which defines Mother Earth as a "living system comprised of the interrelated, independent and complementary, indivisible community of all life systems and living beings that share a common destiny."[84] This law asserts Mother Earth's rights to life, to water, to diversity of life, to clean air, to balance, to restoration, and to life free of contamination, while also specifying the state's obligations to protect it and establishing the position of ombudsman of Mother Earth.[85] Similarly, New Zealand in 2012 granted the Whanganui, the nation's third-longest river, rights of personhood, and appointed a joint council made up of the crown and in part by the local indigenous community, giving the community standing to represent the river's interests.[86] The river will be recognized as a person under the law "in the same way a company is, which will give it rights and interests," the Minister of Treaty

---

[80] Radcliffe 2012.

[81] Hailwood 2005, 201.

[82] Mason 2009, 293.

[83] Curtin 2003, 293.

[84] Bolivia's Law of the Rights of Mother Earth 2010, as cited in Luttenberger and Luttenberger 2012, 35.

[85] Luttenberger and Luttenberger 2012.

[86] Luttenberger and Luttenberger 2012; Shuttleworth 2012.

Negotiations told a statewide newspaper.[87] These nations' actions, alongside Ecuador, lend to volumes of debate on the expansion and reframing of ecological citizenship.

Citizen or Commodity?

Up until now, this chapter has focused on a reconceptualization of ecological citizenship based upon Ecuador's constitution. The rights to nature language certainly brings to light a new framework for the non-human world's inclusion into our moral and political community while also challenging our existing dominant social paradigms toward the "other." Now we must ask how likely it is that the practical engagement of treating nature as citizen rather than commodity will be successful in a country so heavily dependent on an extractive natural resource economy. Is it impossible for the reframed citizenship described above to exist without an alternate development paradigm? Prior to the passage of Ecuador's constitution, President Rafael Correa proposed an internationally binding agreement to leave oil in the ground in exchange for international contributions of at least half of the extraction revenues the state would have otherwise received.[88] Initially heralded as an "ecological win" for the Amazon, the Yasuní-ITT Initiative — as it was later dubbed — now faces failure after Correa in late 2013 announced he would abandon the plan and instead pursue oil extraction in less than one percent of Yasuní National Park.[89] Instead of a subject of rights, nature, it appears, remains a stock of resources despite the constitution's adoption.

Critics say that rather than offering universal protection of nature as rights-bearing subjects, Ecuador's constitution now represents ultimately greater state control over nature. Tanasescu goes on to describe the "double personality" the rights of nature face:

> On the one hand, ...[nature's rights] are presented as a moral principle of inclusion, akin to liberating the enslaved (literally so for their proponents). On the other hand, they appear to be a subtle weapon against a state that is unwilling to concede some principles that the indigenous in particular hold dear.[90]

While indigenous communities wanted a "clear right" to oppose or approve development projects, the government fought for (and won) maintaining state ownership of all natural resources — meaning "the state can decide to exploit any natural resources it deems to be of national importance, as long as it consults the affected communities, without having an actual obligation to listen."[91] Indeed, the 2008 constitution even furthered the state's powers over natural resource development and management, in addition to providing for the nationalization of

---

[87] Shuttleworth 2012.
[88] Espinosa 2013.
[89] Hill 2013.
[90] Tanasescu 2013, 850.
[91] Tanasescu 2013, 850.

extractive industries.[92] While the state's first "rights for nature" case in court favored a local community over the municipal government for dumping construction materials into the Rio Vilcabamba, enforcing the sentence has been "incomplete and careless."[93]

Ecuador's second "rights for nature" case exemplifies the state's increased powers under the new constitution and its enforcement capabilities. In 2011, the national government's interior minister sought an injunction against foreign-owned illegal mining operations polluting rivers in the north and won the case "for the protection of the rights of nature and the people."[94] The Ecuadorian court ordered the mining activities cease and for the army to move into the territories and seize or destroy — by use of explosives — all of the personal property used for mining.[95] At the same time, critics of the government point to increased gold and copper exploration with the state holding majority shares.[96] Also underway is increased oil exploration in the Amazon region carried out by state-owned entities or China Petrochemical Corporation, on behalf of the Ecuadorian state-owned Petroecuador.[97] Offering an even more biting critique of Correa's administration, Whittemore argues Correa's government has never demonstrated a "sincere intention or ability" to uphold or enforce the "rights to nature" amendments over economic gains.[98] After at least two decades of denationalizing development, Ecuador's constitution has effectively brought development back under national control that seeks to conserve price mechanisms over nature's rights.[99] From these reports it becomes unfortunately apparent that creating a legal space for nature means little when its rights remain subject to market forces rather than responsibilities or intrinsic value.

Conclusion

Despite the increasing attention that has been given to the idea of ecological citizenship over the past two decades, very little government policy addresses rights and obligations owned beyond the human species. Ecuador's constitutional action is extraordinary in its incorporation of an indigenous worldview into national politics, beyond the dominant and anthropocentric status quo, and it prompts a reconsideration of how we theorize and include the natural world in citizenship studies. In this chapter, I have argued against the normative liberal and civic republican framework, concluding instead that Ecuador's provision inspires a new ethical-political orientation that calls for more inclusive conceptualizations of citizenship and a deeper recognition of the natural world's value akin with the country's growing indigenous political and social influence. I have proposed two revised characteristics of ecological citizenship: a shifting of emphasis from

---

[92] Arsel 2012.

[93] Global Alliance for the Rights of Nature 2012. See more about enforcement challenges in the case from Daly 2012.

[94] Daly 2012, 65.

[95] Daly 2012; Tanasescu 2013.

[96] Arsel 2012.

[97] Arsel 2012.

[98] Whittemore 2011, 659.

[99] Arsel 2012.

individualism to the collective community and an inclusion of humans' obligations to the natural world.

Ecuador's constitution certainly could transform legal treatment of the environment if its amendments were to be enforced.[100] But at the time of writing there is little evidence to suggest a decommodification of nature in a country so heavily dependent on its exploitation. Political studies provide numerous examples of countries in the developing world that draft laws or constitutional provisions that promise more than existing institutional establishments can deliver. Some scholars assume that certain political systems, particularly liberal democracies, cannot carry out the aims associated with progressive ecological enactments,[101] while others predict the rise of eco-authoritarian regimes that will politically force citizens to mitigate environmental damage.[102] In early-2014 the world's first Ethics Tribunal of the Rights of Nature and Mother Earth was held in Ecuador as a permanent platform for hearing and judging cases from across the globe. Among the cases accepted by the Tribunal are the Chevron-Texaco pollution case in Ecuador, the Yasuní-ITT oil project, persecution of Ecuadorian environmental activists under the new constitution and Ecuadorian mining. Citing humans' responsibility to the natural world, the Tribunal claims that it "is ironic that Ecuador is abandoning its leadership and failing to respect its own constitution."[103] A reframing and expansion of ecological citizenship is necessary now for decision-making about the future of our planet and all its inhabitants. As Wildcat rightly concludes: "We must consider how our lives and the life of the planet might be better served realizing that politics based on inalienable rights might be empty, indeed poverty stricken, without a necessary complementary recognition of our inalienable responsibilities."[104]

Notes

I would like to thank delegates of the Sixth International Conference on Environmental, Cultural, Economic, and Social Sustainability for their helpful comments on an earlier version of this article; Dr. Peter Jacques of the University of Central Florida for his advice and encouragement; and my husband Lee Mullon for his critique of this work.

References

Alvaro, Mercedes and Daniel Gilbert. 2013. Ecuador Court Affirms and Halves Chevron Judgment. *The Wall Street Journal*, November 12, B1.

Amazon Watch. 2014. First International Tribune on Rights of Nature. Available at http://amazonwatch.org/news/2014/0121-first-international-tribunal-on-rights-of-nature, accessed February 11, 2014.

---

[100] Whittemore 2011.

[101] Barry 2001; Eckersley 2004.

[102] Heilbronner 1974; Kennedy 1993.

[103] Amazon Watch 2014.

[104] Wildcat 2009, 131.

Andolina, Robert. 2003. The Sovereign and its Shadow: Constituent Assembly and Indigenous Movement in Ecuador. *Journal of Latin American Studies* 35(4): 721-750.

Arsel, Murat. 2012. Between "Marx and Markets"? The State, the "Left Turn" and Nature in Ecuador. *Tijdschrift voor Economische en Sociale Geografie* 103(2): 150-163.

Barry, John. 2001. Justice, Nature and Political Economy. *New Political Economy* 30(3): 381-394.

Barry, John. 2002. Vulnerability and Virtue: Democracy, Dependency and Ecological Stewardship. In *Democracy and the Claims of Nature*, edited by B.A. Minteer and B. P. Taylor, 133-152. Oxford: Rowman and Littlefield.

Bell, Derek. 2005. Liberal Environmental Citizenship. *Environmental Politics* 14(2): 179-194.

Bruckerhoff, Joshua J. 2008. Giving Nature Constitutional Protection: A Less Anthropocentric Interpretation of Environmental Rights. *Texas Law Review* 86(3): 615-646.

Cepak, Michael L. 2012. Strange Powers: Conservation, Science, and Transparency in an Indigenous Political Project. *Anthropology Today* 28(4): 14-17.

Christoff, Peter. 1996. Ecological Citizens and Ecologically Guided Democracy. In *Democracy and Green Political Thought: Sustainability, Rights, and Citizenship*, edited by B. Doherty and M. de Geus, 151-169. London: Routledge.

Clark, Nigel, and Nick Stevenson. 2003. Care in the Time of Catastrophe: Citizenship, Community and the Ecological Imagination. *Journal of Human Rights* 2(2): 235-246.

Consejo Nacional de Planificación [Ecuador]. 2009. *Plan Nacional de Desarrollo 2009-2013*. Quito: SENPLADES.

Curry, Patrick. 2000. Redefining Community: Towards an Ecological Republicanism. *Biodiversity and Conservation* 9: 1059-1071.

Curtin, Deane. 2003. Ecological Citizenship. In *Handbook of Citizenship Studies*, edited by Engin F. Isin and Bryan S. Turner, 289-304. London: Sage.

Daly, Erin. 2012. The Ecuadorian Exemplar: The First Ever Vindications of Constitutional Rights of Nature. *Review of European Community and International Environmental Law* 21(1): 63-66.

de la Cadena, Marisol. 2010. Indigenous Cosmopolitics in the Andes: Conceptual Reflections Beyond "Politics". *Cultural Anthropology* 25(2): 334-370.

Dean, Hartley. 2001. Green Citizenship. *Social Policy & Administration* 35(5): 490-505.

Depledge, Michael H. and Cinnamon P. Carlarne. 2008. Environmental Rights and Wrongs. *Environmental Science and Technology* 42(4): 990-994.

de-Shalit, Avner. 2000. *The Environment Between Theory and Practice*. Oxford: Oxford University Press.

Dobson, Andrew. 2003. *Citizenship and the Environment.* Oxford: Oxford University Press.

Dobson, Andrew, and Derek Bell. 2006. Introduction. In *Environmental Citizenship*, edited by Andrew Dobson and Derek Bell, 1-17. Cambridge, MA: The MIT Press.

Dosh, Paul, and Nicole Kilgerman. 2009. Correa vs. Social Movements: Showdown in Ecuador. *NACLA Report on the Americas* 42(5): 21-40.

Eckersley, Robyn. 2004. *The Green State: Rethinking Democracy and Sovereignty*. Cambridge, MA: The MIT Press.

Espinosa, Cristina. 2013. The Riddle of Leaving the Oil in the Soil – Ecuador's Yasuní-ITT Project from a Discourse Perspective. *Forest Policy and Economics* 36: 27-36.

Gabrielson, Teena. 2008. Green Citizenship: A Review and Critique. *Citizenship Studies* 12(4): 429-446.

Galeano, Eduardo. 2008. Nature is Not Mute. *The Progressive* 72(4): 19.

Gerlach, Allen. 2003. *Indians, Oil, and Politics: A Recent History of Ecuador*. Wilmington, DE: Scholarly Resources, Inc.

Global Alliance for the Rights of Nature. 2012. Vilcabamba River Case Law: 1 Year After. Available at http://therightsofnature.org/rights-of-nature-laws/vilcabamba-river-1-year-after/, accessed February 11, 2014.

Goodwin, Robert. 1992. *Green Political Theory*. Oxford: Oxford University Press.

Hailwood, Simon. 2005. Environmental Citizenship as Reasonable Citizenship. *Environmental Politics* 14(2): 195-210.

Hayward, Tim. 2000. Constitutional Environmental Rights: A Case for Political Analysis. *Political Studies* 48(3): 558-572.

Heilbronner, Robert. 1974. *An Inquiry Into the Human Prospect*. New York: Harper and Row.

Hill, David. 2013. Why Ecuador's President is Misleading the World on Yasuni-ITT. *The Guardian,* October 15. Available at http://www.theguardian.com/environment/andes-to-the-amazon/2013/oct/15/ecuador-president-misleading-yasuni, accessed April 7, 2014

Holder, Cindy L. and Jeff J. Corntassel. 2002. Indigenous Peoples and Multicultural Citizenship: Bridging Collective and Individual Rights. *Human Rights Quarterly* 24(1): 126-151.

Jagers, Sverker C. 2009. In Search of Ecological Citizenship. *Environmental Politics* 18(1): 18-36.

Jones, Ashby. 2014. International Law 101: Why the $9.5BN Chevron Ruling Has (So Far) Led to Nothing. *The Wall Street Journal*, March 4. Available at http://blogs.wsj.com/law/2014/03/04/why-the-9-5-billlion-ruling-against-chevron-has-so-far-amounted-to-nothing/, accessed April 7, 2014.

Kearney, Amanda and John J. Bradley. 2009. 'Too Strong to Ever Not Be There': Place Names and Emotional Geographies. *Social and Cultural Geography* 10(1): 77-94.

Kennedy, Paul. 1993. *Preparing for the Twenty-First Century*. London: Harper Collins Publishers.

Kimerling, J. 1994. The Environmental Audit of Texaco's Amazon Oil Fields: Environmental Justice or Business as Usual? *Harvard Human Rights Journal* 7: 199-224.

Krauss, Clifford. 2014. Big Victory for Chevron Over Claims in Ecuador. *The New York Times*, March 4, B1-B4.

Latta, Alex P. 2007. Locating Democratic Politics in Ecological Citizenship. *Environmental Politics* 16: 377-393.

Leopold, Aldo. [1949] 1966. *A Sand County Almanac: With Essays on Conservation from Round River.* New York: Ballantine Books.

Ley Orgánica de Biodiversidad of Ecuador, 2009. Asamblea Nacional Memorando No. PAN-FC-09-125, dated Nov. 12, 2009. Available at http://downloads.arqueoecuatoriana.ec/ayhpwxgv/leyes/Proyecto_Ley_Organica_de_la_Biodiversidad.pdf, accessed February 1, 2010.

Luttenberger, Lidija Runko and Axel Luttengerger. 2012. Earth-Centric Approach in Environmental Politics. *Scientific Journal of Maritime Research* 26(1): 27-44.

Maniates, Michael F. 2001. Individualization: Plant a Tree, Buy a Bike, Save the World? *Global Environmental Politics* 1(3): 31-52.

Marshall, T. H. 1992. *Citizenship and Social Class.* In *Citizenship and Social Class*, edited by T. Bottomore, 1-51. London: Pluto Press.

Mason, Andrew. 2009. Environmental Obligations and the Limits of Transnational Citizenship. *Political Studies* 57: 280-297.

Mato, Daniel. 2008. Transnational Relations, Culture, Communication and Social Change. Trans. Emeshe Juhasz-Mininberg. *Social Identities* 14(3): 415-435.

McGregor, Deborah. 2004. Coming Full Circle: Indigenous Knowledge, Environment, and Our Future. *American Indian Quarterly* 28(3/4): 385-410.

Melo-Escrihuela, Carme. 2008. Promoting Ecological Citizenship: Rights, Duties and Political Agency. *ACME* 7(2): 113-134.

Mychalejko, Cyril. 2008. Ecuador's Proposed New Constitution Gives Rights to Nature. *Upside Down World*, September 25. Available at http://upsidedownworld.org/main/ecuador-archives-49/1494-ecuadors-constitution-gives-rights-to-nature, accessed April 7, 2014.

Naess, Arne. 1973. The Shallow and the Deep, Long-Range Ecology Movement: A Summary. *Inquiry* 16: 95-100.

Nozick, Robert. 1974. *Anarchy, State, Utopia.* Oxford: Blackwell.

Paehlke, Robert. 2004. Ethics, Green Citizenship and Globalization. Unpublished paper presented at the Annual Meeting of the International Studies Association, Quebec, March 17-20.

Quintero Lopez, Rafael. 2008. *La Constitución del 2008 : Un Análisis Político.* Quito: Abya-Yala.

Radcliffe, Sarah A. 2012. Development for a Postneoliberal Era? *Sumak kawsay*, Living Well and the Limits to Decolonisation in Ecuador. *Geoforum* 43: 240-249.

Regan, Tom. 1988. *The Case for Animal Rights.* London: Routledge.

Santacruz, Silvia. 2009. The Other L.A., *Forbes*, November 20. Available at http://www.forbes.com/2009/11/20/lago-agrio-ecuador-chevron-texaco-opinions-contributors-silvia-santacruz.html, accessed April 7, 2014.

Shelton, Dinah L. 2006. Human Rights and the Environment: What Specific Environmental Rights Have Been Recognized? *Denver Journal of International Law and Policy* 35(1): 129-171.

Shuttleworth, Kate. 2012. Agreement Entitles Whanganui River to Legal Identity. *The New Zealand Herald*, August 30. Available at http://www.nzherald.co.nz/nz/news/article.cfm?c_id=1&objectid=10830586, accessed April 7, 2014.

Singer, Peter. 1977. *Animal Liberation*. New York: Avon Books.

Smith, Gar. 2009. In Ecuador, Trees Now Have Rights. *Earth Island Journal* 23(4): 15.

Smith, Kimberly K. 2006. Natural Subjects: Nature and Political Community. *Environmental Values* 15(3): 343-353.

Smith, Mark J. 1998. *Ecologism: Towards Ecological Citizenship*. Buckingham: Open University Press.

Soveroski, Marie. 2007. Environmental Rights Versus Environmental Wrongs: Forum Over Substance? *Review of European Community and International Environmental Law* 16(3): 261-273.

Stone, Christopher D. 1972. Should Trees Have Standing? Toward Legal Rights for Natural Objects. *University of Southern California Law Review* 42: 450-482.

Tanasescu, Mihnea. 2013. The Rights of Nature in Ecuador: The Making of an Idea. *International Journal of Environmental Studies* 70(6): 846-861.

The Pachamama Alliance. 2008. Pachamama and Ecuador's Constitutional Assembly Explore Creation of an Environmental State. Available at http://www.pachamama.org/news/pachamama-and-Ecuador%E2%80%99s-constitutional-assembly-explore-creation-of-an-environmental-state, accessed March 31, 2009.

Valdivia, Gabriela. 2007. The 'Amazonian Trial of the Century': Indigenous Identities, Transnational Networks, and Petroleum in Ecuador. *Alternatives* 32: 41-72.

Walsh, Catherine. 2009. *Interculturalidad, Estado, Sociedad: Luchas (De)Coloniales de Nuestra Época*. Quito: Universidad Andina Simon Bolivar – Abya Yala.

Whittemore, Mary Elizabeth. 2011. The Problem of Enforcing Nature's Rights Under Ecuador's Constitution: Why the 2008 Environmental Amendments Have No Bite. *Pacific Rim Law and Policy Journal* 20(2): 659-691.

Wildcat, Daniel R. 2009. *Red Alert!: Saving the Planet with Indigenous Knowledge*. Golden, CO: Fulcrum Publishing.

Part VI: Citizenship

Editors' Note: One of the earliest and most influential environmental philosophers was the American forester and environmentalist Aldo Leopold (1887-1948). His most famous work, *A Sand County Almanc*, was publishd one year after his death in 1949. Probably the most cited line from the book - and one that is considered foundational in the study of envuironmental ethics - is "A thing is right when it tends to preserve the integrity, stability, and beauty of the biotic community. It is wrong when it tends otherwise." A longer quote from the same book, which challenges us to rethink our notions of community and citizenship, is:

> The land ethic simply enlarges the boundaries of the community to include soils, waters, plants, and animals, or collectively: the land.
>
> This sounds simple: do we not already sing our love of and obligation to the land of the free and the home of the brave? Yes, but just what and whom do we love? Certainly not the soil, which we are sending helter-skelter downriver. Certainly not the waters, which we assume have no function except to turn turbines, float barges, and carry off sewage. Certainly not the plants, of which we exterminate whole communities without batting an eye. Certainly not the animals, of which we have already extirpated many of the largest and most beautiful species. A land ethic of course cannot prevent the alteration, management, and use of these 'resources,' but it does affirm their right to continued existence in a natural state.

Aldo Leopold (1949)[1]

---

[1] Leopold, Aldo. [1949] 1966. *A Sound County Almanac: With Essays on Conservation from Round River*. New York: Ballantine Books, 239-240.

Chapter 19: Disruptive Social Innovation for a Low-carbon World

Samuel Alexander

Introduction

It is becoming increasingly clear that small, incremental changes to the way humans use and produce energy are unlikely to catalyze a transformation to a low-carbon civilization, at least, not within the ever-tightening time frame urged by the world's climate scientists. In September 2013, the Intergovernmental Panel on Climate Change (IPCC) published its fifth report,[1] in which it estimated that the world's carbon budget – that is, the maximum carbon emissions available if the world is to have a good chance keeping global warming below 2 degrees – is likely to be entirely used within the timeframe of 2028-2038, based on current trends. If "business as usual" continues, the trends indicate that we may be facing a future that is 4 degrees hotter, or more.[2] It is not clear to what extent civilization is compatible with such a climate.

This calls for an urgent and committed re-evaluation of dominant strategies for transitioning beyond fossil fuels. If there is any hope for rapid decarbonization today, it surely lies, at this late stage, in movements, innovations, or technologies that do not seek to produce change through a smooth series of increments, but through an ability to somehow "disrupt" the status quo and fundamentally redirect the world's trajectory toward a low-carbon future.

The phrase "disruptive innovation" will be used in this chapter to refer to rapid and far-reaching societal change that is provoked by the abrupt emergence of a social movement, technology, business model, or confluence of such phenomena. This usage draws loosely on the work of Clayton Christensen, who coined the term disruptive innovation to describe times when commercial enterprises develop new business models or technologies that rapidly change the market in ways that are both unexpected and game-changing.[3] In Christensen's

---

[1] Intergovernmental Panel on Climate Change 2013.
[2] World Bank 2012; Christoff ed. 2013.
[3] Christensen 1997.

work, a disruptive innovation is contrasted with a "sustaining innovation", which is less about changing the game and more about competing more effectively in the same game.

While there may be a certain irony to using terminology from commercial discourse to refer to socio-technical innovations that could potentially shake the very energy basis of the global economy, the language of disruptive innovation aptly describes the extent and speed at which any transition to low-carbon world must proceed. With the carbon budget shrinking as business as usual persists – to say nothing of the myriad other ecological crises worsening by the day – a progression of sustaining innovations seems unlikely to affect the changes necessary. The task is too urgent; the extent of change needed, too great.

As implied above, disruptive innovations can take place within various spheres of life: social, economic, technological, institutional, and political. In order to transition to a low-carbon world, it is likely that a coordinated confluence of innovations from all such spheres will be required to produce deep systemic and structural change. This paper focuses on the *socio-cultural* sphere. Without denying the importance of other spheres of transformative change, there are reasons to think the socio-cultural sphere may be of particular importance in driving the transition to a low-carbon world.

The socio-cultural domain may have special disruptive potential due to the fact that other spheres of innovation can be understood as *tools* or *means*, whereas the socio-cultural sphere can be understood to be the source of *goals* or *ends*. This difference is important because until there is a culture driven by the values and vision of a low-carbon world, available tools or means for societal change (e.g. legislation, technology, capital, etc.) are likely to be misdirected, and perhaps even be employed in counter-productive ways. In much the same way as the tool of "fire" can have a positive or negative impact on our lives, depending on how it used and how much of it there is, the tools of technology, business, and politics can advance or inhibit the transition to a low-carbon society, depending on the social values and desires that shape their implementation and development. For these reasons, the socio-cultural sphere can be considered fundamental, in the sense that it provides the *ends* towards which available *means* are directed.

This point deserves some elaboration. The nature and development of technology in a society, for example, will take different forms depending on the social context and the dominant social values which drive innovation. If the primary end of research and development is profit-maximization, not the desire for a low-carbon world, it follows that technological innovations are just as likely to inhibit, rather than facilitate, a low-carbon transition (e.g. fracking technologies). A similar dynamic exists with respect to business and politics. Until there is a socio-cultural context that *incentivizes* or *demands* economic or political change in the direction of a low-carbon world, the tools of business and public policy are unlikely to be sources of disruptive innovation, but at most sources of sustaining innovation. More likely still, they will merely serve to reify and entrench the status quo.

Another way to think about the importance of the socio-cultural sphere is in terms of sequencing; that is, in terms of what order various innovations may need to take place on the path to a low-carbon world. By the time business, politics, and technology are capable of disrupting the status quo, it may be that a revolution in social values would need to have already taken place, in order to have driven such innovation in the first place and been receptive to it. After all, it is no good developing a cheap and efficient solar panel if few people are interested in renewable energy; just as an effective carbon policy will not be the foundation of a successful political campaign until the social conditions are ripe for its acceptance. Again, this is not meant to downplay the undeniable importance of technological, economic, and political innovations on the path to a low-carbon society. A coordinated, multi-faceted approach is both necessary and desirable. But insofar as technology, business, and politics are a reflection of the culture in which they are situated, it would seem that disruptive innovation in the socio-cultural sphere may need to be the prime mover which would then enable or ignite further disruptive innovations in other spheres of life.

Significantly, the socio-cultural sphere is also the domain where individuals acting as private citizens have most agency. We may not feel like we have much influence over the decisions of our members of parliament, or the decisions of big business or other global institutions, but within the structural constraints of any society there nevertheless resides a realm of freedom through which citizens and communities can resist and oppose the existing order and make their influence felt. However small those acts of opposition (or renewal) might seem in isolation, when they form part of a large social movement, their cumulative impact can reshape society "from below" and ultimately form a tidal wave of revolutionary significance, washing away the old world, or aspects of it, and clearing space for the new. A brief glance at the history of social movements shows this to be true.

Of course, to suggest that technology, business, and politics are a merely reflection of culture is a contestable and, in many ways, an overly simplistic proposition. Public policy, for example, rather than always being shaped by culture in a uni-directional way, sometimes takes the lead in societal development and is influenced by forces other than culture. The same can be said of the spheres of business and technology, both of which *shape* culture as they are *shaped by* culture, in a dialectical fashion. Nevertheless, it would be fair to state that any transformative politics, technology, or business model needs to be complemented, and probably preceded by, a co-relative transformation in the socio-cultural sphere. This suggests that citizens must carefully consider not only what cultural or social conditions would best facilitate the realization of a low-carbon world, but also what role social or cultural movements might have to play in producing those conditions.

The purpose of this paper, then, is to review contenders, so to speak, for the category of most innovative social movements working toward a low-carbon world, movements which are potentially "disruptive" in the sense outlined above. Different citizens would surely make different choices and see different degrees of potential in today's various social movements for a low-carbon world. Be that as it may, the movements reviewed below do seem to jump out somewhat as obvious contenders, not only for what they already are, but more importantly, what they are promising, or even threatening, to become. The analyses offered are linked insofar as they suggest that individuals, as engaged citizens, have various ways to initiate deep change by acting decisively themselves, without waiting for change to be driven by politicians, business, or technological advancement. In an age that often seems to be coloured by political paralysis, economic path-dependency, and techno-fetishism, highlighting the transformative power we have as citizens can provide real, albeit tentative, grounds for thinking that a low-carbon world can indeed emerge from the grassroots – if only we choose it.

Potential Contenders for Disruptive Social Innovation

The Fossil Fuel Divestment Campaign

The analysis will begin with the fossil fuel divestment campaign, which was initiated late in 2012 by climate activist, Bill McKibben, and his team at the campaigning network 350.org. Since launching, this campaign has taken on a life of its own, as nascent social movements tend to do. The disruptive potential of this campaign lies in how directly it challenges the *financial foundations* of the fossil fuel industry, without which the industry could not support itself or develop new projects. Find a way to remove the financial lifeblood of the industry – that is, find a way to remove or minimize shareholder investment in fossil fuels – and the industry would inevitably wither away by the very same logic of capital that currently sustains it.

Motivated by this possibility, McKibben and his team organized a campaign for fossil fuel divestment, which calls on individuals, communities, institutions, and governments to withdraw or "divest" their financial support from the fossil fuel industry with the ultimate aim of crippling it.[4] Without investment, the fossil fuel industry cannot exist; without the fossil fuel industry, the primary cause of climate change is eliminated. The genius of this campaign lies both in its simplicity and its directness, and the movement seems to be growing in momentum.

In the US, 380 college campuses have committed to divestment,[5] with successful divestment having already been achieved in nine universities and colleges, 22 cities, and 10 religious organizations, with further campaigns under way in Canada, the United Kingdom, Sweden, Finland, India, Bangladesh, Australia and New Zealand.[6] While the campaign encourages individuals to divest wherever possible, the main focus is on larger institutions and organizations where the real financial power lays, especially banks, pension schemes, universities, churches, and governments. A 2013 report from Oxford University concludes that this is the fastest growing divestment campaign in history.[7]

Significantly, the transformative potential of divestment as a strategy for deep societal change is not without precedent. As the fossil fuel divestment website notes:

> There have been a handful of successful divestment campaigns in recent history, including Darfur, Tobacco and others, but the largest and most impactful one came to a head around the issue of South African

---

[4] McKibben 2012.
[5] Conifino 2013.
[6] Go Fossil Free 2014a.
[7] Ansar, Caldecott, and Tilbury 2013.

Apartheid. By the mid-1980s, 155 campuses—including some of the most famous in the country—had divested from companies doing business in South Africa. 26 state governments, 22 counties, and 90 cities, including some of the nation's biggest, took their money from multinationals that did business in the country. The South African divestment campaign helped break the back of the Apartheid government, and usher in an era of democracy and equality.[8]

One of the most interesting and promising elements to the fossil fuel divestment campaign is how it mixes financial self-interest with environmental and humanitarian ethics. The ethical side of the divestment campaign is obvious enough: fossil fuel emissions are the primary cause of climate change,[9] thus a moral case can easily be made that people and institutions should not be investing in, or profiting from, an industry that is in the process of destabilizing the climate with potentially devastating social and environmental consequences.[10] Just as it would be unethical to profit from the slave trade or from the sale of ivory, it is ethically dubious to profit from the cause of climate change.

This ethical defense is arguably grounds enough for divesting from fossil fuels, but the fascinating thing about the divestment campaign is how to frame a supplementary argument based on self-interest, and perhaps this is where the real revolutionary potential of the campaign lies. McKibben argues that citizens or institutions concerned about *their own financial assets* should immediately divest from fossil fuels because there is a "carbon bubble" waiting to burst.

The carbon bubble hypothesis is based on the notion of a carbon budget, which represents the estimated amount of carbon emissions the atmosphere could safely absorb (where "safely" means keeping temperatures under 2 degrees from pre-industrial levels, which is the target internationally agreed to in the Copenhagen Accord of 2009).[11] Put simply, McKibben and others argue that the carbon bubble exists because the amount of fossil fuels already discovered far exceeds the world's carbon budget. More precisely, it is estimated that embedded carbon in existing global fossil fuel reserves lies in the vicinity of 2795Gt, but the world's carbon budget is only around 565Gt.[12] What this means is that, if the world is to stop temperatures rising above the threshold of 2 degrees, around 80% of the fossil fuels already discovered simply cannot be burned and must remain in the ground. A carbon bubble exists, therefore, because currently fossil fuel shares are priced on the assumption that all reserves will be produced. It follows that any serious response to climate change is going to burst the carbon bubble by turning a vast amount of fossil fuel reserves into "stranded assets" of little or no value.

The divestment campaign is calling on people to recognize both the ethics and the economics of this situation, and withdraw their shares in the fossil fuel industry before the bubble bursts and the value of their shares implode. Not only do investors have a financial self-interest to do so, a broad-based divestment from

---

[8] Go Fossil Free 2014b.
[9] Intergovernmental Panel on Climate Change 2013.
[10] World Bank 2012; Christoff ed. 2013.
[11] Carbon Tracker Initiative (undated); Intergovernmental Panel on Climate Change 2013.
[12] Carbon Tracker Initiative (undated); McKibben 2012.

fossil fuels could crash the industry by destroying its investment base, which, as noted, is a leading aim of the campaign.

Even if the campaign does not manage to bankrupt the industry economically, the campaign may nevertheless advance a critically important goal of stigmatizing the fossil fuel industry as a primary enemy of climate stabilization, thereby bankrupting it politically and socially. Historically, stigmatization has been an important function of divestment campaigns.[13]

While it could be argued that the divestment campaign makes the mistake of thinking that market mechanisms can provide a solution to climate change, this potential indictment can easily be reconceived as a defense of the strategy: divestment uses the existing mechanisms of capitalism to undermine what is arguably capitalism's defining industry. If ever there was an Achilles' Heel to the fossil fuel industry, the divestment campaign just might be it.

Of course, it is not enough merely to divest from the fossil fuel industry; it is equally important to reinvest in a clean energy economy. This additional reinvestment strategy provides further grounds for thinking that the divestment campaign could be of transformative significance in bringing about a post-carbon or low-carbon world.

Transition Initiatives

If the fossil fuel divestment campaign is one of the most promising social movements *opposing* and *undermining* the carbon-based society, the Transition Towns movement is arguably one of the most promising and coherent social movements focused on *building an alternative* society. This movement burst onto the scene in Ireland in 2005, and already there are more than one thousand Transition Towns around the world, in more than forty countries, including Australia. Given that this movement is explicitly seeking to mobilize communities for a low-carbon future, it is important to consider what exactly defines a Transition Town and evaluate the extent to which this movement has disruptive potential.

The fundamental aims of the movement are to respond to the twin challenges of peak oil and climate change by decarbonizing and delocalizing the economy through a community-led model of change based on permaculture principles.[14] In doing so, the movement runs counter to the dominant narrative of globalization, and instead offers a positive, highly localized vision of a low-carbon future, as well as an evolving roadmap for getting there through grassroots activism. While this young and promising movement is not without its critics,[15] there are some, such as Ted Trainer, who argue that if civilization is to make it into the next half of the century in any desirable form, "it will be via some kind of Transition Towns process."[16]

According to the movement's co-founder, Rob Hopkins, the strategy and vision of Transition is based on four key assumptions:[17]

---

[13] Ansar, Caldecott, and Tilbury 2013.

[14] Hopkins 2008; Holmgren 2002.

[15] James 2010, 14-5.

[16] Trainer 2009, 11.

[17] Hopkins 2008, 134.

1. That life with dramatically lower energy consumption is inevitable, and that it's better to plan for it than to be taken by surprise;
2. That our settlements and communities presently lack the resilience to enable them to weather the sever energy [and economic] shocks that will accompany peak oil [and climate change];
3. That we have to act collectively, and we have to act now;
4. That by unleashing the collective genius of those around us to creatively and proactively design our energy descent, we can build ways of living that are more connected, more enriching, and that recognise the biological limits of our planet.

The rationale for engaging in grassroots activity is that "if we wait for governments, it'll be too little, too late. If we act as individuals, it'll be too little. But if we act as communities, it might just be enough, just in time."[18] According to some commentators, this approach represents a "pragmatic turn"[19] insofar as it focuses on *doing* sustainability here and now. In other words, it is a form of "DIY politics,"[20] one that does not involve waiting for governments to provide solutions, but rather depends upon an actively engaged citizenry.

This approach is particularly relevant in Australia, where Tony Abbott's conservative government is showing no signs of progressing the nation toward a low-carbon future, meaning that any movement toward such a future may have to be driven "from below." Therein lies the promise and coherency of the Transition Movement: in an era of political paralysis, it seems that the *only* path beyond fossil fuels is one led by communities acting locally, and in that regard the Transition Movement is leading the way. Of course, whether grassroots movements for a low-carbon world ultimately march under the banner of "transition" is of little importance; what is necessary and important is that people do not wait for governments to act.

The paradigm shift of Transition is articulated around notions of "decarbonization" and "relocalization" of the economy. What this means in practice is complex, but the overarching idea is that decarbonization is necessary and desirable for reasons of peak oil and climate change, and given how carbon-intensive global trade is, decarbonization implies delocalizing economic processes. As well as this, another central goal of the movement is to build community resilience, a term which can be broadly defined as the capacity to withstand shocks and the ability to adapt after disturbances.[21]

Notably, crisis in the current system is presented not as a cause for despair but as a transformational opportunity, a change for the better that should be embraced rather than feared. [22] Consequently, the vision presented by the Transition Movement is very positive, one that is "full of hope"[23] for a more "nourishing and abundant future."[24] Hopkins, who is by far the most prominent

---

[18] Hopkins 2013, 45.
[19] Barry and Quilley 2008, 2.
[20] Barry and Quilley 2009, 3.
[21] Hopkins 2008.
[22] Hopkins 2011, 45.
[23] Bunting 2009.
[24] Hopkins 2008, 5.

spokesperson for the movement, plays a crucial role in promoting such an optimistic message, while at the same time acknowledging the extent of the global problems and asserting there is no guarantee of success.[25] By doing so, Hopkins skilfully walks a delicate line: he openly acknowledges the magnitude of the global predicament, but quickly proceeds to focus on positive, local responses and action. Whether his positivity is justifiable is an open question – some argue that it is not[26] – but it is nevertheless proving to be a means of inspiring and mobilising communities in ways that "doomsayers" are unlikely to ever realise.

As promising as the Transition Movement may be, there are crucial questions it needs to confront and reflect on if it wants to fully realise its potential for deep societal transformation. Firstly, some critics argue that the movement suffers, just as the broader environmental movement arguably suffers, from the inability to expand much beyond the usual middle-class, generally well-educated participants, who have the time, security, and privilege to engage in social and environmental activism. While the Transition Movement is ostensibly inclusive, this self-image requires examination in order to assess whether it is as inclusive and as diverse as it claims to be, and what this might mean for the movement's prospects. Can it "scale up" sufficiently? Secondly, there is the issue of whether a grassroots, community-led movement can change the macro-economic and political structures of global capitalism "from below" through re-localization strategies, or whether the movement may need to engage in more conventional "top down" political activity if it is to have any chance of achieving its ambitious goals. Other critics argue that the movement is insufficiently radical in its vision. Does the movement need to engage more critically with the broader paradigm of capitalism, its growth imperative and social norms and values that constrict the imagination? Is building local resilience within the existing system an adequate strategy? And does the movement recognize that decarbonization almost certainly means giving up many aspects of affluent, consumer lifestyles? This is not the forum to offer answers to these probing questions,[27] but engaging critically with these issues could advance the debate around a movement that may indeed hold some of the keys to transitioning to a just and sustainable, low-carbon world.

It may be that the practical reality of the Transition Movement has been "over-hyped" to some extent, but what seems clear is that a low-carbon world will never emerge unless there is an engaged citizenry that speaks up and gets active. Currently, the Transition Movement is the most promising example of such a social movement, mobilizing communities for a world beyond fossil fuels, and meeting with some success.

Collaborative Consumption and the Sharing Economy

The term "collaborative consumption" has emerged as one of the socio-economic buzzwords of recent times, with *Time* magazine listing it as one of the big ideas that will change the world.[28] Surprisingly, perhaps, collaborative consumption is in many ways an alternative name for "sharing", although as the prime website

---

[25] Hopkins 2011, 17.

[26] See, for example, Smith and Positano 2010.

[27] Alloun and Alexander 2014.

[28] Walsh, 2011.

dedicated to this concept notes, it is "sharing reinvented through technology."[29] But if human beings have been sharing their wealth, possessions, and skills (to varying extents) throughout history, what role could collaborative consumption play in the transition to a low-carbon world? And to what extent could something as mundane-sounding as sharing have disruptive potential?

While the term was coined decades ago,[30] collaborative consumption only began entering the popular lexicon over the last few years, primarily through the work of Rachel Botsman and Roo Rogers, who define this emerging practice as: "Traditional sharing, bartering, lending, trading, renting, gifting, and swapping, redefined through technology and peer communities."[31] The innovation here is that people are using online forums and technologies to offer or acquire access to things without necessarily buying or selling them; instead, they often share, hire, or gift them in more or less informal ways, facilitated by online peer communities which make it easy to list or search for available goods and services. In economic parlance, the transaction costs of sharing or trading are markedly reduced through the use of the internet, making it more efficient than ever to connect formal or informal sharers and traders.

Examples of collaborative consumption are many, varied, and expanding. A representative example is the upsurge in car-sharing businesses, which involve either a central business purchasing limited cars that are then used by a community of people (e.g. Zipcar, Flexicar and GoGet), or alternatively, the central business can facilitate peer-to-peer car sharing (e.g. Car-Next-Door). The genius here was in recognizing that many, if not most, cars sit idle for a huge portion of the day or week, opening up space to utilize them more efficiently through sharing access. If a person can easily hire a neighbor's car for an hour or two when needed, this means less need for a personal car. For the same reason, bike hire has also taken off in various cities around the world, often facilitated by local or national governments. Organizations like Lyft facilitate ride-sharing.

There is also a variety of websites that facilitate sharing with or without monetary exchanges, such as the Sharetribe, Streetbank, or Open Shed, which also function to make better use of existing resources. For example, if there are easy ways to facilitate sharing online, not everyone on the street needs a lawnmower or a jigsaw (since they tend to sit idle), thereby minimizing the need for superfluous production and consumption. Other websites, such as Freecycle, rather than facilitating sharing or trading, simply facilitate the gifting of unwanted or superfluous things, thereby reducing the flow of waste to landfill. One of the real success stories of collaborative consumption has been Air BnB, which allows people to list a room or rooms in their home as short or long term accommodation for travelers, providing people with an alternative to hotels and backpacker accommodation.

What these examples show is that collaborative consumption is often more about *access* to goods and services than *ownership*. Botsman and Rogers argue that this new form of consumption behavior and entrepreneurship has the ability to radically transform business, cultures of consumption, and the environmental

---

[29] Collaborative Consumption 2014.

[30] Felson and Spaeth 1978, 614-24.

[31] Botsman and Rogers 2011, xv.

movement, with potentially deep implications on various aspects of life.[32] This transformative potential remains even when the exchange of money is involved, as is the case with prominent websites such as Craigslist, Ebay, and Gumtree. These types of websites still provide an efficient means of allocating or reallocating goods and services beneath the surface of the traditional economy. If a household finds itself with a surplus couch, table, or set of curtains, there are now numerous channels that are available to connect such goods with people who need them, through the click of a button.

There are three broad categories of collaborative consumption: Product Service Systems, Redistribution Markets, and Collaborative Lifestyles:[33]

- *Product Service Systems* refer to the switch from an ownership model of consumption toward a usage or access model. Thus, people pay for, or get the benefit of a product, without owning it.
- *Redistribution Markets* facilitate the redistribution of goods from where they are not needed to any place or person where they are needed. These markets have always existed, but current technologies, especially online social networks, are fuelling this type of collaborative consumption.[34] These markets can either involve gifting, sharing, bartering, or more conventional trading, and they are challenging traditional business and consumption methods. According to Botsman and Rogers "redistribution is the fifth 'R' - reduce, recycle, reuse, repair and redistribute."[35]
- *Collaborative Lifestyles* are less about sharing tangible assets, and more about sharing things like time, space, and skills – again, facilitated by online forums.

It is difficult to deny the potential of these types of collaborative exchange to challenge dominant cultures of wasteful and excessive consumption. Especially in affluent societies, where a vast amount of goods lie idle and unused, there seems to be huge potential for avoiding further production of goods by facilitating the efficient reallocation of existing goods through sharing, barter, and trade. Sharing and redistribution via these methods clearly provide a path to reduced carbon emissions, by minimizing the need for continuous production, and they seem to have the added benefit of promoting community interaction at the nexus of social and economic life. Of course, there is the further incentive of self-interest: many people are drawn to collaborative consumption for the obvious reason that it can save money and hassle, or even make money.

Nevertheless, this potentially disruptive innovation has various risks that ought to be borne in mind too. For example, one of the obvious benefits for individuals who consume collaboratively is reduced costs; with less need to purchase a commodity, money is saved by hiring or borrowing only when needed. But this gives rise to a risk of a "rebound effect."[36] That is, if sharing saves an individual money, that arguably provides that person with increased funds to

---

[32] Botsman and Rogers 2011.
[33] Botsman and Rogers 2011.
[34] van de Glind 2013.
[35] Botsman and Rogers 2011, 73.
[36] Herring and Sorrell 2009.

purchase other things, potentially negating the environmental benefits of collaborative consumption. Similarly, by providing cheaper access to goods and services, collaborative consumption could actually increase consumption. For example, the cheap accommodation provided through Air BnB could make carbon-intensive travel more financially affordable, again negating the potential environmental benefits of sharing. What this suggests is that, if collaborative consumption is to help catalyze the transition to a low-carbon world, these new mechanisms of exchange must be accompanied by an ethics of sufficiency,[37] which is to say, an ethics that consciously use collaborative consumption as a means of reducing the impact of one's consumption, rather than as a means of maximizing consumption. Otherwise, collaborative consumption could just as easily promote rather than undermine consumerist cultures.

The Voluntary Simplicity Movement: Reimagining the Good Life

Throughout history there have been individuals, communities, and subcultures that have expressed doubts about the ethics, and even the desirability, of materialistic lifestyles and value-orientations.[38] However, in the present era of chronic environmental degradation, climate change, and burgeoning population, social movements exploring alternatives to Western-style consumer lifestyles are becoming increasingly relevant, insofar as they offer a direct and coherent response to such global problems. If the economy is in ecological overshoot, it follows that the global consumer class must reduce consumption; by doing so the wealthiest segments of the population also leave more resources for those living in destitution.[39] Of course, reducing consumption at the personal or household level is unlikely to be a *sufficient* response to social and ecological problems, but many argue that a cultural rejection of high-impact consumer lifestyles is a *necessary part* of the transformation needed to achieve a just and sustainable world.[40]

Indeed, climate scientist Kevin Anderson has recently been receiving considerable attention for arguing that climate stabilization requires that most people in the wealthier parts of the world must consume, not just differently and more efficiently; they must actually consume *less*.[41] This is not a strategy or approach that many climate scientists or other sustainability advocates have either recognized or been brave enough to acknowledge publicly, but Anderson does not shy away from the radical implications of the numbers. It is worth unpacking Anderson's forceful emissions-based justification for consuming less, because this is a theoretical innovation with potentially disruptive social and political implications.

Anderson's justification for reducing consumption can be summarized quite briefly. In the Copenhagen Accord of 2009, the international community agreed to take the actions necessary to stop temperatures rising 2 degrees above pre-industrial levels. In order to meet this goal (and keep within the estimated carbon

---

[37] Princen 2005.
[38] Vanenbroeck 1991; Kasser 2002.
[39] Vale and Vale 2013.
[40] Alexander ed. 2009; Trainer 2010; Burch 2013.
[41] Anderson 2012; Anderson 2013.

budget that goal implies), Anderson shows that the wealthier nations will need to reduce their emissions by around 8-10% a year, if they are to leave the poorer parts of the world a fair share of the carbon budget.[42] The great challenge this presents, however, is that economists claim that emissions reductions above 3% or 4% p.a. are incompatible with economic growth.[43] It follows that higher reductions of 8% or 10% p.a. will necessitate giving up economic growth as a national goal and embracing "degrowth" policies that deliberately initiate a process of "planned economic contraction."[44] In terms of lifestyle implications, this evidence-based response to climate change would mean that people in wealthy nations would have to "cut back very significantly on consumption."[45]

This argument is supported by the fact that it is likely to take much more than a decade to really scale up renewable energy and reduce emissions significantly through *energy supply* transitions.[46] While it is necessary, of course, to transition to renewable energy supply, Anderson concludes that the only way to reduce emissions sufficiently in the short-to-medium term is to greatly reduce *energy demand* by consuming and producing less goods and services.[47] This goes directly against other climate and broader environmental analyses, most of which continue to insist that increases in consumption and economic growth in the rich world are compatible with environmental health, climatic stability, and social justice.[48]

Against this backdrop, the significance and disruptive potential of the Voluntary Simplicity Movement becomes apparent. This movement can be understood broadly as a diverse and loosely-knit social movement made of up people who are resisting high consumption lifestyles and who are seeking, in various ways, a lower consumption but higher quality of life alternative.[49] In practice, this way of life might involve growing organic food or supporting local farmers' markets, harvesting rainwater, mending or making clothes, cycling or walking rather than driving, avoiding airflight, limiting work hours, co-housing, purchasing second-hand or "fair trade", progressively reducing energy consumption, and generally minimizing waste and all superfluous purchases. Although participants in this anti-consumerist movement find justification and motivation in a wide range of personal, social, ecological, and even spiritual grounds,[50] Anderson's new emissions-based case for consuming less arguably provides the movement with its most compelling and urgent justification.

The largest empirical analysis of the Voluntary Simplicity Movement shows that there could now be as many as 200 million people in the developed regions of the world exploring, to varying degrees, lifestyles of reduced and restrained consumption.[51] This signifies an emerging social movement of potentially

---

[42] Anderson 2013; Anderson and Bows 2011, 20-44.
[43] Stern 2007.
[44] Alexander 2012a, 349-368.
[45] Anderson 2012.
[46] Anderson and Bows 2011.
[47] Anderson 2013.
[48] Grantham Institute for Climate Change 2013; United Nations 2012.
[49] Alexander ed. 2009.
[50] Burch 2013.
[51] Alexander and Ussher 2012, 66-86.

transformative significance, especially if it were ever to radicalize and organize itself with political intent. Notably, that same empirical study showed that the movement was developing both a "group consciousness" and a "political sensibility", features which are arguably necessary for any social movement if it is to use its collective power in influential ways. As more people are exposed to the type of reasoning unpacked by Kevin Anderson – that is, as more people see that responding to climate change actually requires consuming less – the Voluntary Simplicity Movement could well grow in size and influence, perhaps with surprising speed.

Interestingly, the justification for embracing lifestyles of voluntary simplicity does not begin and end with ecological or humanitarian arguments. In recent decades there has been a huge amount of literature exploring the relationship between income and subjective wellbeing, and the results undermine the culturally entrenched assumption that "money buys happiness."[52] Although the empirical debate is not over, the weight of evidence strongly suggests that money and material wealth is important at low levels of income, but once basic material needs for food, shelter, clothing and so forth have been met, money has fast diminishing marginal returns. In other words, beyond the basic needs threshold, the things that really contribute most to human wellbeing are not monetary or material, but instead things like socializing, creative activity, meaningful work, and other *non-material* sources of meaning and satisfaction. This literature is arguably a ticking time bomb for consumer culture, because if more people came to see that consumerist lifestyles are not a reliable path to a happy and meaningful existence, they would presumably give up the consumerist lifestyle and seek happiness and meaning in realms other than consumption. Although this culture shift might be motivated primarily by self-interest, clearly it would have beneficial social and ecological implications. The point is that an overwhelming case is developing for people to explore post-consumerist lifestyles of reduced or restrained consumption, suggesting that the conditions for a cultural revolution are ripe.

It is also worth acknowledging a new and controversial analysis presented by David Holmgren, co-originator of the permaculture concept, which provides further grounds for thinking that the Voluntary Simplicity Movement could have disruptive potential.[53] Voluntary simplicity has always been an implicit feature of the permaculture worldview, insofar as permaculture is about designing a way of life that minimizes waste in order to work with nature rather than against nature.[54] But Holmgren recently placed voluntary simplicity at the center of his thinking, and arrived at a provocative theory of change that has received a vast amount of online attention.

Always doubtful of the prospects of convincing politicians to lead the necessary transition to a low-carbon world, Holmgren has grown increasingly skeptical that any mass movement at the social level is going to produce significant change either. Accordingly, his pessimism has driven him to conclude that the best we can hope at this late stage is to deliberately "crash" the existing fossil-fuel-based system and build a permaculture alternative as the existing

---

[52] Alexander 2012b, 2-21.
[53] Holmgren 2013.
[54] Holmgren 2002.

system deteriorates. His provocative theory, to oversimplify, is that if a new, relatively small social movement of anti-consumers were able to radically reduce their consumption, this reduction in demand for commodities could destabilize the global economy, which is already struggling. More precisely, Holmgren hypothesizes that if merely 10% of people in a nation could reduce their consumption by 50%, this could signify a 5% reduction in total demand, which, although small, would likely cause havoc with any growth-based economy. It is important to emphasize that Holmgren does not romanticize the process of collapse; he acknowledges the worrying risks his strategy poses. First and foremost, it is unpredictable in its consequences. Nevertheless, he argues that whatever risks his strategy poses, there are greater risks – both socially and environmentally – in letting the existing system continue to degrade planetary ecosystems. What is most interesting about Holmgren's strategy is that it does rely on a mass movement. He believes that a relatively small but radical anti-consumerist movement could be a truly disruptive force.

Whether one agrees with Holmgren's strategy or not, it provides further grounds for seeing the Voluntary Simplicity Movement as a social movement of potentially transformative significance. If growth-based economies require cultures of insatiable consumers to function, this suggests that such economies could be fundamentally transformed if enough people withdrew their support and instead embraced lifestyles of voluntary simplicity. Furthermore, it should be clear enough that any transition beyond fossil fuels is going to have hugely significant lifestyle implications, and voluntary simplicity – the idea of living more with less – is arguably the most coherent and attractive way of "reimagining the good life" beyond a fossil-fuel-based economy. Indeed, one of the most radical acts of liberation and opposition in a consumer culture is the "great refusal" to consume more than one needs. Admittedly, this is not an organized movement that draws attention to itself, and it is not attached to a "buzz word" that excites much media attention. But it does seem to be bubbling under the surface of the dominant culture of consumption, threatening to expand.

Redefining Progress through Alternative Indicators to GDP

If social movements based on notions of transition, voluntary simplicity, collaborative consumption and permaculture provide coherent and attractive means of reconceptualizing life at the personal, household, and community levels, the emergence of alternative indicators to Gross Domestic Product (GDP) present themselves as an important way to reconceptualize the meaning of macro-economic progress on the path to a just and sustainable, low-carbon world. While this subject, at first instance, might seem to be more of an economic innovation than a socio-cultural innovation, the fact that these alternative indicators seem to be receiving increased socio-cultural attention and support is the "movement" which, it will be suggested, has disruptive potential. As noted in the introduction, economic, political, and technological innovations are unlikely to be disruptive until there is a culture that desires them, or is at least receptive to them, and there are reasons to think that support for alternative indicators are on the rise. As leading advocates for alternative indicators have recently noted, "The chance to dethrone GDP is now in sight."[55] In this section we explore why might this be significant.

Economic growth is conventionally defined as a rise in GDP. These national accounts first emerged in the 1930s with the onset of the Great Depression, and developed significantly during World War II to assist with planning. But it was really in the post-war era that GDP came to prominence, not only in the United States but also increasingly around the world.[56] Almost immediately international comparisons of GDP per capita were made as a way of assessing the relative progress of nations, and today almost all governments around the world consider growth in GDP to be their overriding objective.[57] Not only is growth widely considered the best means of keeping unemployment at bay, growth is also considered the best means of providing individuals and governments with more economic power to purchase those things they need or desire most. The underlying assumption is that growth in GDP is always good, such that a bigger economy is always better.

The critical flaw in this macroeconomic paradigm lies in the fact that that GDP is by no means a holistic measure of a nation's progress, a point made decades ago by pioneering ecological economists Herman Daly and John Cobb.[58] GDP is merely a measure of the total market activity of a nation over a given period – a measure that makes no distinction between market activity that contributes to wellbeing and activity that does not. For example, GDP treats market expenditure on guns, anti-depressants, and cleaning up oil spills, no differently from expenditure on education, solar panels, and bicycles. All market activity is considered good. But obviously the type of market activity influences a society's wellbeing, not just its extent, so GDP is a very crude measurement of progress, at best.

Furthermore, GDP says nothing at all about the level or nature of *non-market* activity in a society, such as community engagement, health, or the functioning of ecosystems; nor does GDP say anything about the distribution of wealth in a society.[59] That last point on inequality is important in light of recent evidence showing that economies that have broader distributions of wealth do better on a whole host of social indicators.[60] For present purposes, however, it is the absence of ecological factors within the GDP accounts that signify their greatest failing. It is no good having a growing, carbon-intensive economy if such growth undermines the ecosystems (including climate systems) upon which wellbeing fundamentally depends.

What all this means is that GDP should not be used as a proxy for national progress, because it totally overlooks these critically important factors. Indeed, as Robert Kennedy famously noted, GDP measures "everything except that which makes life worthwhile."[61] Furthermore, treating GDP as a proxy for progress

---

[55] Costanza, Kubiszewski, Giovannini, Lovins, McGlade, Pickett, Ragnarsdóttir, Roberts, Vogli, and Wilkinson 2014, 283-285.
[56] Collins 2000.
[57] Purdey 2010; Hamilton 2003.
[58] Daly and Cobb 1989.
[59] Stiglitz, Sen, and Fitoussi 2010.
[60] Wilkinson and Pickett 2010.
[61] Costanza et al 2014, 283.

obviously encourages governments to shape their policies to maximize GDP, even if that means degrading the environment, destroying communities, and increasing inequalities of wealth. In other words, fetishizing GDP can lead nations to seek growth of the economy, even if that would have negative impacts on the overall wellbeing of a nation.

Wanting to provide a much more nuanced and comprehensive indicators of the overall progress, Daly and Cobb pioneered the development of the Index of Sustainable Economic Welfare (ISEW).[62] This index and others like it – such as the Genuine Progress Indicator (GPI), the Happy Planet Index (HPI), and the Bhutanese notion of Gross National Happiness (GNH) – take into consideration important social and ecological factors that GDP simply does not reflect.[63] For example, the ISEW and GPI begin with total private consumption expenditure and then make deductions for such things as resource depletion, pollution, income inequalities, crime, loss of leisure and defensive expenditures, and make additions for such things as public infrastructure, volunteering, and domestic work. The aim is to measure, as accurately as possible, the overall wellbeing of a nation, including its sustainability, not just its total market activity.

The results from such indexes tend to show that despite steady growth in GDP over recent decades, the genuine progress of many developed nations has been stagnant or even in decline.[64] Put otherwise, the results indicate that growth has stopped contributing to wellbeing in most parts of the developed world and now may even be causing the very problems that it is supposed to be solving, suggesting that many developed nations have entered a phase of "uneconomic growth."[65] If this is so, it would mean that developed nations should stop treating GDP growth as the primary answer to societal problems and instead develop policies that more directly advance wellbeing. Such policies could include eliminating poverty, broadening the distribution of wealth, and protecting the environment, with GDP growth being a goal of lesser importance.

Importantly, some governments and institutions around the world are beginning to take these alternative indicators very serious, led by the Bhutanese government, which has been shaping policy based on notions of Gross Domestic Happiness since 1972. During his presidency of France, Nicholas Sarkozy commissioned three prominent economists – Amartya Sen, Joseph Stiglitz, and Jean-Paul Fitoussi – to examine alternatives to GDP measures; the commission concluded that GDP was grossly inadequate as a measure of progress and that alternatives were necessary.[66] In the past three years the US states of Vermont and Maryland have adopted the GPI as a measure of progress, and have implemented policies specifically aimed at improving it. [67] These types of politico-economic developments have arguably been made possible by a growing cultural dissatisfaction with conventional measures of progress based on GDP, evidenced by the explosion of interest in alternative indicators.

---

[62] Daly and Cobb 1989.

[63] Lawn 2006; Lawn and Clarke 2008; Costanza et al 2014.

[64] Kubiszewski, Costanza, Franco, Lawn, Talberth, Jackson, and Aylmer 2013, 57-68.

[65] Daly 1999.

[66] Stiglitz, Sen and Fittoussi 2010.

[67] Costanza et al 2014.

The disruptive potential of this shift in thinking lies in the fact that "[w]hat we measure affects what we do."[68] If alternative indicators continue to take root in the public consciousness, this may make it much easier to frame the transition to a low-carbon world as something that is genuinely in a nation's interest, even if it is not a policy that maximizes GDP.

Conclusion

Any transition beyond fossil fuels is going to need a coordinated, multifaceted effort between various spheres of life: social, economic, technological, institutional, and political. This paper has reviewed various innovations in the socio-cultural sphere – innovations that have the potential to disrupt the current trajectory and rapidly reorient the world toward a low-carbon future. Just as the socio-cultural sphere will need the support of disruptive innovations in other spheres of life, it is also likely that even the innovations within the socio-cultural domain will need to be coordinated in order to provide mutual support, if they are to fulfill their transformative potentials. Could it be that the socio-cultural conditions for rapid transformation are almost here? While it remains difficult to be confident, this review suggests that there are genuine grounds for hope. And if, as individuals and communities, we have the power to initiate the transition to a low-carbon world, it is only a small step further to conclude that that we must also have the responsibility.

References

Alexander, Samuel, ed. 2009. *Voluntary Simplicity: The Poetic Alternative to Consumer Culture.* Whanganui, New Zealand: Stead and Daughters.

Alexander, Samuel. 2012a. Planned Economic Contraction: The Emerging Case for Degrowth. *Environmental Politics* 21(3): 349-368.

Alexander, Samuel. 2012b. The Optimal Material Threshold: Toward an Economics of Sufficiency. *Real-World Economics Review* 61: 2-21.

Alexander, Samuel, and Simon Ussher. 2012. The Voluntary Simplicity Movement: A Multi-National Survey Analysis in Theoretical Context. *The Journal of Consumer Culture* 12(1): 66-86.

Alloun, Esther and Samuel Alexander, 2014. The Transition Movement: Questions of Diversity, Power, and Affluence. *Simplicity Institute Report* 14g. Available at http://simplicityinstitute.org/wp-content/uploads /2011/04/TransitionMovement.pdf, accessed August 1, 2014.

Anderson, Kevin. 2012. An Interview with Kevin Anderson. *Transition Culture,* November 2.

Anderson, Kevin. 2013. Avoiding Dangerous Climate Change Demands De-Growth Strategies from Wealthier Nations. Available at http://kevinanderson.info/blog/avoiding-dangerous-climate-change-demands-de-growth-strategies-from-wealthier-nations/, accessed February 1, 2014.

---

[68] Stiglitz, Sen and Fitoussi 2010, xvii.

Anderson, Kevin, and Alice Bows. 2011. Beyond "Dangerous" Climate Change: Emission Scenarios for a New World. *Philosophical Transactions of the Royal Society* 369: 20-44.

Ansar, Atif, Ben Caldecott, and James Tilbury. 2013. *Stranded Assets and the Fossil Fuel Divestment Campaign: What Does Divestment Mean for the Valuation of Fossil Fuel Assets.* Stranded Asset Programme, Oxford University. Available at http://www.smithschool.ox.ac.uk/research/stranded-assets/SAP-divestment-report-final.pdf, accessed February 1, 2014.

Barry, John, and Stephen Quilley. 2008. Transition Towns: "Survival", "Resilience" and Sustainable Communities – Outline of a Research Agenda. *Advances in Ecopolitics* 2: 14-37.

Barry, John, and Stephen Quilley. 2009. The Transition to Sustainability: Transition Towns and Sustainable Communities. In *The Transition to Sustainable Living and Practice (Advances in Ecopolitics Volume 4),* edited by John Barry and Liam Leonard, 1-28. Bingley: Emerald Publishing.

Botsman, Rachel, and Roo Rogers. 2011. *What's Mine is Yours: The Rise of Collaborative Consumption.* New York: HarperCollins.

Bunting, Madeleine. 2009. Beyond Westminster's Bankrupted Practices, a New Idealism is Emerging. *The Guardian,* June 1. Available at http://www.theguardian.com/commentisfree/2009/may/31/reform-transition-a-new-politics, accessed February 1, 2014.

Burch, Mark. 2013. *The Hidden Door: Mindful Sufficiency as an Alternative to Extinction.* Melbourne: Simplicity Institute.

Carbon Tracker Initiative. Undated. *Unburnable Carbon: Are the World's Financial Markets Carrying a Carbon Bubble?* Carbon Tracker Initiative. Available at http://www.carbontracker.org/wp-content/uploads/downloads/2011/07/Unburnable-Carbon-Full-rev2.pdf, accessed February 19, 2014.

Christensen, Clayton. 1997. *The Innovator's Dilemma: When New Technologies Cause Great Firms to Fail.* Boston: Harvard Business Press.

Christoff, Peter. ed. 2013. *Four Degrees of Global Warming.* London: Taylor and Francis.

Collaborative Consumption. 2014. Available at http://www.collaborativeconsumption.com/, accessed February 1, 2014.

Collins, Robert. 2000. *More: The Politics of Economic Growth in Post-War America.* Oxford: Oxford University Press.

Conifino, Jo. 2013. Bill McKibben: Fossil Fuel Divestment Campaign Builds Momentum. *The Guardian,* October, 31. Available at http://www.theguardian.com/sustainable-business/bill-mckibben-fossil-fuel-divestment-campaign-climate, accessed February 1, 2014.

Costanza, Robert, Ida Kubiszewski, Enrico Giovannini, Hunter Lovins, Jacqueline McGlade, Kate Pickett, Kristin Ragnarsdóttir, Debra Roberts, Roberto Vogli, and Richard Wilkinson. 2014. Time to Leave GDP Behind. *Nature* 505: 283-285.

Daly, Herman. 1999. Uneconomic Growth in Theory and Fact: (The First Annual Feasta Lecture), 26 April. Available at http://www.feasta.org/documents/feastareview/daly.htm, accessed February 1, 2014.

Daly, Hermann, and John Cobb.1989. *For the Common Good: Redirecting the Economy Toward Community, Environment, and a Sustainable Future.* Boston: Beacon Press.

Felson, Marcus and Joe Spaeth. 1978. Community Structure and Collaborative Consumption: A Routine Activity Approach. *American Behavioural Scientist* 21: 614-24.

Go Fossil Free. 2014a. Commitments. Available at http://gofossilfree.org/commitments/, accessed February 1, 2014.

Go Fossil Free. 2014b. Frequently Asked Questions. Available at http://gofossilfree.org/faq/, accessed February 1, 2014.

Grantham Institute for Climate Change. 2013. Halving global CO2 by 2050: Technologies and Costs. Available at http://www3.imperial.ac.uk/climatechange/publications/collaborative/halving-global-co2-by-2050, accessed February 1, 2014.

Hamilton, Clive. 2003. *Growth Fetish.* Crows Nest, NSW: Allen & Unwin.

Herring, H. and S. Sorrell (eds). 2008. *Energy efficiency and sustainable consumption: Dealing with the rebound effect.* Basingstoke: Palgrave Macmillan.

Holmgren, David. 2002. *Permaculture: Principles and Pathways beyond Sustainability.* Hepburn: Holmgren Design Services.

Holmgren, David. 2013. Crash on Demand: Welcome to the Brown-Tech Future. *Simplicity Institute Report 13c.*

Hopkins, Rob. 2008. *The Transition Handbook: From Oil Dependency to Local Resilience.* White River Junction, VT: Chelsea Green Publishing.

Hopkins, Rob. 2011. *The Transition Companion: Making Your Community More Resilient in Uncertain Times.* White River Junction, VT: Chelsea Green Publishing.

Hopkins, Rob. 2013. *The Power of Just Doing Stuff: How Local Action Can Change the World.* Cambridge: UIT/Green Books.

Intergovernmental Panel on Climate Change. 2013. *Climate Change 2013: The Physical Science Basis – The Fifth Assessment Report.* Cambridge: Cambridge University Press.

James, Chris. 2010. Transition and Raw Resources. *Arena Magazine* 104: 14-15.

Kasser, Tim. 2002. *The High Price of Materialism.* Cambridge MA: MIT Press.

Kubiszewski, Ida, Robert Costanza, Carol Franco, Philip Lawn, John Talberth, Tim Jackson, and Camille Aylmer. 2013. Beyond GDP: Measuring and Achieving Global Genuine Progress. *Ecological Economics* 93: 57-68.

Lawn, Philip. 2006. *Sustainable Development Indicators in Ecological Economics.* Cheltenham: Edward Elgar.

Lawn, Philip, and Michael Clarke. 2008. *Sustainable Welfare in the Asia-Pacific: Studies Using the Genuine Progress Indicator.* Cheltenham: Edward Elgar.

McKibben, Bill. 2012. Global Warming's Terrifying New Math. *Rolling Stone,* July 19, 2012. Available at http://www.rollingstone.com/politics/news/global-warmings-terrifying-new-math-20120719, accessed February 1, 2014.

Princen, Thomas. 2005. *The Logic of Sufficiency.* Cambridge, MA: MIT Press.

Purdey, Stephen. 2010. *Economic Growth, the Environment, and International Relations: The Growth Paradigm.* New York: Routledge.

Smith, Joseph, and Sandro Positano. 2010. *The Self-Destructive Affluence of the First World: The Coming Crisis of Global Poverty and Ecological Collapse.* New York: Edwin Mellen.

Stern, Nicholas. 2007. *The Economics of Climate Change.* Cambridge: Cambridge University Press.

Stiglitz, Joseph, Amartya Sen, and Jean-Paul Fitoussi. 2010. *Mis-Measuring our Lives: Why GDP Doesn't Add Up.* New York: The News Press.

Trainer, Ted. 2009. Transitioning to Where? *Arena Magazine* 102: 11-12.

Trainer, Ted. 2010. *The Transition to a Sustainable and Just World.* Sydney: Envirobook.

United Nations. 2012. The Future We Want. A/RES/66/88. Available at http://daccess-dds-ny.un.org/doc/UNDOC/GEN/N11/476/10/PDF/N1147610.pdf?OpenElement, accessed February 1, 2014.

Vale Robert, and Barbara Vale. 2013. *Living within a Fair Share Ecological Footprint.* London: Earthscan.

van de Glind, Pieter. 2013. *The Consumer Potential of Collaborative Consumption* (Masters thesis), Utrecht University. Available at http://dspace.library.uu.nl/handle/1874/280661, accessed February 1, 2014.

Vanenbroeck, Goldian. ed. 1991. *Less is More: An Anthology of Ancient and Modern Voices Raised in Praise of Simplicity.* Vermont: Inner Traditions.

Walsh, Bryan. 2011. Today's Smart Choice: Don't Own. Share. *Time*, March 17. Available at http://content.time.com/time/specials/packages/article/0,28804,2059521_2059717_2059710,00.html, accessed February 1, 2014.

Wilkinson, Richard and Kate Pickett. 2010. *The Spirit Level: Why Greater Equality Makes Societies Stronger.* London: Penguin.

World Bank. 2012. *Turn Down the Heat: Why a 4 Degree Warmer World Must be Avoided.* Washington: World Bank.

Chapter 20: Human Insecurity through Economic Development: Educational Strategies to Destabilize the Dominant Paradigm

Alexander K. Lautensach and Sabina W. Lautensach

In a 2010 public address Jane Goodall, the famous primatologist and activist, proposed that the current global environmental malaise was the result of economic considerations too often winning over environmental ones for too many years. In this she reflected the widely popularized view that economic and environmental concerns were somehow positioned at loggerheads in a planet-wide quest for ideological dominance. Moreover, an overwhelming amount of evidence suggests that economic ideals tend to be favored by decision-makers at every level. This view culminated in announcements of the "death of environmentalism." [1] Those announcements ignored the fact that environmentalism manifests in numerous and diverse variants, such as biocentric concerns for individual life forms, ecocentric concerns for ecosystems, and "shallow green" concerns about the welfare of future human generations and its dependency on healthy environmental support structures. [2] They allege widespread despair among "environmentalists" about imminent ecological Armageddon and about the fact that too many opportunities keep being missed,[3] without acknowledging that the implications of ecological deterioration will compromise all human endeavors indiscriminately - surely grounds for worry to any reasonable person.

Much less interesting to the average journalist seems to be the semantic and conceptual commonalities between economics and ecology – the simple yet incredibly complex quest for understanding how humanity's home works, and for applying that knowledge to promote the flourishing of the human species. In other words, people who care about the flourishing of economies and people who care about the functioning of our environmental support structures share a concern about the sustainable management and distribution of scarce resources.

---

[1] Shellenberger and Nordhaus 2005.
[2] Curry 2011.
[3] Turner 2007.

Given this significant consensus, how deep does that apparent disagreement really go? If it is in fact superficial, and the consensus extensive, what are the conceptual possibilities for a reconciliation or compromise? Considering how much heat and noise are produced by the friction of the dispute, and how little it accomplishes in terms of addressing the urgent problems humanity faces, should such reconciliation not receive the utmost priority? And how could such a theoretical compromise inform pedagogical innovations that are designed to promote the transition to sustainable living?

To address those questions we will first explicate what shared assumptions contribute to the conceptual consensus between the pro-economics and the pro-environment views. This will allow us to more precisely circumscribe what turns out as the two main areas of contention between the two views: ecocentrism/biocentrism and cornucopianism, respectively. We shall then propose some educational strategies that make use of the consensus areas and that resolve the conflict in ways that ultimately would make it easier for learners at all levels to adopt more sustainable ways of living as responsible global citizens. The worsening global environmental crisis has now elevated that educational goal to the highest priority.[4]

The Extent of Consensus between the Pro-economics and Pro-environment Views

As mentioned above, it seems plausible to assume that most citizens who are concerned about the well-being of their national economies and about global economic development also care about the sustainable management of scarce resources and about the goals of improving the lot of the world's poorest and leaving a better world to future generations. Likewise, most people who advocate the conservation and protection of ecosystems, species diversity, and natural resources probably count among their primary motives a similar concern for the sustainable flourishing of humanity. The two sides might employ different language but at heart their moral goals in the long term tend to overlap significantly.

We suggest that besides those common moral goals, the two sides also share the following basic assumptions.

- Most scarce resources are bound to become even scarcer as population numbers continue to increase. This applies to renewable resources such as timber as much as to non-renewable ones such as petroleum.[5]
- Some resources that so far have been considered abundant in many parts of the world, such as fresh water, are likely also to become scarcer as our numbers and our consumption increase.[6]

---

[4] Ehrlich and Ehrlich 2010; Lautensach 2010. Among the abundant literature on the various manifestations of the crisis, a much smaller subset acknowledges the significance of education in its causation and mitigation.

[5] Grant 2005.

[6] Meadows et al. 2004.

- Efforts towards the sustainable production and management of those resources can mitigate the negative impacts those scarcities are exerting on human well-being and human security.[7]
- The flourishing of economies at the regional, national and global levels allows human societies a greater range of options for sustainable resource management, it can buy them time for urgent policy reforms, and it alleviates human misery in general but especially as it is caused by scarcities.

These assumptions formed the basis for the eight Millennium Development Goals of the United Nations, designed to "end extreme poverty worldwide" by 2015.[8] The Goals, formulated in 2005, manifest the extent of global consensus on those four assumptions listed above. Another indication was the increasing popularity of human security as a conceptual basis for international regimes and initiatives.[9] The 1995 establishment of the Human Security Network, in association with the UN, presents a landmark for the increasing recognition of human security by the international community. In order to establish the central role of human security in manifesting the consensus between pro-economics and pro-environment views, we shall briefly summarize the concept.

Human security in its most comprehensive interpretation subsumes the above mentioned four assumptions. Often expressed as the two ideals of "freedom from fear" and "freedom from want",[10] the concept focuses on the security of the individual citizen as opposed to conventional security that focuses first and foremost on the security of the state.[11] Inasmuch as threats to human security arise from the scarcity of certain resources, efforts to promote human security are designed to operate on two fronts: to ensure a minimum per capita level of supply of the critical resource, and to ensure its more equitable distribution across the population.[12]

Instead of defining human security as the freedom from fear and want, some analysts have found it more expedient to focus on the actual and potential sources of insecurity. This allows them to eliminate unreasonable, unjust and counterproductive demands from the scope of targets – demands that are often formulated by security providers rather than by the victims of insecurity. The title of this chapter reflects that problem-focused orientation.

Surveys of the sources of insecurity suggest that a comprehensive definition of human security needs to include four broad areas that we refer to as the "four pillars" of human security.[13] They include the traditional area of military/strategic security of the state; economic security, particularly as defined by heterodox models of sustainable economies; health-related security, defined by epidemiological context and the complex determinants of community health and health care priorities; and environmental security that models the complex

---

[7] Homer-Dixon 1999.
[8] UN Millennium Project 2005, 1.
[9] Spady and Lautensach 2013.
[10] United Nations 2000.
[11] Griffin 1995; Hampson et al. 2002.
[12] Meadows et al. 2004.
[13] Lautensach 2006.

interactions between human populations and their ecological support structures, the source and sink functions of their host ecosystems. Environmental security is defined as security from "critical adverse effects caused directly or indirectly by environmental change."[14]

Elsewhere we have elaborated how each pillar can contribute to our understanding of the sources of human insecurity and how it enables us to mitigate their effects. [15] The four pillar model agrees with other similarly multidisciplinary models such as the UNDP's seven dimensions of human security (economic, food, health, environmental, personal, community, and political security). [16] The central role of human security as a principle for international consensus building and policy formation has received increasing recognition from the United Nations and other international organizations, as well as from many national governments. This extent of political recognition indicates a certain conceptual consensus between pro-economics and pro-environment views. While by itself each view favors one of the pillars over the rest, advocating human security in its comprehensive form and making it a reality for most of humanity requires a balanced attention to all four pillars and a reconciliation between economism and environmentalism. This raises the questions how the apparent conflict between the two originated in the first place, and how that conflict could be resolved. We shall address those questions presently.

Resolving the Conflict between Pro-economics and Pro-environment Views

The preceding section might have given the impression that the clamorous conflict between the two views was somehow immaterial or based on nothing more than misunderstandings. That would indicate a clear agenda for reconciliation. Unfortunately it is not going to be that easy, because of two major areas of disagreement that remain outside the common ground of consensus outlined above.

A distinct proportion of calls for environmental protection and conservation are informed by moral concerns that, unlike the rest, do not centre primarily on the well-being of humans or the future flourishing of humanity. One reason for such non-anthropocentric motives is a biocentric concern for the well-being and continued existence of individual non-human life forms or of non-human species. Another reason can be an ecocentric valuation of ecosystems over any individual species or organism; such ethics are predicated on the assumption that ecosystems have higher moral standing than the sum of their constituent populations or individuals. Both biocentric and ecocentric ethics come in numerous variations and combinations but they all share the fundamental premise that *Homo sapiens* do not, and should not, take centre position and top priority in the moral universe.[17] It is this non-anthropocentric premise that often attracts the accusation of being "anti-human", "ecofascist", "elitist", or even antisocial. Critics, many of

---

[14] Barnett 2007, 5.
[15] Lautensach 2006.
[16] UNDP 1994, 24-33.
[17] Curry 2011, 57.

them of the pro-economy persuasion, tend to claim the moral high ground of humanitarianism and denounce such sentiments as misanthropic and hence morally reprehensible.

We do not intend to engage here in a metaethical comparative analysis of alternative environmental ethics, although we do believe, along with others, that some forms of ecocentrism can emerge as a superior ethic from such a comparison.[18] In the context of this chapter it is more important to note that public debates about differences in values are rarely conducted at the level of a metaethical comparison and more often assume the form of rhetorical mudslinging involving pejorative attributes such as the ones mentioned above. Moreover, most conventional economic models rely on the ideal of individuals as rational actors (erroneously, in our view) and downplay or ignore the role of values (and their cultural underpinnings) as determinants of human behavior. This often leads their advocates to claim a "rational", "objective" (and hence superior) position and to charge their opponents with naïve idealism. All in all, the area of value differences remains treacherous territory in the conflict between pro-economy and pro-environment views, one that remains rather intractable to attempts at reconciliation. What tends to be ignored by both warring factions is the fact that calls for environmental protection are by no means necessarily contingent on non-anthropocentric ethics but often amount to no more than "enlightened self interest."[19]

The other major intractable area concerns a peculiar factual belief, widespread among pro-economy advocates, that the growth of economies and populations is not subject to any physical limits. This belief has been referred to as cornucopianism.[20] We refer to this belief as peculiar because it is primarily voiced by people, including several Nobel laureates, who otherwise frequently invoke such ideals as realism, objectivity, and scientific substantiation.[21] But when faced with the (very scientifically grounded) objection that in fact positively *nothing* in the physical universe can grow without limits they tend to accuse the critic of "Malthusianism" without engaging in a logical argument (which they are likely to lose).[22] Advocates of economic growth often invoke the "axiom of infinite substitutability" which claims that any resource, once it becomes sufficiently scarce, will be substituted by another resource that is less scarce, *ad infinitum*.[23] However, this merely restates the cornucopian assumption in a different form.

We have presented a more detailed critique of cornucopianism on conceptual, empirical, and consequentialist grounds elsewhere, as have others.[24] What makes cornucopianism such a remarkable phenomenon is not just the extent to which it confounds any attempt at scientific justification, nor the astounding zeal with which many of its adherents attempt to defend it with pseudo-scientific arguments. Remarkable is its popularity and unquestioned adoption by millions in

---

[18] The full argument is given in Lautensach 2009a and Lautensach 2010.

[19] Jamieson 2008.

[20] Ehrlich and Holdren 1971.

[21] For example, see Simon 1990 or North 1995.

[22] For an example of this fallacious counterargument see Sen 1998, 205.

[23] Pilzer 1990; Pearce and Warford 1993.

[24] Lautensach 2010; Coale 1974; Rees 2004; Nadeau 2009.

an age of economic recession, a popularity that still allows highly educated individuals to advocate it in public at no risk to their professional reputation.[25] More than any other single concept, cornucopianism prevents mainstream economists from making positive contributions towards meeting the challenge of the human predicament.[26] Politicians at all levels still refer to economic growth as a good in itself.[27] More significantly, cornucopianism has hampered efforts at the policy level to promote human security for the world's poorest, diverting attention from conceptions of development that do not depend on growth.[28]

We believe that, besides the confusion over ethical differences, cornucopianism is the main reason why the conflict shows such little sign of being laid to rest. In recent years it has tended to recede somewhat into the implicit realm, at least as far as academic discourse is concerned. Its main premise has become obfuscated with the advent of terms such as "sustainable growth", "medium term sustainability", and "sustainable development."[29] Other analysts claim that all shortages can be solved through the equitable distribution of resources, thus implicitly endorsing cornucopianism without relying on it.[30] It seems unlikely that any macroeconomic model could escape the cornucopian fallacy as long as it does not explicitly take into account the constraint by physical limits to growth.

Promoters of the dominant ideology of progress who recognized at least implicitly some vague ecological limits attempted a synthesis, as in the Brundtland commission's concept of sustainable development.[31] Its dictum of "development that meets the needs of the present without compromising the ability of future generations to meet their own needs" neatly omits the reference

---

[25] Ehrlich and Ehrlich 1990, 164. The authors (p.291) quote the inaugural lecture of W. Beckerman, an economist at University College, London. Also, Nobel laureate Robert Solow claimed that "the world can, in effect, get along without natural resources" (quoted in McNeill 2000, 375), a claim that he subsequently retracted (Solow 1993). The open advocacy of an evidently fallacious position by academics points to a curious problematique in the current education system. It will be addressed later in this essay.

[26] The origins of cornucopianism lie in imperial China and mercantilist Europe, even though it gained in popularity only after the Great Depression of the 1930s (McNeill 2000). Apparently the cornucopian delusion dates back to the physiocrat school of economics which emerged in the 18th century before the laws of thermodynamics were discovered (Schumpeter 1954). This does not, of course, excuse the fact that during the past two centuries mainstream economists have found no reason to update their dogma. Small wonder that Lutzenberger (1996) was moved to call classical economics "the dimwitted science." Wackernagel and Rees (1996, 41) refer to cornucopianism as "flat-earth economics" and Hazel Henderson (1999) called it "a kind of brain damage." Herman Daly (1993) has termed sustainable growth an "impossibility theorem."

[27] The phenomenon of an instrumental value becoming reinterpreted over time as an end value has been referred to as goal displacement (Wenz 1988).

[28] Lautensach and Lautensach 2013a; Costanza et al (2014) have documented how such models of truly sustainable development rest on alternative measures of human wellbeing instead of GDP.

[29] In fact, on closer analysis many initiatives conceived under the aim of "sustainable development" appear to be neither very sustainable nor do they constitute development in any proper sense (Lautensach and Lautensach 2013a).

[30] See, for example, Sen 1998, and Sen 1999.

[31] WCED 1987, 43.

to growth but leaves open to interpretation to what extent "basic needs" could be met without further economic growth. The viability of such a synthesis of sustainability and development relies on one's definition of development. Without sounding too dismissive we can assert that human individuals and human societies have near infinite potential for qualitative improvement. If we define growth as qualitative development (as in the concept of "personal growth") we can avoid any collisions with physical limits. What renders the conventional concept of sustainable development questionable is the extent to which most of its advocates seem to adhere to the notion of development as quantitative growth, primarily of economies. They cannot convince without abandoning that notion, nor can the conflict between pro-economics and pro-environment interests be settled until they do so.[32]

It seems that in the midst of this discursive posturing between the two camps, sustainability has emerged as a patch of ideological no man's land that is claimed by both sides via divergent definitions. The divergence refers to the extent to which one's definition respects the basic ecological principles that govern the existence of all living species, specifically the physical limits to a population's growth and to its utilization of resources. The most widely used definition of sustainability, captured by the Brundtland Report's dictum quoted above, does not mention ecology. A widely quoted alternative, the 2000 UNESCO-sponsored Earth Charter, referred to respect for nature, universal human rights, economic justice, and a culture of peace.[33] Although this circumscription of sustainability describes more accurately the necessary condition for how human needs can be met sustainably, it still does not address the question of limits. The meaning of sustainability has become even fuzzier with the efforts of sociologists, economists and anthropologists to introduce diverse definitions within their respective disciplines, while private industry and political organizations contributed further to the confusion by spinning the term to meet their interests.

On the pro-environment side, sustainability tends to be defined in more straightforward terms: The activities of a community are sustainable if they convert resources into waste no faster than the community's ecological support systems can convert waste into resources. This simple version, attributed to Steve Goldfinger,[34] captures the essence of more complex definitions such as the one by Lemons, "the continued satisfaction of basic human physical needs, such as food, water, shelter, and of higher-level social and cultural needs, such as security, freedom, education, employment, and recreation", along with "continued productivity and functioning of ecosystems."[35] The rationale of this definition requires a brief explanation.

As with all living species, the activities of human populations take place within ecosystems, and are supported by them. Ecosystems consist of communities of species and their non-living physical environment; they serve as environmental support structures for human enterprises that deliver raw materials and energy and recycle wastes. By linking sustainability to ecological cycles and their capacities the ecological definition provides the sufficient condition for

---

[32] This argument is presented in greater detail in Lautensach and Lautensach 2013a.
[33] Earth Charter Initiative 2000.
[34] Chambers et al. 2000, 2.
[35] Lemons 1996, 198.

sustainability, not just the necessary one of human needs being met. This focus on what is scientifically sufficient seems rather important if we are to have any chance at attaining sustainability.

We propose that because of their more scientifically convincing framework, the ecological definitions of sustainability circumscribe most others. [36] In comparison, the Brundtland definition of sustainability seems rather less useful because of its lack of conciseness, inattention to metaethical objections and its neglect of fundamental ecological limitations. The survival of our species for the several million years that a relatively successful species could hope for[37] appears to depend on our attaining a state of equilibrium with the biosphere in order to prevent drastic population crashes with disastrous consequences for our civilization. A state of equilibrium could be attained by a society in which population, use of resources, disposal of waste, and environment would be in a sustainable balance.[38]

The logical preference for the ecological definition of sustainability plays a decisive role in the reconciliation between the pro-economics and pro-environment advocates. We work on the assumption that vast majorities on both sides are interested in pursuing a sustainable future for humanity and its increasingly unified civilization. Accepting the ecological definition of sustainability, and with it the reality of ecological limits to growth, necessitates that the pro-economics side abandon cornucopianism. Of the four pillars of human security, environmental security emerges as the "ultimate security,"[39] forming the base on which the remaining pillars stand.[40] Economic security cannot be effectively pursued without environmental security already being in place. Thus, the reconciliation demands that economics as a discipline is revised, abandoning those neoclassical concepts that remain irreconcilable.[41] Only healthy environmental support structures make it possible for economies to flourish in the long term; along with the equitable distribution of assets and power that condition provides the most reliable recipe for sustainable human security.

But the focus on sustainability also requires that the pro-environment side refrain from invoking inflammatory ecocentric arguments and recognize that the sustainable pursuit of *human* interests often suffices as an argument for conservation, because it depends on the sustainable flourishing of environmental support structures. If that should move the media to refrain from referring to

---

[36] Reasons for the primacy of environmental security over other dimensions of security are shown in Lautensach and Lautensach 2013b. Because of its scientific soundness the "concentric circle" model of a hierarchical arrangement of security dimensions has gained acceptance over the once popular "triple bottom line" model (Boulding 1993). Furthermore, the elucidation of concrete boundaries that describe an ecologically safe operating space for humanity has helped to link environmental security with sustainability (Griggs et al. 2013). Notwithstanding the primacy of environmental sustainability, many human endeavors are also subject to sociopolitical and cultural dimensions of sustainability.

[37] Pimm 2001, 202.

[38] Tickell 1997; Rees 2004.

[39] Myers 1993.

[40] Lautensach and Lautensach 2013b.

[41] Specific revisions are laid out in Mander 2007. Following this line of argument Sara Parkin (2010) coined the term "reconciliation economics".

anyone who expresses a concern about sustainability as an "environmentalist", so much the better. In many respects the long-term interests of the human species conflate with what could arguably be construed as the interests of supporting ecosystems.[42] This reconciliation means that "environmentalists" escape the charge of misanthropy, and "pro-economists" can no longer be accused of running roughshod over nature. We believe that neither the media industry nor advertising nor legislation can reliably accomplish such fundamental changes in human thinking, but that education can.[43] We shall now turn to the question how this reconciliation can be accomplished through educational strategies.

Helping Learners Deconstruct the Golden Calf

The Judaeo-Christian scriptures describe the dismay experienced by Moses when on his return from his communion with God on Mount Sinai he found the Israelites worshipping a fabricated god in the form of a golden calf statue. Our discussion of the path toward reconciliation between pro-economy and pro-environment advocates led to the conclusion that they both need to become advocates of sustainability. In itself that seems simple enough as few reasonable people could be expected to dislike sustainability, or even not to express an interest in its attainment.[44] However, we have also seen that two obstacles stand in the way of that reconciliation; the unscientific belief in unlimited growth, and the unnecessary advocacy of ecocentric priorities. Addressing those obstacles must, therefore, be the aim of our educational strategies, with cornucopianism as their main target – the false god fabricated on the basis of a distorted vision.

Educationists have long blamed the hidden curriculum in public education for the transmission of ideological content that serves to reproduce and perpetuate dominant power relationships.[45] The hidden curriculum has been defined as "the values, norms and beliefs that are transmitted to students through the underlying structure of meaning in both the formal content as well as the social relations of school and classroom life."[46] It includes assumptions and ideals that are not spelled out in curriculum documents and that are not mentioned explicitly in the classroom or in any learning material. Yet its significance cannot be overstated. A large contingent of affective learning outcomes (dispositions) are transmitted through it, as is other ideological content such as implicit assumptions, priorities, judgments, prejudices and expectations. The hidden curriculum is considered the main instrument through which education systems worldwide reproduce ideologies, cultures, and societal power structures.

---

[42] Curry 2011; this conflation also refutes the tediously simplistic division between a purportedly environmentally sensitive political left and an economically sensitive political right.

[43] For an explicit presentation of arguments why educational strategies represent the most promising means towards changing a society's dominant environmental ethics, see Lautensach 2010. A comprehensive analysis of the numerous manifestations and ideological roots of our "war against nature" is given in Hawkins 2013.

[44] The exception are so-called narrow anthropocentrists: Lautensach 2010.

[45] Contenta 1993; McLaren 1994; Tekian 2009; Eisler 2000; Lautensach 2013c.

[46] Giroux and Penna 1979, 22.

Louis Althusser described an ideology as "a 'representation' of the imaginary relationship of individuals to their real conditions of existence" and a "system of thought or consciousness" – the sum of explanations that help us make sense of the world around us.[47] We suggest that cornucopianism, as well as the general view that prioritizes economics,[48] fits that description. People cling to it because it helps them justify their ways of life, their ambitions and aspirations. Cornucopianism forms an essential component in the dominant ideology of progress.[49] Getting adult citizens to unlearn such a treasured belief will be a daunting undertaking. Preventing young learners from falling prey to it by reshaping their formal education seems easier in comparison.

We propose that educational measures to protect young learners from the cornucopian delusion should involve, firstly, rendering explicit and analyzing its ideological nature.[50] To this end, statements from prominent individuals invoking the necessity of economic growth (and/or population growth) are collected and then critically analyzed in a collaborative approach by learners. A useful example might be a UN Secretary General's statement about global development.[51] Critical questions might include what resources would be required for such development, where they might come from, how much growth might be implied, and what the potential limiting factors might be. As the capstone question the learner might ask whether the originator of the statement was likely to be aware of those critical issues and for what reasons he/she might have made the statement regardless of those uncertainties. This first stage can be used to prepare the learners for either of the other two.

During the second stage, the learner becomes familiar with growth processes in the natural world (e.g. of a population of rabbits, or cancer cells), as well as in artificial systems (e.g. of computer users), including economies. Each process is analyzed for its limiting factors and how those factors nudge the growth curve towards a steady-state equilibrium, represented as a sigmoid function over time. This remedial instruction in basic scientific literacy has been described as part of "closing the culture gap."[52] The learner is now able to revisit the examples from the first stage and state with some scientific authority why the invoked growth process cannot possibly persist indefinitely.

As a third stage, learners are encouraged to practice distinguishing between statements of value and statements of fact. This is not easy for some young learners and the analytical skills required need to be developed gradually, using numerous practice examples. Gradually the learners will develop their own

---

[47] Quoted in Felluga 2003, 1.

[48] This view was referred to by Curry (2011) as economism.

[49] Lautensach 2010; Ehrlich and Ehrlich 2010.

[50] This has been suggested by Tekian (2009) for medical education. Sara Parkin (2010, 73) asked the following critical question to deconstruct cornucopianism: *"What is growing, where, for whom, and at what cost?"*

[51] In his introduction of the UN's Millennium Development Goals, Kofi Annan (2005, 8) presented a "shared vision of development" that depended on a further increase of global economic activity, despite the fact that the global economy has been in self-destructive "overshoot" mode since the mid-1980s.

[52] Ehrlich and Ehrlich 2010.

algorithm for distinguishing between the two kinds of statements. They will also understand that most statements constitute combinations of the two kinds.

At this point the learner should be able to focus on the topic of sustainability. The framing of the topic depends of course on the grade level, but a skilled teacher could adapt the topic to any level between upper primary and university. One possible line of investigation is the discussion of ecological footprints and how they can be used to illustrate regional sustainability, ecological overshoot, exploitation of disempowered populations, and ways to address overshoot.[53] Another possibility lies in the discussion of historical case examples of cultures using unsustainable practices, experiencing overshoot, and consequently collapsing, and of potential ways in which their demise could have been prevented – and how it was in fact prevented in other cases.[54] Discussing the situation and implications of global overshoot provides convincing grounds for recognizing the urgency of the situation and for engaging with its implications.[55]

The ecological footprint concept shows a particularly rich potential for diverse learning activities.[56] It can illustrate a population's consumption level, its extent of sustainability, individual life style choices, and international trade relations. For example, the national footprints of countries whose economies are referred to as "flourishing" – i.e. rapidly growing – can be compared to the footprints of so-called "poor", "underdeveloped" countries whose economies show little growth.[57] Invariably the learner will discover that it is the latter countries that are practicing sustainability while the former expand largely by exploiting other regions and compromising human security.[58] This realization contrasts with ubiquitous calls by political leaders for "development", "growth", and "progress". The resulting cognitive dissonance provides a fruitful ground for discussion. The students will realize to what great extent political decisions are driven by ideology and how small a role scientific reasoning plays in it.

A challenge for the teacher will be to not let despair and apathy take over the discussion. Numerous case examples are available that illustrate how people have begun to initiate the transition towards sustainability in their region.[59] Grassroots initiatives – everyday practices from the six R's (reduce, reuse, recycle, refuse, rethink, repair) to choosing a wiser mode of transportation, determining one's personal footprint and then deliberately reducing it – all represent areas where young people can experience empowerment. Their invaluable advantage at that point will be that they do not yet have to unlearn or reconcile with the substantial amount of internalized counterproductive ideological baggage that tends to encumber many older citizens.

---

[53] Lautensach 2009b.

[54] Jared Diamond's *Collapse* and other works provide numerous case studies: Diamond 2005; Ludwig et al. 1993.

[55] Imhoff et al. 2004; Ehrlich and Ehrlich 2013.

[56] Wackernagel and Rees 1996; Lautensach 2009b.

[57] Dietz, Rosa and York 2009; Rees 1996.

[58] This analysis is described in Lautensach and Lautensach 2013a. It refutes a highly publicised claim by the World Economics Forum, as do The Ecologist and Friends of the Earth 2001.

[59] Turner 2007.

The most challenging direction or extension that this learning process might be taken towards is the topic of "development", as attempts to debate this subject may become (and have, in the authors' experience) an occasion for heated tempers, shouting matches and disillusionment among learners. The reason is not only that several conflicting definitions exist for the term. The agenda of development attracts people with strong humanitarian ideals, who feel that saving lives and eliminating poverty (however it may be defined) should be given the utmost priority among all global development policies. Having learned about footprints and the extent of global overshoot (50 percent above the maximum sustainable impact in 2008)[60] they will figure out what little sustenance an individual person would receive under a global regime of completely equitable distribution, and how much less that will amount to as the world population continues to grow or if the overshoot were suddenly reduced.[61] One way to address this situation, to prevent it from deteriorating and to derive some pedagogical benefit might be to turn the discussion to ethics.

Turning to ethics represents a significant course change in the pedagogical agenda. It means that the learner no longer focuses primarily on the universe that "is" and faces instead a multiplicity of universes that "ought to be." It might help to point out the fact, however, that despite that logical is/ought divide all descriptive accounts contain implicit value statements. Turning to ethical aspects merely means that those statements can be explicated and discussed.

In the public school system of British Columbia, as well as those of other Canadian provinces and many other countries, students are generally not exposed much to the practice of openly discussing values; religious schools are an exception. Nor do most teacher training programs prepare teachers sufficiently for handling the subject safely and effectively. This makes ethics a challenging subject, one that places both the student and the teacher at risk.[62] Moreover, opinions diverge widely on how much of a role teachers and education systems should be allowed to play in the area of value education.[63] Yet under the unprecedented imperatives presented by the global environmental crisis the potential benefits of such a role tend to outweigh the potential harms.[64] Moreover, the ethics of responsible citizenship have always been part of the educational agenda in schools.

Among the three major schools of ethics – virtue ethics, deontology and utilitarianism – plenty of arguments are found that can be used in classroom discussions to substantiate various points of view on economism and environmentalism. Other schools of axiological thought can provide further contributions, resulting in a wide diversity of moral arguments on any given issue. For example, as a result of the learning outcomes described above, the student will realize that objections to cornucopianism can be raised for consequentialist reasons (e.g. because it informs harmful behavior) as well as on deontological grounds (e.g. because it condones injustice). However, it is

---

[60] WWF 2012, 9, 40.
[61] Cribb 2010; Myers et al. 2002; Roberts 2008. For further discussions of these projections, see Lautensach and Lautensach 2013b, chapter 20.
[62] Freakley and Burgh 2000.
[63] Stiggins 2008; Anderson and Bourke 2000; Gutmann 1999.
[64] For a detailed argument to support this claim, see Lautensach 2010.

important that such a pluralistic approach does not become mired in moral relativism. That is to say, learners will benefit more if they are directed towards metaethical comparisons of the kind that allow them to decide which ethical views are preferable over others given certain fundamental moral priorities that we all agree on, such as sustainability, absence of suffering, and human rights. Specifically, students should be encouraged to ask such questions as "In what ways can economism or environmentalism promote justice?" or "How can each view result in harmful behavior?" – leading to an evaluative decision as to which ethic can serve humanity better in the long term.

Conclusion

We suggest that one outcome of such deliberations will be that the learner not only becomes convinced, through their own train of analytical thoughts, of the scientific fallacies underlying cornucopianism and economism but also becomes conscious of the fact that those beliefs constitute part of an ideology both powerful and objectionable from most ethical perspectives. Once the learner realizes that ideology can never be avoided in the classroom, the need for a judicious and well-reasoned comparative evaluation of ideologies will become obvious. As a result, the merits of many forms of "environmentalism" will become quite evident, especially under a utilitarian perspective. A similar comparison of anthropocentrism and ecocentrism also reveals that ecocentrism as an ethic is more conducive towards sustainability. [65] However, under the pedagogical project of ideological critique that metaethical discussion can be omitted in favor of the simple pragmatic cost-benefit consideration that ecocentric claims are not necessary for a convincing argument for sustainability. With that, both obstacles towards reconciliation are addressed, and the specific pedagogical aim of this chapter is achieved.

Although from many journalists' perspectives reconciliation seems less attractive than conflict, the successful education of responsible citizens relies on this reconciliation. The values and obligations of citizenship are linked to the normative principles of human security, which in turn relies on sustainable environmental security. This realization should pave the way towards a public consensus on reintroducing ethics into formal curricula. The goals of that curricular revision should include the deconstruction of the cornucopian ideal and the inclusion of an ethic of sustainability among the values of responsible citizenship.

Much more controversial will be the answers to questions about the moral merit of various development strategies, such as the plights for eliminating poverty or fighting global hunger. Those answers will depend largely on which school of ethics is applied. For example, preventing the deaths of thousands from starvation can be defended as virtuous (the Aristotelian position), or it can be criticized for contributing to the risk of the future starvation of the survivors' descendants (the utilitarian position). Students will come to the important realization that many of the most important moral questions have multiple "right" answers, none entirely satisfactory and all containing some arguable merit. Yet

---

[65] Curry 2011; Lautensach 2009a; 2010.

some answers clearly seem preferable to others. Our own classroom experiences suggest that the benefits from this realization are worth considerable investments in time and effort. To be sure, such an open advocacy of moral pluralism and pragmatism carries risks, and it may present the teacher with additional challenges. Yet, we believe that a pluralist-pragmatic orientation in the classroom (as well as in wider society) bears considerable advantages in our quest for developing an acceptable value base that makes it possible for citizens, and humanity as a whole, to live sustainably.[66] We reserve that argument for another occasion.

References

Anderson, Lorin W. and Sid F. Bourke. 2000. *Assessing Affective Characteristics in the Schools* (second edition). Mahwah, NJ: Lawrence Erlbaum.

Annan, Kofi. 2005. Executive Summary. In *Larger Freedom: Towards Development, Security, and Human Rights for All*. New York: United Nations. Available at http://www.un.org/en/ga/search/view_doc.asp?symbol=A/59/2005, accessed March 16, 2014.

Barnett, Jonathan. 2007. Environmental Security and Peace. *Journal of Human Security* 3 (1): 4-16.

Boulding, Kenneth E. 1993. The Economics of the Coming Spaceship Earth. In *Valuing the Earth: Economics, Ecology, Ethics*, edited by Herman Daly and Kenneth Townsend, 297-310. Cambridge MA: MIT Press.

Callicott, J. Baird. 1999. *Beyond the Land Ethic: More Essays in Environmental Philosophy*. Albany, NY: State University of New York Press.

Chambers, Nicky, Craig Simmons and Mathis Wackernagel. 2000. *Sharing Nature's Interest: Ecological Footprints as an Indicator of Sustainability*. London: Earthscan Publ. Ltd.

Coale, Ansley J. 1974. The History of the Human Population. *Scientific American* 231 (3): 40-51.

Contenta, Sandro. 1993. *Rituals of Failure: What Schools Really Teach*. Toronto: Between the Lines.

Costanza, Robert, Ida Kubiszewski, Enrico Giovannini, Hunter Lovins, Jacqueline NcGlade, Kate E. Pickett, Kristin Vala Ragnarsdottir, Debra Roberts, Roberto De Vogli & Richard Wilkinson. 2014. *Nature* 505: 283-285.

Cribb, Julian. 2010. *The Coming Famine: The Global Food Crisis and What We Can Do to Avoid It*. Collingwood, Australia: CSIRO Publishing.

Curry, Patrick. 2011. *Ecological Ethics: An Introduction* (second edition). Cambridge, UK: Polity Press.

---

[66] Some environmental ethicists even consider moral pluralism irreconcilable with sustainability; Callicott 1999. Our claim is based on a consequentialist comparison of the pluralist-pragmatic approach with moral monism or unreflected relativism. The latter is the default approach in many secular school systems while the former is favored by religious institutions. For similar discussions, see Curry 2011 and Stone 1987.

Daly, Herman E. 1993. Sustainable Growth: an Impossibility Theorem. In *Valuing the Earth: Economics, Ecology, Ethics*, edited by Herman Daly and Kenneth Townsend, 267-274. Cambridge MA.: MIT Press.

Diamond, Jared. 2005. *Collapse: How Societies Choose to Fail or Succeed*. London: Viking Penguin.

Dietz, Thomas, Eugene Rosa and Richard York. 2009. Environmentally Efficient Well-being: Rethinking Sustainability as the Relationship Between Human Well-being and Environmental Impacts. *Human Ecology Review* 16: 114-123.

Earth Charter Initiative. 2000. Values and Principles to Foster a Sustainable Future. Available at http://www.earthcharterinaction.org/content/pages/Read-the-Charter.html , accessed March 23, 2014.

Ehrlich, Paul R. and John P. Holdren. 1971. The Impact of Population Growth. *Science* 171: 1212-1217.

Ehrlich, Paul R. and Anne H. Ehrlich. 1990. *The Population Explosion*. New York: Simon and Schuster.

Ehrlich, Paul R. and Anne H. Ehrlich. 2010. The culture gap and its needed closures. *International Journal of Environmental Studies* 67(4): 481-492.

Ehrlich, Paul R. and Anne H. Ehrlich. 2013. Can a collapse of global civilisation be avoided? *Proceedings of the Royal Society (Biological Sciences)* 280(1754): 1-9.

Eisler, Reanne. 2000. *Tomorrow's Children: A Blueprint for Partnership Education in the 21st Century*. Boulder, CO: Westview Press.

Felluga, Dino. 2003. Modules on Althusser: On Ideology. *Introductory Guide to Critical Theory*. Lafayette, IN: Purdue University. Available at http://www.purdue.edu/guidetotheory/marxism/modules/althusserideolo gy.html , accessed December 23, 2010.

Freakley, Mark and Gilbert Burgh. 2000. *Engaging with Ethics: Ethical Inquiry for Teachers*. Katoomba NSW Australia: Social Science Press.

Giroux, Henry A. and Anthony N. Penna. 1979. Social education in the classroom: The dynamics of the hidden curriculum. *Theory and Research in Social Education* 7 (1): 21–42.

Grant, Lindsey. 2005. *The Collapsing Bubble: Growth and Fossil Energy*. Santa Ana, CA: Seven Locks Press.

Griffin, Keith. 1995. Global Prospects for Development and Human Security. *Canadian Journal of Development Studies* XVI (3): 359-370.

Griggs, David, Mark Stafford-Smith, Owen Gaffney, Johan Rockström, Marcus C. Öhman, Priya Shyamsundar, Will Steffen, Gisbert Glaser, Norichika Kanie and Ian Noble. 2013. Sustainable Development Goals for People and Planet. *Nature* 495 (March 21): 305-307.

Gutmann, Amy. 1999. *Democratic Education*. Princeton, NJ: Princeton University Press.

Hampson, Fen O., Jean Daudelin, John Hay, Holly Reid and Todd Marton. 2002. *Madness in the Multitude: Human Security and World Disorder*. Oxford: Oxford University Press.

Hawkins, Ronnie. 2013. Our War Against Nature. In *Human Security in World Affairs: Problems and Opportunities*, edited by Alexander K. Lautensach and Sabina W. Lautensach, 227-248. Vienna: Caesarpress.

Henderson, Hazel. 1999. *Beyond Globalisation: Shaping a Sustainable Global Economy*. New York: Kumarian Press.

Homer-Dixon, Thomas. 1999. *Environment, Scarcity, and Violence*. Princeton, NJ: Princeton University Press.

Imhoff, Marc, Lahouari Bounoua, Taylor Ricketts, Colby Loucks, Robert Harriss and William T. Lawrence. 2004. Global patterns in human consumption of net primary production. *Nature* 429 (6994): 870-873.

Jamieson, Dale. 2008. *Ethics and the Environment: An Introduction*. Cambridge: Cambridge University Press.

Lautensach, Alexander K. 2006. Expanding Human Security. *Journal of Human Security* 2 (3): 5-14.

Lautensach, Alexander K. 2009a. The Ethical Basis for Sustainable Human Security: A Place for Anthropocentrism? *Journal of Bioethical Inquiry* 6 (4): 437-455.

Lautensach, Alexander. 2009b. Teaching Values Through the Ecological Footprint. *Green Theory and Praxis: Journal of Ecopedagogy* 5(1): 153-168.

Lautensach, Alexander K. 2010. *Environmental Ethics for the Future: Rethinking Education to Achieve Sustainability*. Saarbruecken, Germany: Lambert Academic Publishing.

Lautensach, Alexander K. and Sabina W. Lautensach. 2013a. Why "Sustainable Development" is Often Neither: A Constructive Critique. *Challenges in Sustainability* 1(1): 3-15.

Lautensach, Alexander K. and Sabina W. Lautensach, eds. 2013b. Introduction. In *Human Security in World Affairs: Problems and Opportunities*, xix-xxxviii. Vienna, Austria: Caesarpress.

Lautensach, Alexander K. 2013c. Shaping the hidden curriculum in education: A strategy towards sustainability. *Journal of Teaching and Education* 2(4): 119-129.

Lemons, John. 1996. Afterword: University Education in Sustainable Development and Environmental Protection. In *Earth Summit Ethics: Toward a Reconstructive Postmodern Philosophy of Environmental Education*, edited by J. Baird Callicott and Fernando J.R. da Rocha, 193-217. Albany, NY: State University of New York Press.

Ludwig, Donald, Ray Hilborn and Carl Walters. 1993. Uncertainty, Resource Exploitation, and Conservation: Lessons from History. *Science* 260: 17-36.

Lutzenberger, Jose. 1996. Science, Technology, Economics, Ethics, and Environment. In *Earth Summit Ethics: Toward a Reconstructive Postmodern Philosophy of Environmental Education*, edited by J. Baird Callicott and Fernando J.R. da Rocha, 23-45. Albany, NY: State University of New York Press.

Mander, Jerry, ed. 2007. Manifesto on Global Economic Transitions. International Forum of Globalization and Institute for Policy Studies and Global Project on Economic Transitions. Available at http://www.global-vision.org/papers/IFG-manifesto.pdf, accessed March 23, 2014.

McLaren, Peter. 1994. *Life in Schools: An Introduction to Critical Pedagogy in the Foundations of Education* (second edition). New York: Longman.

McNeill, John R. 2000. Ideas Matter: A Political History of the Twentieth Century Environment. *Current History* 99 (640): 371-382.

Meadows, Dennis, Jorg Randers and Donella Meadows. 2004. *Limits to Growth: The 30-Year Update.* White River Junction, VT: Chelsea Green Publishing Co.

Myers, Norman. 1993. *Ultimate Security: The Environmental Basis of Political Stability.* New York: W.W. Norton and Co.

Myers, Norman, Richard Norgaards and Jorgen Randers. 2002. Tracking the Ecological Overshoot of the Human Economy. *Proceedings of the National Academy of Sciences* 99: 9266-9271.

Nadeau, Robert L. 2009. Brother, Can You Spare Me a Planet? Mainstream Economic Theory and the Environmental Crisis. *Sapiens* 2 (1): 71-78.

North, Richard. 1995. *Life on a Modern Planet: A Manifesto for Progress.* Manchester: Manchester University Press.

Parkin, Sara. 2010. *The Positive Deviant: Sustainability Leadership in a Perverse World.* London: Earthscan.

Pearce, David W. and Jeremy J. Warford. 1993. *World Without End: Economics, Environment and Sustainable Development.* Oxford: World Bank and Oxford University Press.

Pilzer, Paul Z. 1990. *Unlimited Wealth: The Theory and Practice of Economic Alchemy.* New York: Crown Publishers.

Pimm, Stuart L. 2001. *The World According to Pimm: A Scientist Audits the Earth.* New York: McGraw-Hill.

Rees, William E. 1996. Revisiting Carrying Capacity: Area-based Indicators of Sustainability. *Population and Environment* 17: 195-215.

Rees, William E. 2004. Waking the Sleepwalkers – Globalisation and Sustainability: Conflict or Convergence. In *The Human Ecological Footprint*, edited by Ward Chesworth,1-34. Guelph, Canada: University of Guelph.

Roberts, Paul. 2008. *The End of Food.* New York: Houghton Mifflin Co.

Schumpeter, Josef. 1954. *History of Economic Analysis.* London: George Allen.

Sen, Amartya. 1998. Population: Delusion and Reality. In *Applied Ethics: A Multicultural Approach*, edited by Larry May, Shari Collins-Chobanian and Kai Wong, 195-215. Upper Saddle River, NJ: Prentice Hall.

Sen, Amartya. 1999. *Development as Freedom.* Oxford: Oxford University Press.

Shellenberger, Michael and Ted Nordhaus. 2005. *The Death of Environmentalism: Global Warming Politics in a Post-Environmental World.* Available at http://www.thebreakthrough.org/images/Death_of_Environmentalism.pdf, accessed March 23, 2014.

Simon, Julian L. 1990. *Population Matters.* New York: Transaction Publications.

Solow, Robert. 1993. An Almost Practical Step Toward Sustainability. *Resources Policy* : 162-172.

Spady, Donald and Alexander K. Lautensach. 2013. Why Human Security Needs Our Attention. In *Human Security in World Affairs: Problems and Opportunities*, edited by Alexander K. Lautensach and Sabina W. Lautensach, 17-33. Vienna, Austria: Caesarpress.

Stiggins, Rick. 2008. *An Introduction to Student-Involved Assessment for Learning* (fifth edition). Upper Saddle River, NJ: Pearson.

Stone, Christopher D. 1987. *Earth and Other Ethics: The Case for Moral Pluralism*. New York: Harper and Row.

Tekian, Ara. 2009. Must the hidden curriculum be the "black box" for unspoken truth? *Medical Education* 43: 822–823.

The Ecologist and Friends of the Earth. 2001. Keeping Score. *The Ecologist* 31 (3): 44-47.

Tickell, Sir Crispin. 1997. God and Gaia. *Resurgence* 185 (November/December): 46-49.

Turner, Chris. 2007. *The Geography of Hope: A Tour of the World We Need*. Toronto: Random House Canada.

United Nations. 2000. *We the Peoples: The Role of the United Nations in the 21st Century*. New York: United Nations.

UNDP (United Nations Development Programme). 1994. *Human Development Report: New Dimensions of Human Security*. New York and Oxford: Oxford University Press.

UN Millennium Project. 2005. *Investing in Development: A Practical Plan to Achieve the Millennium Development Goals*. New York: United Nations.

WCED (World Commission on Environment and Development). 1987. *Our Common Future*. Oxford: Oxford University Press.

Wackernagel, Mathis and William E. Rees. 1996. *Our Ecological Footprint: Reducing Human Impact on the Earth*. Oxford: John Carpenter.

Wenz, Peter. 1988. *Environmental Justice*. Albany, NY: State University of New York Press.

WWF (World Wildlife Fund). 2012. *Living Planet Report 2012*. Available at http://wwf.panda.org/about_our_earth/all_publications/living_planet_rep ort/ , accessed March 23, 2014.

Chapter 21: Home: An Alternative Time-spatial Concept for Sustainable Development

Elin Wihlborg and Per Assmo

Introduction

Today environmental problems and solutions are often described in global terms, which require international solutions of a technical and managerial character. But even if such problems need international attention, the environmental problems that we face actually originate as a result of locally behaviors and actions. Hence if we are to address global environmental change then we must start to address those problems that exist at a local level, in the everyday lives of people; the home.

The on-going debate about sustainable development largely takes its point of departure from the view presented in the World Commission on Environment and Development (WCED) report *Our Common Future*, where the concept is described as "development that meets the needs of the present without compromising the ability of future generations to meet their own needs."[1] This general broad definition can be seen as a form of international guidance for policies to balance economic and social systems and ecological conditions by focusing on a triangle in which sustainable development is formed by the three P's, namely people (the social dimension), planet (the environmental dimension), and profit/prosperity (the economic dimension). The WCED's perspective of sustainable development has to a large extent become the conventional view where the environmental and social aspects of economic growth should be included in decision-making processes.[2] By integrating a household perspective on sustainable development there are openings to integrated sustainable development strategies at the local as well as at the global level. People do not intentionally destroy the local environment, which forms their local livelihood. The way we organize our daily livelihoods and use natural resources is largely determined by economic institutions and social structures, which are political constructions of the society.

---

[1] World Commission on Environment and Development 1987, 43.
[2] European Commission 2002; Seghezzo 2009; Barkemeyer et al 2014.

The conventional sustainability triangle displays society and the environment as separate "pillars", which are dependent upon the third economic pillar. This view largely limits the explanatory power of development to economic reasoning, in which "technocentric" economically structured approaches are regarded as the best suited to solve environmental as well as social developmental problems.[3] Hence, to reach sustainable development, which integrates the ecological, social and economic dimensions, there is a need to focus the analysis on the everyday lives of people at the local level,[4] and in particular on the actual users of resources.[5] Hence, there is a need to look for a new conceptual toolbox that can enable us to understand the complexity of local livelihoods, and thereby find new alternative structures and policies to implement societal processes of change. This chapter articulates such an approach by elaborating on the meanings of the home and its implications on the possibilities for sustainable development.

The chapter builds on a time-geographical approach by looking out from the local time-spatial context; the *home*. We aim to explore the integrated aspects of everyday life in the home, which constitute the constraints and opportunities of people's livelihood, and discuss its implications on the opportunities for local sustainable development.

By creating a conceptual meaning of the home, we can distinguish between and discuss the distinct and integrated implications of the home in three dimensions: as a physical *dwelling* of resource management, the *household* as an economic unit, and the *family* as a social unit. The chapter discusses and outlines an alternative analytical time-spatial approach that acknowledges the interplay of social, economic and ecological constraints and opportunities in people's everyday life. The time-spatial approach highlights the importance of creating different policy instruments that can enhance sustainable ways of living in the home, and thereby in society as a whole.

Perspectives of Sustainable Development

Sustainable development is a normative and complicated concept. Here we will follow the basic ideas of the WCED report, since our aim here is not to critique sustainable development as a concept but, rather, to focus on its local implications in the home. Most approaches towards environmental problems focus on policies and strategies of a technical and managerial character.[6] The concept of *ecological modernization* is a commonly used perspective of this technocentric orientation, which argues that almost no radical changes of the present capitalist system are needed.[7] Even if the ecological modernization perspective is criticized,[8] most other approaches to sustainable development consider economic growth as a precondition for obtaining the economic capabilities to tackle environmental

---

[3] Christen and Schmidt 2012.

[4] Soderholm 2010.

[5] Shove and Walker 2010; Spaargaren and van Vliet 2000; Skill 2008; Hägerstrand et al 2009; Assmo and Wihlborg 2010.

[6] Dryzek and Schlosberg 2005; Hopwood et.al. 2000.

[7] Langhelle 2000; Norton 2005; Hajer and Wagenaar 2003.

[8] Langhelle 2000.

problems.[9] Most conventional policies and strategies emphasizing sustainable development are to a large extent trapped in technocentric oriented market economy models. This kind of ecological modernization thinking enhances strategies that advocate general technical and managerial institutional solutions. Such a one-sided focus on production, institutions and structures lacks a focus and understanding of the actual users of resources; the people. This can be tackled differently if analysis focuses more explicitly on the home.

The applicability and usefulness of the conventional concept of sustainable development are limited due to different reasons. The sectoral focus on separate pillars of development tends to impede a more integrated understanding of the processes of change in society. The conventional analyses, policies and strategies applied in the conventional view of sustainable development often focus on economic, environmental or social factors, one at a time.[10] Hence, these models fail to connect the different factors into a more all-integrated understanding and analysis, which can highlight the complex processes of change in time and space.[11] Furthermore, the division and focus on separate pillars also undermine the focus and understanding of the actual actors in the process; the people. Thus there is a need to look for a more embracing approach that includes the processes of everyday life in the home.

To reverse the conventional perspectives and search for an alternative toolbox we will open a time-geographical approach that takes its starting point as the local level, launching an approach to focus on the home as a hub for sustainable development. In so doing we will also try to embrace the complex relationships of opportunities and constraints among the three pillars: people, environment and economy.

Sustainable Development in a Time-spatial Approach

Societies can use nature in order to survive and prosper, but history shows that societies also have the power to exceed nature's carrying capacity, which eventually can lead to environmental and societal collapse.[12] How a society responds to constraints and opportunities depends on the physical resources, economic structures, social institutions and cultural values that exist in that particular society. Basically, the political organization develops and forms the structure of a society as well as the economic and social arrangements of that society. When we view society and its economic and social arrangements as a political construction, it is important to place the focus of our analyses on the basis of the local society; the home.

If we begin our analysis of the daily activities of people as the crucial starting point for resource management – sustainable development – we may turn the analysis almost up-side down. Inspired by Hägerstrand's classical conceptualization of time and space in theories of time-geography, we will

---

[9] Joas 2001; Barry 2005.
[10] Bell and Morse 2008.
[11] Eckerberg and Joas 2004.
[12] Diamond 2005.

develop our elaboration of the home into a model on how sustainability in homes can be formed.

Time-geography can mainly be seen as an ontological perspective, focusing on how to identify "all" processes taking place in sections of time-space. Time-geography is not explicitly a conventional theory – providing tools for analysis and explanations – but it provides a coherent methodological approach of a notation system that contextualizes processes of change in time and space (including place).[13] Hereby the approach also carries an ontological interpretation of the inter-connectedness of "everything" that makes analyses inclusive.

Hägerstrand, the founding father of time-geography, stressed the importance of a contextual "all-embracing" perspective, arguing that the world can be viewed as pockets with mixed assortments of beings and processes that share a common existence in space and time.[14] Time-geography thereby provides a platform from which it is possible to observe and describe certain elements of reality, without losing touch with the total context.[15] The basic notion in classical time-geography is the co-existence of time and space. The fundamental meaning of "taking place" actually points at the power of the process, which always takes place in a specific physical space during a specific period of time.[16] In classical time-geography the human utilization of the physical world is in focus, closely linked to socio-cultural and mental dimensions. Hence, a classical time-geographical perspective is not really a subject specific theory in a classical conventional academic sense, but rather a way of connecting and relating to the different worlds in describing the reality. Being so, a time-geographical perspective provides an alternative structure of development thinking, which attempts to consolidate the spatial and temporal perspectives of different academic disciplines into a more concrete analytical platform.[17]

People's daily life is formed by time-spatial opportunities and constraints in time and the resources made available hereby. The time-spatial approach is hence an attempt to form a broader conceptual framework capable of contextualizing the connections and relations between nature and human actions for knowledge examination, analysis and synthesis. There are two main concepts in time-geography opening for analysis of the home from a time-geographical perspective: the constraints of actors; and the pockets of local orders they form. These are now examined in turn.

Constraints

A basic time-geographical concept that shows how and why processes in time-space develop as they did is the concept of constraints. Hägerstrand identified three main forms of constraint: capacity constraints, coupling constraints and authority constraints. *Capacity* and *coupling constraints* are primarily connected to the capacity of the individual. Capacity constraints focus on an individual's actual capacity to act in the time-spatial setting where she is. Coupling constraints

---

[13] Hägerstrand 1953; 1970; 1985; Hägerstrand et al 2009.
[14] Hägerstrand 1985; Hägerstrand et al 2009.
[15] Lenntorp 1999; Thrift 2005; Krantz 2006.
[16] Hägerstrand 1970; Lenntorp 1999.
[17] Lenntorp 1999; Ellegård and Wihlborg 2001; Krantz 2006.

refer to the relations of the individual and focus on everything that limits her co-ordination with other people and physical artefacts. *Authority constraints* have a more structural and social origin and include everything that has a power to steer the actor's actions. Constraints of authority can be expressed through restrictions by laws and regulations (formal institutions), but could also be viewed through discourses, norms and culture (informal institutions) that are related to the attitudes and values that exist in a community.[18] By identifying the actor's constraints in a given situation it is also possible to identify her potential resources. The constraints are time-spatial specific and framed into pockets of local orders.

Pocket of Local Order: An Analysis of Time-spatial Practices

The concept of a *pocket of a local order* can be seen as a response to the criticism that the time-geographical approach was more descriptive than analytical. Giddens made that clear, but also identified the time-geographical constraints as an approach that enables analysis, in particular of power.[19] Hägerstrand addressed this by introducing the concept of pocket of local order, to open for elaborations and analyses of how constraints are managed and resources are made available in a specific pocket of time space to fulfill specific projects.[20]

A pocket of local order is defined as a distinct section of time-space in which the actors form a specific order to conduct specific projects. The pocket of local order concept thus aims to capture the interplay of actors and structure to analyze local practices.[21] Resources in the pocket can be human actors, artifacts and natural resources. The activities conducted in the time-spatial pocket of local order define and re-produce the order. Hereby, some pockets are maintained to make activities routinenized and embedded into a setting of resources. The home is a typical pocket of local order. It is an order having the power to frame and define how resources are made available and how constraints are overcome in everyday practices.[22] In the next section our elaboration on the home in relation to sustainability will be guided by this conceptualization.

A Three Dimensional Time-spatial Contextualization of Everyday Life: The Home

The place and content of everyday life is unique for every individual and it develops over time. It builds on the experiences and knowledge we gain throughout our lives, through everything we hear and see, but also where and when we have gained these experiences. Everyday life takes place within physical contexts and takes time. To understand the time-spatial contextualization of everyday life and thereby also sustainable development the analysis takes its point of departure in the local arena; the home. The home is seen as the

---

[18] Hägerstrand 1985; Hägerstrand et al 2009.
[19] Giddens 1984.
[20] Hägerstrand 1985.
[21] Hägerstrand 1985; Ellegård and Vilhelmson 2004; Assmo and Wihlborg 2010.
[22] Ellegård and Vilhelmson 2004.

embracing concept constituted by individuals in close collaboration with other individuals, organizations and structures.[23]

We will here argue that the home can be viewed and analyzed in three integrated conceptual dimensions as: the *dwelling* (related to the physical), the *household* (related to economic resource management), and the *family* (related to the social and emotional). This three-folded approach has an interdisciplinary grounding. Many local-oriented studies of sustainable development focus on specific topics, activities, or attitudes. However, along the arguments described in the previous section, such studies often have a rather narrow approach focusing on specific phenomena of environmental, social or economic character. Analysis of what is taking place in the home appears in a range of disciplines. In relation to sustainability issues of the home, we can identify studies that take off from disciplines like psychology, human ecology, economics and sociology. For example, studies in psychology pay attention to the environmental attitudes and norms individuals have and their relationship to behavior.[24] In human ecology and household economics there are examples that examine the home as a node in life cycle analysis of grocery consumption and the generation of household waste.[25] In microeconomics and sociology we can find studies focusing on the division of labor, unpaid household work and how costs for different activities are evaluated by householders.[26] In sociology there are studies, which examine the use and access of households with regards to technology and socio–technical systems.[27] All these dimensions are analytical and are mutually interwoven into daily webs of activity.

The Three Dimensions of the Home

The model we propose here is based on three inter-related dimensions of the home: dwelling, household and family (Figure 21.1). The dwelling is based on our need for a place to live, and the physical items and resources that constitute the home. The physical attributes are accessed through the economic householding of the home. We buy, own and manage the physical attributes through the household. There are social values and norms regarding where, when and with whom we conduct certain activities at home. In the home the social relations most often refer to some kind of kinship or family. But these are all related and combined in the home.

---

[23] Giddens 1984; Ellegård and Wihlborg 2001.

[24] Barr 2002.

[25] Shove 2003; Bubolz and Sontag 1993; Spaargaren and van Vliet 2000; Åberg 2000.

[26] Becker 1991.

[27] Shove 2003; Silverstone and Hirsch 1992.

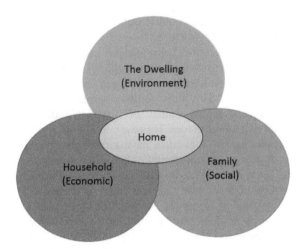

Figure 21.1: The three inter-related dimensions of the home

These three dimensions are fundamental aspects of the everyday lives of people, and form their homes. We will now elaborate on each of these three dimensions of the home in relation to sustainable development.

The Dwelling

The interpretation of the dwelling in this conceptual approach builds on a time-spatial contextualization of activities. Time and space are mutually dependent and inseparable. Every activity, social or natural, takes place somewhere and takes place within a certain period of time. And when taking place, the activity as such has to occur within a specific place. Here there are openings to include "actors" in a broader sense,[28] in the forming of the pocket of local order's dwelling dimension. The expression that an activity is taking place, implies that the activity and the individuals and artifact conducting the activity have the power to actually own or use the place at a specific time.[29]

To regard the home as a dwelling is to view the home as a physical artifact and a space that makes use of physical resources. The location and design of the dwelling define how every day activities are organized. Everyday activities are related both to the property as such and its surroundings, focusing on the basic physical and spatial context. The parameters of the dwelling are not only limited to the house as such but can also be extended to include a larger physical area where the dwelling is located. It is therefore often extremely complicated to divide what is conventionally regarded as the private and the public spheres concerning the politics of the home.

A local-individual focus on people's daily livelihood constraints and opportunities makes the dwelling a key unit of analysis, since it is a place where humans live out their daily lives, interact with each other and where labor is divided. The environmental impact of everyday activities around the dwelling can

---

[28] Latour 2005.
[29] Wihlborg 2011.

be seen at the intersections between what are commonly distinguished as the private–public, micro–macro, and actor–structure spheres.[30] The location and structure of the dwelling defines, for example, energy consumption and demands for transportation in the home.

Everyday life does not only take place in the dwelling. Thus defining the limitations or boundaries as to what can be seen to take place within the dwelling is far from simple. The limits of what is seen as the dwelling are not easily identified. For example, in some contexts, even the private car could be seen as an extension of the dwelling. One can also notice an increasing "domestification" of artefacts and activities such as entertainment, work and education that can occur within the space of the home.[31] The dwelling can also exist as an important space where lifestyle aspirations can be pursued. To conclude, the dimensions of what can be defined as the dwelling of the home indicate that we are physical bodies – even if we participate in the forming of cyberspaces – and that we are integrated within the physical world when taking place.

Householding of the Home

When looking at the home as an economic unit we can make use of the concept of the household. This concept emerges from the analytical description of the household as a space where resources are utilized or, in other words, resource management. Householding as a verb indicates the activity of holding (managing) a home (house), and that is what happens at home.

For Aristotle, the household was a space where the management of relations among "master and slave, husband and wife, father and children" occurred. Aristotle thereby focused on the household as a space of production. Aristotle also pointed out that as states are made up of households, households have to be considered as political actors. In modern societies, however, the focus on the household as a place of consumption rather than production has become central.[32] The concept of the household relates to the production of tools, properties and other valuables for the family in the dwelling.

In early writings on economics, resources were considered as being of a value, much more than just things given a monetary value. Yet resources include much more than what is today conventionally valued in monetary terms and constraints are thus much more than the monetary cost. New Home Economics also includes resources of a non-monetary value in the distribution of values, resources and activities at home.[33] New Home Economics thereby contributes to a wider understanding of the use of resources and the constraints, most often taken for granted in economics as well as most other social science disciplines.

In both the domestic and the public sphere there are several activities that could be made more visible, even if not paid for. These activities are important, and even crucial to sustaining the livelihood of the household. Today, it is necessary to view the household in different ways so as to acknowledge the important role of daily life and its activities. There is a need to analytically

---

[30] Giddens 1984; Dobson 2003; Dobson and Bell 2006; Skill 2008.
[31] Silverstone and Hirsch 1992.
[32] Paterson 2006.
[33] Becker 1991.

separate the concept of the household from the values formed in family relations and the physical implications inherent in the dwelling. The concept of the household is about householding with everything that is constructed in terms of resources and constraints. There are political constructions of the household dimension in the home, defining the limits of what are deemed to be paid and unpaid activities.[34]

The Family as the Social Dimension of the Home

Concerning the concept of the family, the focus is on the social and cultural aspects of the home. The peer groups we interact with at home create the norms, values and ideas, which are represented in language, symbols and texts. These meanings are created in a specific time and spatial/material context.[35] This time-spatial dimension also includes the intergenerational transformation of norms and values, which are commonly referred to in an analysis of norms and values for the purpose of sustainable development.

A definition of what it means to be a family does not have to include the idea that a family should be biological or based on marriage or even romance. It is rather a conceptual idea of people sharing everyday life, and the ability of people to construct values and to create a sense of closeness through their daily practices. In these relations we invest long term social capital through building shared virtues and meanings. This can be seen as a broader form of citizenship. The liberal democratic concept of the citizen and self and, by extension, the citizen is flawed because the idea of independence and self-determination is at odds with the kind of mutual dependent relationship necessary for ecological flourishing – relationships of kinship, mutuality, empathy and care.[36]

The state takes more or less active and legitimate initiatives to construct the family through various means which may include legislation regarding marriage and cohabitation as well as informative campaigns on, for example, eating and drinking habits, and in many states also on values of environmentally sustainable behavior. Family policies have traditionally been seen as legitimate (in the interest of the nation) and have thereby implicitly or even explicitly been in the interest of the state for the purpose of its structure and survival. The fostering of the "good" citizen has been an obligation for the family and there are numerous policies relating to how to promote the ability of a family to fulfil such obligations. An example of a specific and clearly outspoken normative policy is the Australian Pre-Marriage Education Programs which promote ways to encourage families to stay together. The program encourages families to learn what the state considers to be traditional family values. This argument again points at the political construction of the family dimension in the home.[37]

The *family* is a private sphere of relations, but it is at the same time constituted by public arrangements. The understanding of the private is indeed complicated. The private and public are mutually dependent and at the same time formed by each other. The private is constituted by the public and vice versa. The

---

[34] Assmo and Wihlborg 2012.
[35] Douglas 1970.
[36] MacGregor 2006.
[37] van Acker 2003.

private and the public are two sides of the same coin. Citizenship indicates the interface of the public and the private. [38] Thus sustainable relations in a family perspective is formed in the interplay between the private and public spheres.

Sustainable Development in the Home

The consumption and production patterns that take place in the *household* are part of an intricate web of (monetary and non-monetary) resources and decisions that take place in close interplay between the public and private spheres. The *dwelling* is a private expression of power within a place, but publicly arranged by legislation as well as socio-technical systems. Since the private and public are mutually dependent and at the same time form each other, this chapter integrates these aspects into a broader concept of "everyday life."

By dividing an analysis of the home according to the three aspects discussed above, these three aspects can be directly related to the ecological, social and economic dimensions of sustainable development (Figure 21.2). Despite the integration of the three dimensions of the home, an initial division between these dimensions will help us to analyze the transformation towards achieving the objective of increased sustainable development in the home.

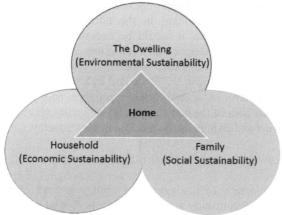

Figure 21.2: How the three dimensions of the home relate to the dimensions of sustainable development

Ecological Sustainability through the Dwelling

The dwelling relates to ecological sustainability, since it refers to the physical part of the home. There are numerous resources that flow through the home. Socio-technical systems as water, the transportation of goods and people, waste and food can be regarded as the physical parts of a home and can be placed into the category of a dwelling. The home is the final destination of several flows of resources in society, like the food chain and energy-production (energy for many industries is used to produce resources for the homes). Thus we argue that an analysis of ecological sustainability should begin by including everyday activities,

---

[38] Barry 2005.

since these activities and the use of resources and materials have an ecological impact.

The design and organization of dwellings have implications for the potential of changing into a more sustainable way of living. Through the design and regulation of dwellings, policies have the potential to enforce and change everyday life style choices. Physical planning could be an efficient, but not always legitimate, policy instrument for governing the location and formation of dwellings. Physical planning is a key policy instrument to lay a ground for more sustainable dwellings and creating openings for sustainability in other terms as well.

Economic Sustainability through the Home

The household dimension of the home is directly related to economic sustainability. In daily life, running a household is often equated with the objective of reaching an acceptable and hopefully sustainable balance of incomes and costs, the overall aim of which is to achieve what family members would regard as a good life.

Large-scale economic sustainability has to be balanced in the same way. However, the household includes resources not recognized as monetarily countable. Economic householding in the home extends beyond large-scale economic sustainability. Thus it would be necessary to consider whether the contemporary economic system could even allow economic sustainability. The ability to reach a level of economic sustainability in the home depends on the political structures that surround the home and its location. There is a demand for an economic structure in societies that promotes and provides sustainable economic relations both in the home and in wider society. Formal politics has the power to construct and change the boundaries of the economy.

Social Sustainability through the Family at Home

The family is a means of analyzing our ability to lead socially sustainable lifestyles and the relationships among individuals. Social norms and values are the basic building blocks of a socially sustainable society. The creation of meanings and values is a common way of creating a broad understanding of social sustainability. What is deemed to be acceptable in a society may be seen as socially sustainable and is a way of defining legitimate behavior.[39]

Thus, information campaigns and changes in norms and values are important for the purpose of creating a sustainable way of living in the home and thereby encouraging sustainable development on a global level. The construction of normality is a critical issue in most post-modern studies, but is also essential for an understanding of the environment. What can be defined as a "normal" home and family is flexible and open to different interpretations of all three aspects of the home.

---

[39] Wihlborg and Skill 2006.

A Time-spatial Approach of Resources and Constraints in the Home

With the time-spatial approach, the different dimensions of the home described above can be brought into an analysis of a life drama. The people are actors that perform in a specific environment. And the roles they play are connected to the common structures of culture, values, rules, institutional frameworks, and power relations, which are set in a particular time-spatial pocket. Hence, the time-spatial structure of societies makes up pre-conditions for the local orders that can be made up in the home by individuals to fulfill their aims and the projects they undertake.[40]

The concept of constraints in a pocket of local order is in this view important, since the different forms of constraints formed at different geographical levels position the constraints and opportunities for people's daily activities. By applying a time-spatial perspective, one can identify and analyze the social (norms, values and structures), physical (dwelling and landscapes) and time (as resource and context) dimensions of the home in an integrated and holistic manner – social-physical-time – in order to understand the broader processes of change in the local society.

The concept of a pocket of local order can be applied on different time-spatial levels. The orders do, however, influence each other. The structural setting and constraints in a larger time-spatial pocket of order may, and most likely will, define and influence the authoritative constraints in a more local pocket of order. This is similar to the hierarchy of policies and political power, but the pocket of local order concept reaches further by including the physical, spatial and temporal implications of power.[41]

Authority constraints restrain and direct time-use for different activities, since people's activities and choices are largely dependent on how the surroundings are structured. Human activities demand structures defining what is appropriate in a certain situation. In a simplified manner one can argue that rules and norms form institutional (formal and informal) authoritative structures and constraints. People (or companies and organizations) therefore act in accordance with the perception of the possibilities and constraints given in that particular time-spatial structure. At the same time, authoritative structures and constraints are not deterministic or static. People form the structures they live in and, thereby, can enable pockets of local order that influence and change existing structures or norms.

With the combination of the dimensions of the home in relation to constraints, a time-spatial analytical approach can physically anchor and integrate social, economic and ecological processes extended over time. Such an analysis may develop from a local time-spatial setting and may focus on the constraints experienced by actors. This analysis can be structured according to the framework shown in Table 21.1.

---

[40] Ellegård and Wihlborg 2001.
[41] Assmo and Wihlborg 2010.

Table 21.1: A time-spatial analytical framework for mapping three types of constraints onto the three dimensions of the home

Home (People)	Capacity Constraints (individual)	Coupling Constraints (relational)	Authoritative Constraints (structural)
Dwelling (Environment)			
Household (Economy)			
Family (Social)			

Through a conceptual analysis of the pockets of local order related to the dimensions of the home one can, therefore, identify the constraints and opportunities for change towards improved sustainability.

Concluding Remarks

This chapter has discussed an alternative analytical tool, a time-spatial approach, which focus on the root of the problem, in the daily livelihood of people; the home. How people respond to environmental, social and economic constraints is dependent on the integration of physical resources, economic institutions, social institutions and cultural values in their particular society. Societies are basically political constructions that are valid and legitimate in a certain time-space pocket.

The construction and setting of the home takes place in the integration of economic, physical and social dimensions. A home also builds on relations with others (the family), the use of place and the resources accessible in the dwelling, and finally, the resources prevalent in the household. Thus, these three dimensions appear to be mutually dependent in practice. By analytically identifying each one and understanding how they are integrated, we can create an analytical approach that will enable us to gain a better understanding of the politics of the home.

The politics concerning sustainable development demands an expansion to the interpretation of what can be considered as legitimate politics. By highlighting and analyzing sustainable development in the different dimensions, there is a potential to develop and clarify why and how policy instruments can relate to aspects of daily life in the home. There is also potential to evaluate changes from a home-impact analysis perspective, as shown here.

The potential for creating policy instruments lies in the everyday approach to actually integrating all aspects of what people do in their daily lives. Since the individual – resource user or citizen – travels through a complex web of mixed spheres of activity, organizations and places during an ordinary day, the individual may encounter many different structures when making up their home. An integrated time-spatial conceptual approach of the dimensions of the home can open up new ways of improving sustainable development that are in all our interests.

References

Assmo, Per, and Elin Wihlborg. 2010. Home: The Arena for Sustainable Development – A conceptual discussion. *The International Journal of Environmental, Cultural, Economic and Social Sustainability*. 6(1) 319-332.

Assmo, Per, and Elin Wihlborg. 2012. *Ydre 2.0*. An Alternative Time-Spatial Approach towards Post-Monetarism. In: *Entrepreneurship, Social Capital and Governance. - Directions for the Sustainable Development and Competitiveness of Regions* edited by Börje Johansson, Charlie Karlsson and Roger Stough, 71-84. London: Edward Elgar Publications.

Åberg, Helena. 2000. *Sustainable Waste Management in Households – From International Policy to Everyday Practice. Experience from Two Swedish Field Studies*. Diss. Göteborg: Acta Universitatis Gothoburgensis.

Barkemeyer, Ralf, Diane Holt, Lutz Preuss, and Stephen Tsang. 2014. What Happened to the "Development" in Sustainable Development? Business Guidelines Two Decades After Brundtland. *Sustainable Development* 22: 15–32.

Barr, Stewart. 2002. *Household Waste in Social Perspective: Values, Attitudes, Situations and Behaviour*. Aldershot: Ashgate

Barry, John 2005. Resistance Is Fertile: From Environmental to Sustainability Citizenship. In *Environmental Citizenship*, edited by Andrew Dobson and Derek Bell, 21-48. Cambridge, London: The MIT Press.

Becker, Gary S. 1991. *A treatise on the family*. Cambridge, MA Harvard University Press.

Bell, Simon and Stephen Morse. 2008. *Sustainability indicators: measuring the immeasurable?* second edition. London: Earthscan Publications Ltd.

Bubolz, Marget M and Suzanne M. Sontag. 1993. Human Ecology Theory. In *Sourcebook of Family Theories and Methods: A Contextual Approach*, edited by Pauline G Boss, William J. Doherty, Ralph LaRossa, Walter R. Schumm and Suzanne K. Steinmetz, 419-447. New York: Plenum.

Christen, Marius and Stephan Schmidt. 2012. A Formal Framework for Conceptions of Sustainability – a Theoretical Contribution to the Discourse in Sustainable Development. *Sustainable Development* 20: 400–410.

Diamond, Jared. 2005. *Collapse – How Societies Choose to Fail or Succeed*. New York: Viking.

Dobson, Andrew. 2003. *Citizenship and the environment*. Oxford: Oxford University Press.

Dobson, Andrew and Derek Bell. 2006. *Environmental citizenship*. Cambridge, MA: MIT Press.

Douglas, Mary. 1970. *Natural Symbols*. London: Barrie and Rockliff.

Dryzek, John S. and David Schlosberg. 2005. (eds) *Debating the Earth: the environmental politics reader*. Oxford: Oxford University Press.

Eckerberg, Katarina, and Marko Joas. 2004. Multi-level Environmental Governance: a concept under stress? *Local Environment* 9(5): 405-412.

Ellegård, Kajsa, and Bertil Vilhelmson. 2004. Home as a Pocket of Local Order: Everyday Activities and The Friction of Distance. *Geografiska Annaler, Series B: Human Geography* 86 (4): 281–296.

Ellegård, Kajsa and Elin Wihlborg. 2001. (eds) *Fånga vardagen. Ett tvärvetenskapligt perspektiv.* Lund: Studentlitteratur.

European Commission. 2002. *The World Summit on Sustainable Development. People, Planet, Prosperity.* Luxembourg: Office for Official Publications of the European Communities.

Hajer, Maarten, and Hendrik Wagenaar (eds). 2003. *Deliberative policy analysis: Understanding governance in the network society.* Cambridge: Cambridge University Press.

Giddens, Anthony. 1984. *The Constitution of Society: Outline of the Theory of Structuration.* Cambridge: Polity Press.

Hägerstrand, Torsten. 1953. *Innovationsförloppet ur korrologisk synpunkt.* Diss. No 25. Lund: Lund University.

Hägerstrand, Torsten. 1970. What about people in Regional Science? *Regional Science Association Papers.* XXIV: 7-21.

Hägerstrand, Torsten. 1985. *Time-Geography: Focus on the Corporeality of Man, Society, and Environment.* (Reprint Science and Praxis of Complexity). New York: United Nations University.

Hägerstrand, Torsten, Kajsa Ellegård, and Uno Svedin. 2009. (eds) *Tillvaroväven.* Stockholm: Formas.

Hopwood Bill, Mary Mellor, and Geoff O'Brien. 2005. Sustainable development: mapping different approaches. *Sustainable Development* 13: 38–52.

Joas, Marko. 2001. *Reflexive Modernisation of the Environmental Administration in Finland: Essays of Institutional and Policy Change within the Finnish National and Local Environmental Administration.* Diss. Åbo: Åbo Akademi.

Krantz, Helena. 2006. Household routines—A time-space issue: A theoretical approach applied on the case of water and sanitation. *Applied Geography* 26: 227–241.

Langhelle, Oluf. 2000. Why Ecological Modernization and Sustainable Development Should Not Be Conflated. *Journal of Environmental Policy and Planning* 2: 303-322.

Latour, Bruno. 2005. *Reassembling the Social. An introduction to Actor-Network Theory.* Oxford: Oxford University Press.

Lenntorp, Bo. 1999. Time-geography – at the end of its beginning. *GeoJournal* 48: 155–158.

Lenntorp, Bo. 2004. Path, Prism, Project, Pocket and Population: An Introduction. *Geografiska Annaler* 86 B: 223-226.

MacGregor, Sherilyn. 2006. *Beyond Mothering Earth: Ecological Citizenship and the Politics of Care.* Vancouver: University of British Columbia Press.

Norton, Bryan G. 2005. *Sustainability. A philosophy of adaptive ecosystem management.* Chicago, IL: The University of Chicago Press.

Paterson, Mark. 2006. *Consumption and everyday life.* New York: Routledge.

Seghezzo, Lucas. 2009. The five dimensions of sustainability. *Environmental Politics* 18(4): 539-556.

Shove, Elizabeth. 2003. *Comfort, Cleanliness and Convenience: the Social Organization of Normality.* Oxford: Berg.

Shove, Elizabeth, and Gordon Walker. 2010. Governing transitions in the sustainability of everyday life. *Research Policy* 39(4): 471-476.

Silverstone, Richard, and Erik Hirsch. 1992. *Consuming Technologies: Media and Information in Domestic Spaces.* London: Routledge.

Skill, Karin 2008. *(Re)creating ecological action space: Householders' Activities for Sustainable Development in Sweden.* Diss. Linköping Studies in Arts and Science, No. 449, Linköping: Linköping University Press.

Soderholm, Patrik 2010. (ed) *Environmental Policy and Household Behaviour: Sustainability and Everyday Life.* London: Earthscan.

Spaargaren, Gert and Bas van Vliet. 2000. Lifestyles, Consumption and the Environment: The Ecological Modernisation of Domestic Consumption. *Environmental Politics* 9(1): 50-77.

Thrift, Nigel 2005. Torsten Hägerstrand and social theory. *Progress in Human Geography.* 29: 337-340.

van Acker, Elisabeth. 2003. Administering Romance: Government Policies Concerning Pre-Marriage Education Programs. *Australian Journal of Public Administration* 62(1): 15-23.

Wihlborg, Elin and Karin Skill. 2006. Legitimate Policies for Sustainable Development and Household Work. *Women and Environment International* 70/71: 35-38.

Wihlborg, Elin. 2011. Makt att äga rum. En essä om tidsgeografisk epistemologi. In *Sammanvävt. Det goda livet i vardagslivs-forskningen,* edited by Jenny Palm and Elin Wihlborg. Linköping: Almlöfs förlag.

World Commission on Environment and Development. 1987. *Our Common Future.* Oxford: Oxford University Press.

About the Editors

David Humphreys is a reader in environmental policy and social sciences programme director at The Open University in the UK where he specializes in environmental politics and policy. He is a member of the editorial board of the journal *Forest Policy and Economics* and a member of the advisory board of the *On Sustainability* knowledge community and corresponding conference and academic journals sponsored by Common Ground. He has written two books on international forest politics, the second of which, *Logjam: Deforestation and the Crisis of Global Governance*, won the International Studies Association's 2008 Harold and Margaret Sprout Award.

Spencer S. Stober is professor of biology and leadership studies at Alvernia University in the US. His research interests include environmental sustainability and the intersection between religion and science. He is a member of the advisory board of the *On Sustainability* knowledge community and corresponding conference and journals sponsored by Common Ground. He recently co-authored *Nature-centered Leadership: An Aspirational Narrative*, with two PhD students (Tracey L. Brown and Sean J. Cullen), published by Common Ground in 2013. He also co-authored a book with Dr. Donna Yarri, Associate Professor of Theology at Alvernia University, entitled *God, Science, and Designer Genes: An Exploration of Emerging Issues in Genetic Technologies*, published by Praeger in 2009.

About the Contributors

Samuel Alexander is a lecturer with the Office for Environmental Programs, University of Melbourne, Australia. He teaches a course in the Masters of Environment called "Consumerism and the Growth Paradigm: Interdisciplinary Perspectives." He is also co-director of the Simplicity Institute (www.simplicityinstitute.org) and research fellow with the Melbourne Sustainable Society Institute. His research interests include post-growth economics, the voluntary simplicity movement, peak oil, and other energy issues. His books include *Entropia: Life Beyond Industrial Civilisation* (2013) and an edited collection called *Voluntary Simplicity: The Poetic Alternative to Consumer Culture* (2009).

Per Assmo is an associate professor in geography at Linkoping University, Sweden. He obtained his PhD in human/economic geography at Göteborg University, Sweden in 1999. Since 2009 he has been a visiting professor at Rhodes University in South Africa. For more than 20 years he has conducted and published research work focusing on development issues in an African as well a European context. In recent years, he has developed a three year Bachelor degree, the International Programme in Politics and Economics (IPPE). He has also conducted research and published articles related to pedagogical aspects of international higher education.

Robert C Carlson, a professor in the School of Engineering at Stanford University for more than four decades, passed away on September 6, 2011. His primary areas of interest in both teaching and research were production and capacity planning, new product development, manufacturing strategy, and sustainable product design, development and manufacturing. Before moving to Stanford, Professor Carlson worked at Bell Labs in Holmdel, N.J., where he was a member of the technical staff in the Operations Analysis and Economic Studies Department.

Christine Dellert is a graduate student at the University of Central Florida, where she is pursuing a degree in political science with a focus on environmental politics. A former journalist, Christine Dellert participated as an observer NGO delegate to the United Nations Framework Convention on Climate Change (UNFCCC) with the youth-focused organization SustainUS. While at the UNFCCC in 2010, she also served as a media coordinator for YOUNGO. The views in Christine Dellert's chapter are the author's own and do not represent any affiliated organizations.

Paul Derby is a professor of anthropology in the sociology program at Castleton State College, Vermont, USA. Professor Derby also heads the Green Campus Initiative at Castleton State College. His research interests include environmental and cultural sustainability in the context of community studies. For fall semesters Professor Derby travels and works with students on community-based sustainable building and cultural heritage projects in New Mexico, USA.

Alan Derbyshire is a senior lecturer in design at the University of Central Lancashire, United Kingdom. His research focuses on developing design strategies that contribute to the notion of "placeness" by reviewing native ecologies and materials and their perceived connection with determined localities. His recent projects involve assessing vernacular ecologies and architecture within specific urban locations, appraising cultural variables, local resources and environmental conditions in order identify innovation and benchmarks of good practice. He was awarded the prestigious Reed and Mallik Medal from the Institute of Civil Engineers for best research paper on the subject of Urban Design in their published journals in 2012.

Elizabeth More Graff is visiting faculty in the School of Design and Construction at Washington State University in Pullman, Washington, US, where she teaches landscape architectural design and design-build. She is a licensed landscape architect with 29 years experience in the professions of the built environment including architectural design, site engineering, and historic preservation. Her life-work is dedicated to qualitative processes of placemaking, systems-thinking community design, art as a catalyst for positive change, natural building, and spiritual and emotional development in relation to sustainable practices and well-being. She currently resides in an adaptively re-used building she co-designed and re-birthed in Moscow, Idaho.

David Grierson is deputy head of the Department of Architecture and chair of the Graduate School of Engineering at the University of Strathclyde Glasgow in Scotland. He has directed the ground breaking Postgraduate Program in Sustainable Engineering since 2004, combining study in specialist, advanced engineering technologies underpinned with training in sustainability. In September 2013 the Cosanti Foundation appointed Dr. Grierson as the first Visiting Professor at Arcosanti, Arizona, USA.

Amzad Hossain is an adjunct research fellow at Curtin University, an action researcher and sustainability scholar whose interests cover simple lifestyle, appropriate technology, ecotourism and values with relevance to social, technological and environmental sustainability. He has two PhDs from Murdoch University and more than 30 refereed publications, including numerous presentations at prestigious international conferences. As a researcher and activist, Amzad divides his time between Australia and Bangladesh where he supervises research students and works for the community.

Popie Hossain-Rhaman completed her PhD at Murdoch University in Perth, Western Australia in the area of education for sustainable development examining the potential contribution of creator-centric Islamic values education. She is currently working as the principal of the Langford Islamic College in Perth and under her leadership the school is striving to achieve high standards both academically and in Islamic studies.

Wolfram Hoefer is an associate professor at the Department of Landscape Architecture at Rutgers, the State University of New Jersey and serves as undergraduate program director for the department. He also serves as co-director of the Rutgers Center for urban environmental sustainability. In 1992 he earned a diploma in landscape architecture from the Technische Universität Berlin and received a doctoral degree from Technische Universtät München in 2000. His research and teaching focus is the cultural interpretation of brownfields as potential elements of the public realm. He is investigating the different cultural interpretations of landscapes by the general public in North America and Europe and how they have an effect on public participation processes as well as professional approaches towards planning and design solutions for adaptive re-use of brownfields.

Glen David Kuecker is a professor of history at DePauw University. He received his PhD in Latin American and global comparative history from Rutgers University, New Brunswick. He co-edited *Latin American Social Movements in the Twenty-first Century: Resistance, Power, and Democracy* (2008). Kuecker's recent work examines how we are going to weather the Perfect Storm of 21st century crises. Most recently, he is exploring the place of cities within the perfect storm, which has led to a focus on understanding eco cities, especially New Songdo City in South Korea. He is founder and manager of an international human rights observation team in Ecuador, which has been active since 2006.

Alexander K Lautensach is an assistant professor at the School of Education of the University of Northern British Columbia in Terrace, British Columbia, Canada. His background includes biology, environmental science, bioethics, and education. In 2010 he published *Environmental Ethics for the Future: Rethinking Education to Achieve Sustainability* (Lambert Academic Publ.). His current research activities focus on human ecology, cross-cultural education, and environmental ethics. His work in human security centers on its health-related and environmental pillars.

Sabina W Lautensach serves on the faculties of the University of Northern British Columbia and M. Gandhi University, Kerala, India, where she lectures in social sciences (global political economy, human security and development studies). She is the founder and current editor-in-chief of the *Journal of Human Security.* As director of the Human Security Institute she coordinates collaborations with colleagues in Canada, Europe, and the Australasian region. She and Dr. Alexander K Lautensach, are the editors of a university graduate level textbook in human security, entitled *Human Security in World Affairs: Problems and Opportunities* (2013). She received her doctorate in anthropology, international relations and fisheries from the University of Otago, New Zealand.

Dora Marinova is a professor of sustainability and deputy director of the Curtin University Sustainability Policy (CUSP) Institute in Perth, Western Australia. She has more than 400 publications and has supervised to successful completion 40 PhD students. Her research interests cover innovation models, including the evolving global green system of innovation, self-reliance and the newly emerging area of sustainometrics related to the modelling and measuring of sustainability. She is a member of the Editorial Board of the *International Journal of Education Economics and Development* (published by Indescience, Switzerland) and *Transformations*: *An Interdisciplinary Journal* (published by EBSCO Publishing, USA).

Kimberly Porter Martin is a professor of anthropology and anthropology program chair at the University of La Verne in Southern California. Her areas of specialization include psychological, cognitive and medical anthropology, gender, Mesoamerica, Polynesia and Western Europe. She is secretary and executive board member for the Southwestern Anthropology Association. Her publications include *Diversity Orientations: Culture, Ethnicity and Race* in cultural diversity in the United States, and *The Evolution of Ethnicity: Integrating Psychological and Social Models* in problems and issues of diversity in the United States, both edited by Larry Naylor and published by Bergin and Garvey.

Michael McIntyre is an associate professor, finance at the Sprott School of Business, Carleton University, Ottawa, Canada. He researches in the areas of governance and financial risk management, the latter with a focus on derivative securities and financial mathematics.

Jennifer E Michaels is a Samuel R. and Marie-Louise Rosenthal professor of humanities and professor of German at Grinnell College in Iowa, US where she has taught courses on language, literature and culture. She received her M.A. degree in German from Edinburgh University and an MA and PhD in German from McGill University in Montreal. She has published four books and numerous articles on German and Austrian literature and culture with a focus on twentieth and twenty-first century literature. She has served on the editorial board of *German Studies Review* and Ariadne Press. She was elected president of the German Studies Association and the Rocky Mountain Modern Language Association and has served as the German Studies Association's representative to the American Council of Learned Societies.

Alisa Moldavanova is an assistant professor of public administration at the political science department at Wayne State University (Detroit, MI). She is a graduate of the School of Public Affairs and Administration at the University of Kansas, where she defended her PhD thesis "Sustainable Public Administration: The Search for Intergenerational Fairness" in 2013. Her research interests include sustainability and intergenerational justice, organization theory, non-profit management, and public service ethics. She is currently doing research on the sustainability of arts organizations in an urban context, and teaching courses on managing public organizations and programs, and human resources management for the Masters of Public Administration program at Wayne State University. Dr. Moldavanova's work has been published in the *International Journal of Sustainability Policy and Practice*.

Steven A Murphy is a professor and a dean at the Ted Rogers School of Management, Ryerson University, Toronto, Canada. His research examines the role of emotions in strategic decision-making, leadership, and other facets of organizational life, and conducts research in the area of senior management dynamics and governance.

Dariush Rafinejad is a consulting associate professor at Stanford University and a core faculty member at the Presidio Graduate School in San Francisco. He has dedicated his current teaching and research to sustainable product development and manufacturing, and renewable energy systems. He has published several case studies, research papers and a book on innovation in product development. Before entering academia, he held senior executive positions in the high tech industry of Silicon Valley for over 25 years. He holds a BSc degree in petroleum engineering from Abadan Institute of Technology, and MSc and PhD degrees in Mechanical Engineering from the University of California, Berkeley.

Elin Wihlborg is a professor in politics at Linkoping University, Sweden and holds a Marie Curie scholarship for research on sustainable e-government (2011-14). She has a Master of social policy from London School of Economics and a PhD in technology and social change. Her research focuses on policies towards households, local planning and policy from inter-disciplinary perspectives.

Jade Wildy is a postgraduate researcher at the University of South Australia. She holds a Masters degree from the University of Adelaide and a Bachelor of visual arts, also from the University of South Australia. Her current research interests are concerned with the ways that the arts can be used to promote positive global environmental change.

Index

CPSIA information can be obtained at www.ICGtesting.com
Printed in the USA
BVOW10s1625281214

380924BV00001B/1/P

9 781612 297262